CELL MEMBRANE TRANSPORT
TRANSPORT
Principles and Techniques

Date Due

CELL MEMBRANE TRANSPORT
Principles and Techniques

by Arnošt Kotyk and Karel Janáček

Laboratory for Cell Membrane Transport
Institute of Microbiology
Czechoslovak Academy of Sciences
Prague, Czechoslovakia

In collaboration with the Staff of the Laboratory

ℙ PLENUM PRESS · NEW YORK—LONDON · 1970

Library of Congress Catalog Card Number 71-107538
SBN 306-30446-5

© 1970 Plenum Press, New York
A Division of Plenum Publishing Corporation
227 West 17th Street, New York, N.Y. 10011

United Kingdom edition published by Plenum Press, London
A Division of Plenum Publishing Corporation, Ltd.
Donington House, 30 Norfolk Street, London W.C. 2, England

Printed in the United States of America

To students of
flow across barriers — not only biological

PREFACE

It is not a particularly rewarding task to engage in writing a book on a subject which is undergoing a rapid and potentially revolutionary development, but, on the other hand, the investigation of transport of substances into and out of cells has reached a stage of maturity or at least of self-realization and this fact alone warrants a closer examination of the subject.

No one will doubt at present that the movement—mostly by selective translocation—of substances, ranging from hydrogen ions to deoxyribonucleic acids, across the cell-surrounding barriers represents one of the salient features of a living cell and that, if we are permitted to go so far, the cessation of the selective transport processes might be considered as the equivalent of cell death. Hardly anybody will question the premise that cell and tissue differentiation within the ontogenetic development of an organism is closely associated with properties of the outer cell face. Perhaps no serious scholar will attempt to refute the concept that membranes with characteristic morphology and composition represent the architectural framework for the whole cell. And probably no experienced biologist will raise objections to the belief that many physiological processes, like nervous impulse conduction and other electrical phenomena of cells and tissues or their volume changes, are associated with membrane-regulated shifts of ions and molecules.

Still, perhaps because all these observations comprise a border discipline *par excellence*, textbooks of biochemistry hardly ever mention the problem of biological transport, textbooks of physiology deal at most with the ion and water movement in the kidney, and textbooks of biophysics (the few there are) treat membrane phenomena and ion distribution in cells rather one-sidedly.

Bringing together all the pertinent information—an awesome task—is thus imperative at this stage of experimental research. However, collating the diversified pieces of evidence within a system has proved to be extremely difficult, apparently owing to our notions of systems acquired either in physics or in chemistry or in biology. It is not intended to be an over-estimation of the magnitude of the work facing the membrane biologist (or chemist or physicist?) if we state that none of the classical frameworks will fit here flawlessly. Therefore, we were compelled to resort to a certain compromise when designing the layout of this book. It includes a physico-chemical treatment arranged according to the degree of complexity of transport models, as well as one based on the type of biological objects examined. We believe that this dualism is not detrimental to the usefulness of the book as each of the sections contains a different type of information and, from the viewpoint of supplying data to the reader, we feel that it is actually of greater practical value.

While we realize that this lack of a unitary system is a handicap for our present effort it will only add to the merit of those who, perhaps in the near future, will be able to propose a neat and orderly classification of transport processes based on their molecular parameters, a challenge to the enzymologist and to the protein chemist.

This is not the first book on biological transport ever written, a monographic treatment of one or more aspects of transport or cell membranes having appeared several times before, the more important and recent ones of these being included in the bibliography at the end of this book. However, this book has been written not only to summarize what we feel to be important experimental evidence and theories attempting to account for it but also to provide the reader with an insight into the technical means that are available for transport studies. And it is this section of the book which we think might be especially useful to those aspiring to study transport phenomena in various types of cells.

It was our intention to include in the text all the physical concepts underlying the more sophisticated phenomena and we did our best to use uniform symbols throughout although certain inconsistencies were unavoidable because of traditionally used symbols in different treatments of the subject (such as the symbol c as well as s used for solute concentration). We attempted to include, wherever appropriate, practical examples to illustrate either the theories or the techniques.

The 26 chapters of this book are divided into 5 sections so as to emphasize the different approaches employed. The first section deals with the morphological and chemical aspects of membranes and it was included to present the student with a physical picture of the object of the subsequent, largely abstract, considerations.

The second section covers most, if not all, of the kinetic concepts and theories involved in membrane transport. It starts necessarily with a chapter on diffusion of substances and migration of ions in a homogeneous medium, to proceed to the transport of substances in anisotropic systems and to the metabolically linked types of transport. The section also includes a brief treatment of special types of transport for which the detailed kinetics are not known, *viz.*, pino- and phagocytosis.

The third section is one which will probably become obsolete most rapidly as it presents the rather meager but fast accumulating evidence on the molecular properties and plausible models of the membrane transport systems.

The fourth section contains technical descriptions of methods and basic equipment used in transport studies, together with practical examples illustrating their application.

The fifth section is a more or less comparative treatment of the transport information as has accrued from studies of different cell types.

The writing of the various chapters of the book was accompanied by frequent mutual consultations and the whole text was revised several times to avoid major inconsistencies and duplicity. If errors and shortcomings are found by readers of the book their criticism should be addressed to A.K. for chapters 1, 3.2., 5.1.2., 5.2., 7, 8, 9, 10, 11, 15, 16, 17, 19, and 26, and to K.J. for chapters 2, 3.1., 4, 5.1.1., 6, 12, 13, 22, 24.1, 24.2., 24.3., and 24.4.

It is a pleasure to acknowledge the factual and moral support for writing this book shown by our colleagues and friends. Among those who actively contributed, greatest credit goes to Dr. Renata Rybová who wrote chapters 18, 20, and 21; to Dr. Jiřina Kolínská who contributed greatly by writing chapters 23, 24.6., 24.7., and 24.8.; to Dr. Milan Höfer who wrote the chapter on mitochondria (25); to Dr. Rudolf Metlička and Mr. Ivan Beneš who wrote chapter 14 on the use of artificial membranes; and to Dr. Stanislav Janda who contributed by sections 23.4.2 and 24.5.

Other collaborators helped us greatly by calling our attention to publications that had escaped our attention (special credit to Dr. Jaroslav Horák) and by discussing the text with us.

We are greatly indebted to Dr. Jiří Ludvík who heads the Laboratory of Electron Microscopy of this institute who supplied a number of original electron microphotographs. Our thanks are due to a number of assistants for typing parts of the manuscript, drawing some of the figures, and photographing others.

It would be most ungrateful not to mention here the debt we owe to

Prof. Arnošt Kleinzeller, now at the University of Pennsylvania, who introduced us into the realm of transport studies and, last but not least, to Prof. Ivan Málek who, as head of the Institute of Microbiology of the Czechoslovak Academy of Sciences, showed profound understanding for the problems of membrane transport when he permitted a whole department of his institute to concentrate on biological transport phenomena.

Prague, July 1969   Arnošt Kotyk
                                                              Karel Janáček
                                            Laboratory for Cell Membrane Transport
                                                 Institute of Microbiology
                                           Czechoslovak Academy of Sciences

CONTENTS

Structural Aspects

1. COMPOSITION AND STRUCTURE OF CELL MEMBRANES

1.1. CELL ENVELOPES AND MEMBRANES

There is perhaps no need for an apology when opening a treatise of cell membrane transport with a consideration of what cell membranes are. Straightforward as this question may appear, it will not elicit the same answer from a cytologist to whom the cell membrane is a rather concrete boundary line (or lines) that he sees or believes to see under his microscope and from a physiologist to whom it represents the barrier separating the cell interior from the external medium which is invested with a number of peculiar permeability properties.

Although the time is not long past when arguments were advanced that there was no definite membrane at the cell boundary but that all the morphological and physiological features could be attributed to the behavior of an interface between protoplasm and aqueous medium, there is now very definite evidence that the membranes surrounding cells are a reality. The advent and development of electron microscopy and appropriate staining techniques supplied impressive documentation for the existence in cells of structures which appeared as two dark bands about 25–40 Å thick, enclosing a light band about 30–60 Å across (Fig. 1.1). This type of structure not only surrounds practically all cells but is seen to occur inside cells and to communicate between the outer cell envelope and a number of cell organelles. Some of these organelles are actually composed of such membraneous structures (Fig. 1.2). It is now recognized that membranes of one type or another may constitute as much as 80% of the dry

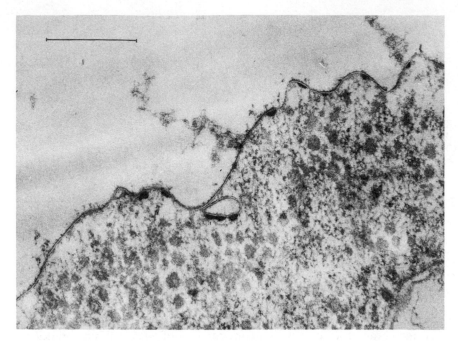

Fig. 1.1. Ultrathin section through *Entamoeba histolytica* showing a unit plasma membrane bounding the cell. Scale line 0.5 μ. (Courtesy of Dr. J. Ludvík, Institute of Microbiology, ČSAV, Prague.)

mass of a cell and display turnover rates of almost 100% per hour, suggesting a significant role in the cell metabolism.

The above observations on the microscopic appearance of membranes date back some 15 years and led Robertson (1955, 1959) to the concept of the "unit membrane" which has represented a unifying theory and has found widespread support. Robertson made use of the fact that heavy-metal fixatives (potassium permanganate, formalin dichromate, osmium tetroxide) stain lipoprotein membranes so as to make them appear as a double dark band in the electron microscope. (This does not mean that the staining by all the above agents is identical. In fact, significant differences may be observed between membranes stained with permanganate and those stained with osmium.) It was shown later that it is the hydrophilic part of the membranes which is stained (an addition of protein to a model system with phospholipid increased the size of the dark bands), the hydrophobic part remaining electron-clear (although at least osmium is known also to react with unsaturated fatty acids; Korn, 1968). The universality of the unit membrane structure which was so convincing by its

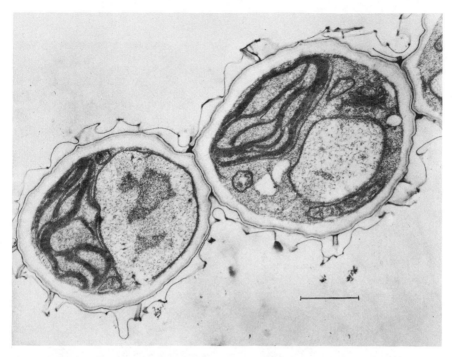

Fig. 1.2. Two cells of the green alga *Scenedesmus quadricauda* showing membraneous lamellae and a heavy cell wall. Scale line 1 μ. (Courtesy of Dr. J. Ludvík, Institute of Microbiology, ČSAV, Prague.)

ubiquity has necessitated a terminological differentiation between the outer cell membrane which is called the *plasma membrane* or is given one of several more specifically coined names, e.g., *plasmalemma*, and the assembly of intracellular unit membranes which are called *cytoplasmic membranes* and *organelle membranes*. Thy cytoplasmic membranes, in their turn, can be divided into rough and smooth, from their microscopic appearance. The *rough* membranes are typically those of the endoplasmic reticulum with ribosomes attached to them (Fig. 1.3), the *smooth* membranes are best represented by those of the Golgi apparatus which appears as a stack of smooth elongate cisternae or lamellae (Fig. 1.4).

The organelle membranes, although preserving much of the unit membrane structure, are most heavily differentiated according to their function. Thus, the nuclear membrane is characterized by 0.05–0.1 μ pores (Fig. 1.5) through which macromolecules can easily pass unless the pores are plugged as seen distinctly in some sections. The mitochondrial membranes (Fig. 1.6) appear to be of a characteristic type, the outer one being the locale of citric

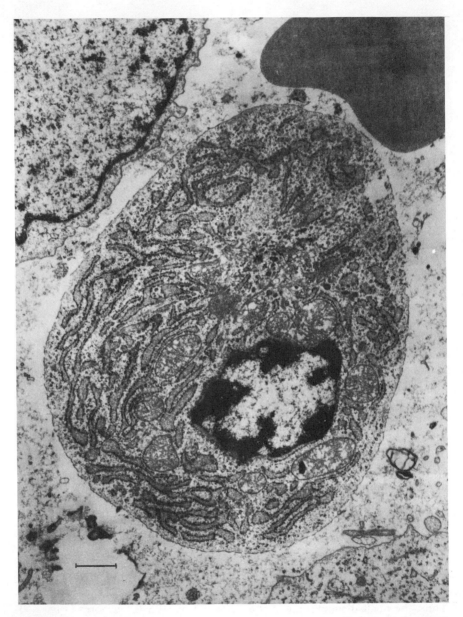

FIG. 1.3. Ultrathin section of a plasma cell of piglet lung showing a richly developed endoplasmic reticulum. The dark-lobed object at upper right is an erythrocyte. Scale line 1 μ. (Courtesy of Dr. J. Ludvík, Institute of Microbiology, ČSAV, Prague.)

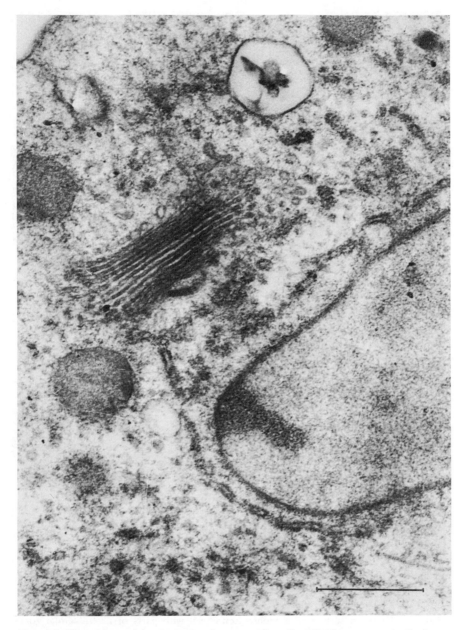

FIG. 1.4. Ultrathin section of a perinuclear part of a cell of *Trichomonas vaginalis* showing the stacked lamellae of a Golgi apparatus. Scale line 0.5 μ. (Courtesy of Dr. J. Ludvík, Institute of Microbiology, ČSAV, Prague.)

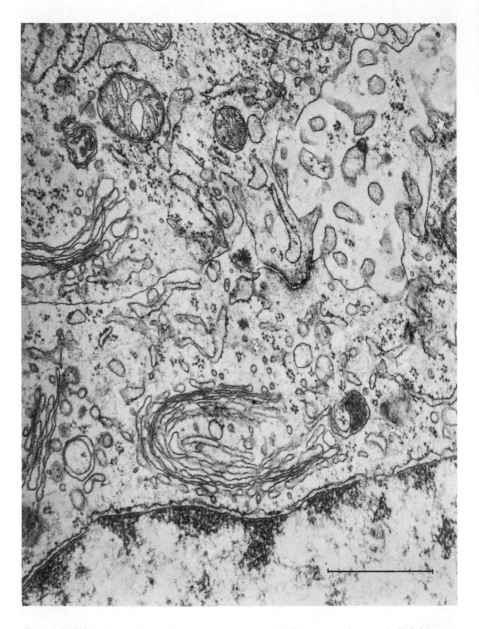

FIG. 1.5. Ultrathin section of a mouse tumor cell with parts of several Golgi zones (lower center and left) and a part of the nuclear membrane. Pores are visible at left (closed by a partition) and at right (with fuzzy material passing through them). Scale line 1 μ. (Courtesy of Dr. J. Ludvík, Institute of Microbiology, ČSAV, Prague.)

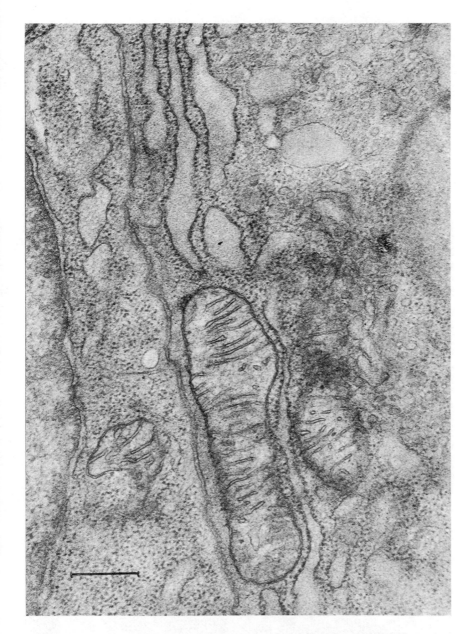

FIG. 1.6. Ultrathin section of a mouse spleen cell with a longitudinally and a transversely cut mitochondrion exhibiting double-membrane cristae. Scale line 0.5 μ. (Courtesy of Dr. J. Ludvík, Institute of Microbiology, ČSAV, Prague.)

acid cycle and fatty acid metabolism enzymes, the inner one, forming the typical cristae, being characterized by prominences (Fig. 1.7) which are likely to contain the whole set of oxidative and oxidative phosphorylation enzymes. Similarly, the membrane surrounding the chloroplasts and other plastids is differentiated better to fulfill its function and is probably made up of various particles and sub-units (Park, 1966).

It is natural that in this book the main emphasis will be placed on the outer, plasma, membrane across which materials from the outside must pass to reach the cell interior and vice versa. Let it be stated here that the plasma membrane need not be the outermost envelope of the cell. In fact,

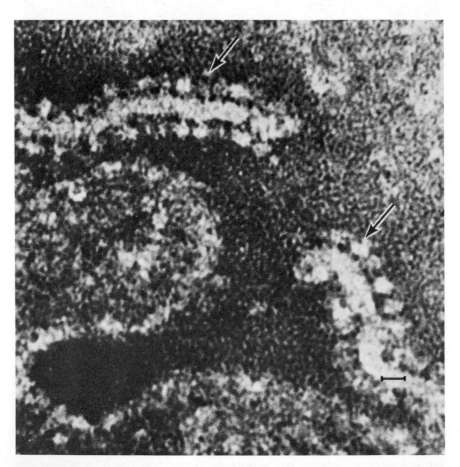

FIG. 1.7. Negatively stained preparation of mitochondrial electron transfer particles (shown by arrows) sitting on the cristae. Scale line 0.01 μ. (Taken with kind permission from Fernández-Morán *et al.*, 1964.)

all freely living cells, most bacteria, yeasts, fungi, and plant cells possess a more or less rigid structure outside their plasma membrane, either a cell wall or a capsule composed of various types of complex polysaccharides but containing also a minority of protein material. However, this supporting structure of cells, although not necessarily devoid of metabolic activity, does not play any significant role in the transport of low-molecular sub-

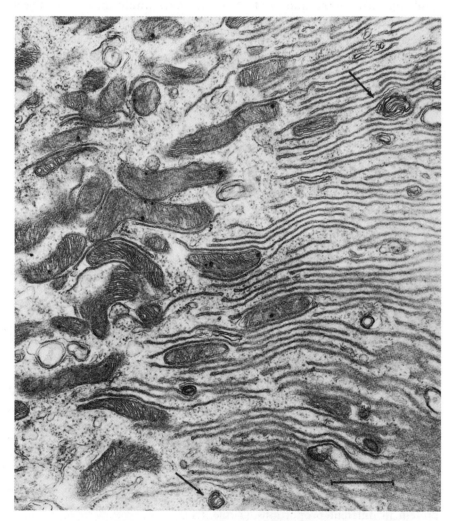

FIG. 1.8. A tangential section of a cell from the anal papilla of a mosquito larva, exhibiting an array of basal membrane folds projecting from the right toward numerous mitochondria. At points marked with arrows the folds appear to be coiled into vesicular structures. Scale line 0.5 μ. (Taken with kind permission from Copeland, 1964.)

stances into and out of cells, as is shown by the transport properties of
cells deprived of this external envelope (yeast *protoplasts*, bacterial *spheroplasts*, *gymnoplasts* in general).

When emphasizing the role of the plasma membrane in the transport
of substances into cells we should not underestimate two possible transport
functions of the intracellular membranes. These membranes are most certainly operative in the translocation of substrates within the cell, a subject
which is at present not amenable to a rigorous examination but which
deserves the utmost attention of biologists and biophysicists. Secondly, the
intimate association of these cytoplasmic membranes with the plasma membrane, on the one hand, and with the cell organelles, on the other (Fig. 1.8),
suggests that these membranes might play an interesting role in the uptake
of nutrients and extrusion of waste products by the cell. Be it as it may,
the outer, plasma, membrane appears to possess all the transport characteristics of the intact cell (experiments done on erythrocytes and their reconstituted stromata or "ghosts") and, hence, our subsequent considerations will be restricted to this outer cell membrane.

1.2. CHEMICAL COMPOSITION OF THE PLASMA MEMBRANE

Early work on the permeation of various compounds into cells indicated the predominantly lipid character of the outer membrane and led
to the design of several membrane models, all of them based on a (mostly)
bimolecular lipid layer forming the permeability barrier. However, considerations of surface tension of cells with respect to water (which was found
to be of the order of 10^0 dyne $\cdot$ cm^{-1} as compared with a value of 10^1 dyne $\cdot$
cm^{-1} for a water/hydrocarbon interface) introduced the idea of a protein
coat lining the lipid layer on each side. Although, as shown by Haydon
and Taylor (1963), the considerations on surface tension were not substan-
tiated, they led to a model which has been widely accepted and which represents the basis of most of the more sophisticated subsequent hypotheses.

The general plausibility of the model was supported by the chemical
analysis of membranes prepared by differential centrifugation of cell homo-
genates which show that the bulk of the membrane material is lipid and
protein in different proportions (Table 1.1).

In addition to these two types of components, most membranes contain
some carbohydrate which may in some cells form an extraneous coat in-
vesting the cell with serological properties and probably playing a role in

Table 1.1. Composition of Some Typical Plasma Membranes

Type of cell	Protein (%)	Lipid (%)	Reference
Micrococcus lysodeikticus	68	23[a]	Salton and Freer (1965)
Bacillus megaterium	70	25[a]	Yudkin (1966)
Pseudomonas aeruginosa	60	35[a]	Norton *et al.* (1963)
Saccharomyces cerevisiae ETH 1022	37	35[b]	Matile *et al.* (1967)
Saccharomyces cerevisiae NCYC 366	49	45	Longley *et al.* (1968)
Human erythrocyte	53	47	Dodge *et al.* (1963)
Avian erythrocyte	89	4	Williams *et al.* (1941)
Ox brain myelin	18–23	73–78	Autilio *et al.* (1964)
Rat liver	85	10	Emmelot *et al.* (1964)
Rat muscle	65	15	Kono and Colowick (1961)

[a] There is no cholesterol in bacterial membranes.
[b] Up to 27% mannan was found in the preparation.

cell interactions. The most important carbohydrate component of membranes is sialic acid which is a group name given to acylated neuraminic acids derived from a pyranose ring

N-acetylneuraminic acid

It is covalently bound with a protein which may, in addition, contain hexoses and amino sugars (Cook, 1968).

Some membrane preparations appear to contain up to 5% ribonucleic acid but it is not quite certain whether the material used could not have been contaminated with, for example, endoplasmic reticulum.

The bulk protein and lipid fractions were analyzed mostly in the isolated membranes of erythrocytes as these can be prepared practically free of foreign material (like haemoglobin) by haemolyzing erythrocytes in 20 mM phosphate buffer at pH 7.65 (Dodge *et al.*, 1963).

Refined techniques, like starch gel or disc electrophoresis of solubilized membrane material, reveal the presence of different proteins, some of them possessing enzyme activity. Animal plasma membranes contain a number of esterases, phosphatases, and peptidases. Erythrocyte membranes, in particular, contain acetylcholinesterase and adenosine triphosphatase and possibly some glycolytic enzymes. Sea urchin egg plasma membrane contains adenyl cyclase. Bacterial membranes were found to contain hydrolases and some dehydrogenases, yeast plasmalemma contains adenosine triphosphatase. Moreover, plasma membranes certainly contain a number of different proteins involved in the transmembrane transport of substances, including the carriers, coupling enzymes, and the like.

The difficulties accompanying the identification of cell membrane proteins were particularly those of solubilization. Different agents were employed to this end, among them aqueous urea, mercaptoethanol, *n*-butanol, Triton X-100, sodium iodide, sodium chloride, ultrasonic treatment, and diversified results were obtained, the maximum of different protein entities released being about twenty.

The lipid fraction of the plasma membranes has been studied somewhat more systematically and has been shown to contain mostly phospholipids, together with cholesterol and glycolipids (cerebrosides). The representation of these compounds is not the same in plasma membranes of different cells and even in the erythrocytes of different species (Table 1.2).

Table 1.2. Lipid Content of Plasma Membranes

Type of cell	Phospholipid	Lecithin	Cephalins	Other	Reference
	Total lipid	(in % of phospholipid)			
Erythrocyte, human	0.58	39	24	37	de Gier and van Deenen (1961)
Erythrocyte, ox	0.56	7	32	61	"
Erythrocyte, pig	0.53	29	35	36	"
Erythrocyte, rabbit	0.58	44	27	29	"
Erythrocyte, rat	0.61	56	18	26	"
Erythrocyte, sheep	0.60	1	36	61	"
Baker's yeast	0.16	23	61	6	Longley *et al.* (1968)

1.3. FINE STRUCTURE OF THE PLASMA MEMBRANE

The above considerations result in two salient features of the plasma membrane: (i) Its microscopic appearance after staining with heavy metals as two dark bands with a light band between them, (ii) its chemical composition as a combination of protein with phospholipid. We will now attempt to combine these two characteristics into a membrane model, plausible both chemically and functionally.

In our first approach we are assisted by the observation that if, say, lecithin is placed in contact with water, it spreads itself over the surface to form a monomolecular film with its hydrophobic (fatty acid) regions out in the air and with its hydrophilic (glycerophosphorylcholine) regions immersed in the aqueous phase. If this film is compressed by a laterally applied force it will buckle to form at first a poorly characterized conglomerate of layers, but, since the phospholipid molecules tend to expose the minimum possible hydrophobic surface to the surrounding water, as the ratio of lecithin to water within a given space increases the phospholipid molecules organize themselves into bimolecular flat structures with the hydrophobic chains pointing inward. This type of structure appears to be energetically most favorable and it is seen to be formed spontaneously if proper conditions exist. This is the basis of what has been called the myelin forms (*cf.* Fernández-Morán, 1962) as they closely resemble the myelin sheath of a small nerve axon.

Such a bimolecular leaflet is formed not only by identical molecules but can equally well be produced if different lipid molecules are present in the mixture, such as lecithin and cholesterol. If certain steric requirements are met the mean area per molecule of such a heterogeneous film is actually less than would be predicted from the average of the two different areas per molecule taken as a sum. All these findings are rather sug-

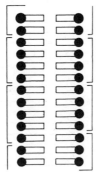

FIG. 1.9. The "paucimolecular" model of cell membrane showing a double layer of lipid molecules facing with their "hydrophilic heads" outward, covered by a protein coat. (From Danielli, 1952.)

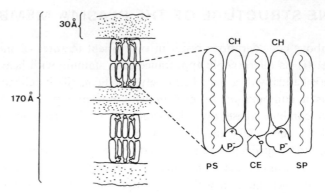

FIG. 1.10. Diagram summarizing the structural features and composition of a peripheral myelinated nerve sheath, based on X-ray diffraction and polarized light data. **PS** Phosphatidylserine, **CH** cholesterol, **CE** cerebroside, **SP** sphingomyelin. (After Finean, 1958.)

gestive of the postulated structure of Robertson's original unit membrane and they are in excellent agreement with the model proposed originally by Danielli and Davson (1935) where a bimolecular lipid leaflet is covered on both sides with a layer of protein (Fig. 1.9). Using Dreiding molecular models, various investigators (e.g., Vandenheuvel, 1966) have suggested a number of plausible arrangements of the various lipid molecules in the bilayer, one of them being shown in Fig. 1.10 (Finean, 1958). The proteins covering the lipid sheet might interact with the lipids by electrostatic forces or by hydrophobic–hydrophobic interaction.

To account for the passive permeability of most plasma membranes to very small molecules the model was extended by Danielli in the early

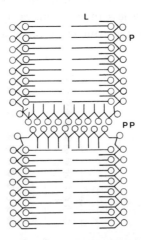

FIG. 1.11. An "aqueous pore" model of the cell membrane according to Stein and Danielli (1956). **L** Lipoid molecule, **P** protein molecule, **PP** polar pore across the membrane.

fifties to include minute hydrophilic pores reaching across the membrane (Fig. 1.11). Actually, some electron micrographs of cell membranes have been interpreted as indicative of the presence of such fine pores (de Robertis *et al.*, 1965) although the evidence is not incontestable.

What has been said so far applies reasonably well to unit membranes of the myelin sheath where it has found support from X-ray analysis. However, substantially more powerful evidence has been obtained during the last few years that speaks against the universality of the bilayer leaflet model.

(i) A uniform unit membrane is unlikely to be present in all cell types because of the variation in the lipid and protein composition and in the thickness observed.

(ii) In many ultra-thin preparations, some of the membranes or their parts appear to show a globular or granular fine structure (Fig. 1.12).

(iii) Refined techniques, like freeze etching (Moor and Mühlethaler, 1963) and high-resolution electron microscopy (Sjöstrand, 1963*b*; Robertson, 1964), aided by special treatment of cell membranes (Dourmashkin *et al.*, 1962) show a typical array of more or less hexagonal units in membranes viewed from top (Fig. 1.13). The diameter of these units lies anywhere between 40 and 150 Å.

(iv) Membranes of the mitochondrial and plastid type cannot be accommodated within the concept of the bilayer membrane, there being always a clear particulate structure present.

(v) The curvature of some membranes (particularly in invaginations) is so great that a bimolecular leaflet would break down under the stress. The birefringence of such curved membranes shows characteristic changes as one proceeds along the length of the membrane, indicating differences in lipid orientation inside the membrane.

(vi) The electrical resistance of lipid bilayer membranes is too great and their permeability to various compounds so low that it is irreconcilable with the situation in living cells.

New light was shed on the problem when models resembling the submicroscopic regular structures of membranes were produced artificially. On mixing lecithin with relatively little water the lecithin molecules organize themselves in a hexagonal array of tubes where water is surrounded by the phospholipid (Fig. 1.14) as indicated again by the fact that the tubes are electron-dense when stained with osmium tetroxide. Addition of some soaps (such as sodium linolenate) to water will produce a reverse pattern where water surrounds the globular tubes of nonpolar material (Luzzati

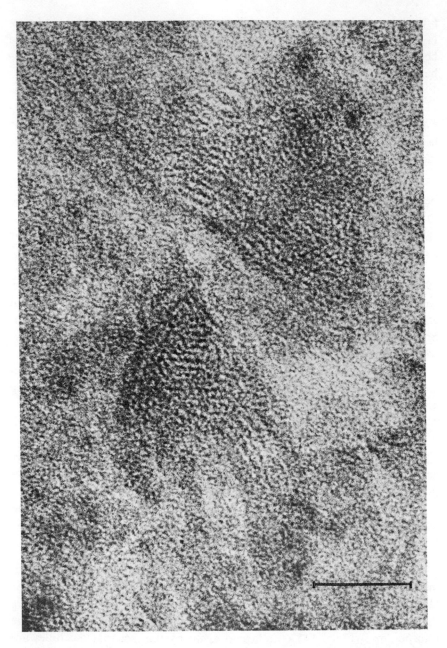

FIG. 1.12. Golgi membranes of exocrine cells of pancreas with regular globular structures oriented along the plane of section. Scale line 0.05 μ. (Reproduced from Sjöstrand, 1964.)

FIG. 1.13. Surface view of the plasma membrane in a frozen-etched cell of *Saccharomyces cerevisiae*. Note the hexagonal arrangement of particles between the invaginations. Scale line 0.5 μ. (Taken with kind permission from Moor and Mühlethaler, 1963.)

Fig. 1.14. A hexagonal phospholipid structure in 3% water. The dense spots are hydrophilic "cylinders" surrounded by lipid. Scale line 0.01 μ. (Taken with permission from Stoeckenius, 1962.)

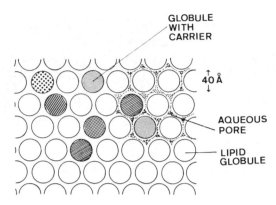

FIG. 1.15. A surface view of the lipid sheet in a micellar membrane. (According to Lucy, 1964.)

and Husson, 1962). Mixtures of lecithin, cholesterol, and saponin will produce, depending on the concentration ratio, structures that can be described as the *hexagonal, helical,* or *stacked-disc* phase, all of these having their counterparts in living objects like viruses, bacterial flagellae, etc. (Lucy and Glauert, 1964). Any of these types of structure which are micellar in nature may account for the deviations from the bilayer model as enumerated above. It can be shown that a micellar membrane composed of 60-Å lipid globules would exhibit roughly 6-Å water-filled pores between the globules, which may be significant for the passive permeability of some hydrophilic molecules. Some of the lipid globules, moreover, may be replaced with proteins of enzymic nature or with transport proteins for different substrates (Fig. 1.15). A sheet of such globules may, but need not, be enclosed on both sides in a continuous protein coat interacting, as in the bilayer model, with the polar groups of the lipid, although, particularly with the mitochondrial structural protein, there seems to be a pronounced hydrophobic interaction between protein and lipid (Fig. 1.16). The arrangement of the globular micelles is thought to be flexible (Lucy, 1964), the adjacent micelles being held together by hydrogen bonds. Interactions with extended protein and/or polysaccharide molecules on the lipid layer surface would account for the stability of the flat sheet of micelles. It was shown by Lenard and Singer (1966) that the protein is about 30% in the α-helical form, the rest in the random-coil conformation, the pattern being similar in widely differing cells.

This type of structure, functionally reminding one of the bilayer-pore model of Stein and Danielli (1956), even if present in a minority as compared with the bilayer model, would readily account for the characteristic

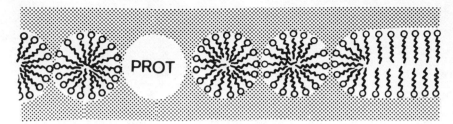

FIG. 1.16. A transverse view of a membrane with lipid globules surrounded by a protein coat. Some of the globules may be occupied by protein (possibly of carrier nature), others may merge with a bimolecular leaflet of lipid (at the right). (According to Lucy, 1964.)

permeability properties shown by cell membranes: (i) Passive diffusion of certain polar compounds (urea, glycol) as well as of nonpolar substances (alkanes, higher alcohols), the former passing through the "interstitial" aqueous matrix, the latter through the lipid globules. (ii) Selective permeability for many compounds (particularly polar, of greater molecular size) made possible through the mediation of special protein systems taking the place of some of the lipid globules.

Lucy's globular micelle model is not the only one that can be reconciled with the existing experimental evidence. Lenard and Singer (1966) suggested a lipoprotein subunit model where both protein and lipid form parts of the membrane surface, the protein, in addition, accounting for enzyme and transport properties (*cf.* also Green *et al.*, 1967) (Fig. 1.17). The common

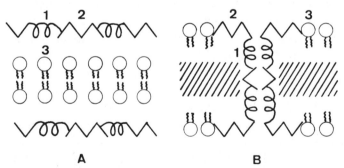

FIG. 1.17. The Danielli–Davson model of cell membrane, modified to include data obtained by optical rotatory dispersion and circular dichroism (**A**). A generalized membrane containing a hydrophilic pore and surface-exposed proteins as well as lipids (**B**). **1** Helical protein, **2** random-coil protein, **3** polar lipid facing with its hydrophilic head outward. The hatched area is assumed to be occupied by nonpolar substances, such as hydrophobic amino acid ends of proteins or lipids. (According to Lenard and Singer, 1966.)

appeal of the subunit and micelle models over the bilayer model lies in the recognition of the role of proteins in the various membrane functions but it is difficult on the present evidence to prefer any one of them over the other.

The granular or subunit structure of plasma membranes has recently received powerful, even if indirect, support from two lines of evidence. One was the series of observations by Engelman and associates (e.g., 1968) who solubilized plasma membranes of *Mycoplasma laidlawii* with dodecyl sulfate to separate lipid–detergent and protein–detergent complexes which resemble in size and composition the membrane subunits observed in some intact membranes. Moreover, on removing the detergent by dialysis and adding Mg^{2+}, the protein and the lipid reaggregate spontaneously to form membranes which resemble greatly the original "unit membranes" of the microorganisms.

The other findings supporting the existence of more or less integral self-contained units were made by Muñoz and co-workers (1968) who isolated particles of Ca^{2+}-activated ATPase from the membrane of *Micrococcus lysodeikticus* and by Harris (1968) who extracted from erythrocyte ghosts one-ring and four-ring particles of protein. Both the ATPase and the protein particles are roughly cylindrical in shape about 100 Å across and might easily fit into the globular micelle model described above.

It thus appears that, although the bimolecular lipid leaflet may be present in some membranes and may invest them with particular stability, the biochemist or membrane physiologist will readily embrace the lipoprotein globule as it offers him an explanation for the diverse phenomena associated with the transport of compounds into cells.

Kinetic Aspects

2. TRANSPORT IN HOMOGENEOUS LIQUID PHASE

2.1. MASS FLOW AND DIFFUSION

2.1.1. Introductory Definitions

Before any theoretical considerations of biological transport are attempted, some physical chemistry of diffusion appears to be in order. Let us begin with the very simple case of transport of electrically neutral particles in a liquid solution of a single substance. Only two components will be present in such a solution: the quantitatively prevailing *solvent* and, in the solvent, the dispersed *solute*. The system considered will be homogeneous in the sense that it will form a single phase and will contain no barriers, such as membranes, and the like.

Let us consider a small volume element inside a solution. How can the molecules of a solute present in this volume be transferred to another place in the solution? In principle, there are two different ways of transporting them. Firstly, the whole volume element may move to a different place. This type of transfer is called *mass* or *bulk* flow. If the volume of the element is dV and its concentration c (say, in $mol \cdot cm^{-3}$) the translocation of the element represents a transfer of $c \cdot dV = dn$ moles of solute. If the volume flow of solution across a surface of unit area proceeds at the rate of

$$\frac{1}{A} \frac{dV}{dt}$$

(e.g., in $cm \cdot sec^{-1}$) the flow of solute across the same area will be

$$\frac{dn}{dt} = c \frac{1}{A} \frac{dV}{dt} \tag{1}$$

Secondly, solute molecules can change their position in space due to their tendency to move away from the volume element through their random thermal movement, prevailing in liquids, and to become randomly distributed in space. This second kind of transfer is called *diffusion*. The random thermal movements being permanent, the diffusion out of the volume element is also permanent, but it depends on the conditions outside this element, whether or not the process will be observable. If the concentration of the solute in the proximity of the element is the same as inside, no concentration changes will occur; the solute particles which leave the volume element will be replaced by identical particles and, unless some of them are isotopically labelled, their exchange will pass unnoticed. If, on the other hand, the volume is surrounded by solution of a different concentration, i.e., if there are concentration gradients in its proximity, a net transfer of solute will take place and the concentration inside the volume element will, in general, change.

It will be seen later that it is only under special and rather artificial conditions that pure mass flow or pure diffusion can be observed even in a homogeneous liquid phase. Most commonly, both occur together. The process of diffusion being permanent in liquids, it does not stop when mass flow takes place and, inversely, as shown by Hartley and Crank (1949), net transfer by diffusion creates pressure differences and hence mass flow. It is only when mass flow takes place in the absence of concentration gradients or at such a rate that diffusion phenomena may be neglected and further in the case of the so-called *self-diffusion* (diffusion of isotopically labelled substances in the absence of gradients in the total, i.e., chemical, concentrations) that these phenomena may be studied in their pure form.

2.1.2. The Laws of Diffusion

It is often expedient to use *formal* or *phenomenological* equations for describing various natural phenomena. The physical meaning of some of these equations is well understood and the values of their coefficients may be deduced from concepts familiar in other branches of physics. For other equations, a satisfactory picture of the described phenomena is still to be found and the coefficients are only empirically determined constants or functions. The formal equations are the first and essential step for a quantitative physical description and whether the meaning of the coefficients is understood or not the formal equations are used in calculations by which the relationships between measured quantities are determined and physical events predicted. A very general equation of this kind, which may serve as a starting point for the description of many transport processes, was derived by T. Teorell (1953):*

★   Flux = Mobility × Concentration × Total driving force   (2)

where *flux* is the "amount of substance which per unit time penetrates per unit area normal to the direction of the transport." This general equation will assume specific forms, describing specific kinds of transport, depending on what kind of driving force (acting on one mole of substance) will be substituted in the equation.

If, for instance, the solute particles are charged, the *gradient* of the *electrical potential* (in other words, the *electrical field*) multiplied by the charge of one mole of solute (Faraday's number times the number of elementary charges per particle) may act as the driving force. The resulting transport is called *migration* of ions in an electrical field and Teorell's equation becomes Ohm's law for a solution.

This example may be of help in finding a suitable driving force for the process of diffusion. There is one situation where the migration of ions proceeds at exactly the same rate as the diffusion of ions in the opposite direction, so that a dynamic equilibrium, the well-known Gibbs–Donnan equilibrium, is established. We are then justified in assuming that under these conditions the driving force in one direction equals the driving force in the opposite direction. Since in the Gibbs–Donnan equilibrium for ions the gradient of electrical potential multiplied by the charge of one mole is equal to the gradient of chemical potential, it follows that the gradient of chemical potential may serve as a driving force in diffusion.

The *chemical potential* μ_i of an uncharged component i of a solution may be defined as the partial molal free energy of this component at constant temperature and pressure or, in other words, as the increase of the free energy of solution, per mole of a component, added in such a minute amount that the composition of the solution is not appreciably altered. In the case of ideal solutions (in which the partial pressures of volatile components are directly proportional to their mole fractions) the chemical potential of the i component can be expressed as the sum of a constant term and a term proportional to the logarithm of the mole fraction of the i component

$$\mu_i = \mu_{i0} + RT \ln x_i \qquad (3)$$

* The symbol ★ will be used to indicate fundamental equations.

For a dilute solution, molar concentration may be used in place of the mole fraction, since $c_w \gg c_i$ and

$$x_i = \frac{c_i}{c_i + c_w} \simeq \frac{c_i}{c_w}$$

Then

$$RT \ln x_i \simeq RT \ln \frac{1}{c_w} + RT \ln c_i$$

and, since $c_w \simeq$ constant, the term $RT \ln (1/c_w)$ may be included in the constant term, so that

$$\mu_i = \mu'_{io} + RT \ln c_i \tag{4}$$

At great dilutions the behavior of all solutions approaches the ideal case; in other cases, *activity* (mole fraction or molar concentration multiplied by a correction factor, called activity coefficient and often empirically determined) is used instead of the mole fraction. In the case of ions the term "chemical potential" may be used for that part of their partial molal free energy which is not related to their electrical charge and can be expressed as above, whereas the complete partial molal free energy is called the *electrochemical potential*. It should be kept in mind that separation of the electrochemical potential into a chemical and an electrical part is meaningful only when measuring, under special conditions, the electrical potential difference rather than the electrochemical potential difference between two media.

Teorell's equation for flux Φ will then take the form

$$\Phi = -Uc \text{ grad } \mu \tag{5}$$

where U is mobility. The minus sign derives from the fact that when the chemical potential increases in the direction that we consider as positive (i.e., when grad μ is a positive quantity) the flux will proceed in the opposite direction, i.e., it will be negative. For the simple case of diffusion in one direction only, say, in the direction of the x-axis, and for a dilute solution, we can write using (4)

$$\Phi = -Uc \frac{\partial \mu}{\partial x} = -Uc \, RT \frac{\partial \ln c}{\partial x} \tag{6}$$

and since, from differential calculus,

$$\frac{d \ln y}{dx} = \frac{1}{y} \frac{dy}{dx}$$

$$\Phi = -RTU \frac{\partial c}{\partial x} \tag{7}$$

Flux Φ is the number of moles (n) transported per unit area (A) per unit time

$$\Phi = \frac{1}{A} \frac{dn}{dt}$$

The expression RTU may be called the diffusion coefficient D and hence

★
$$\frac{1}{A} \frac{dn}{dt} = -D \frac{\partial c}{\partial x} \qquad (8)$$

This equation represents Fick's first law of diffusion defined by the German physiologist A. E. Fick (1829–1901) as an analogy to another famous formal equation, Fourier's law, describing the flow of heat.

If the process of diffusion is not restricted to one direction, Fick's law becomes

$$\frac{1}{A} \frac{dn}{dt} = -D \ \text{grad} \ c = -D \left(i \frac{\partial c}{\partial x} + j \frac{\partial c}{\partial y} + k \frac{\partial c}{\partial z} \right) \qquad (9)$$

A *physical interpretation* of the diffusion coefficient for solute particles that are large in comparison with molecules of the solvent and spherical in shape was presented by Einstein (1908):

$$D = \frac{RT}{N} \frac{1}{6\pi\eta r} \qquad (10)$$

where N is Avogadro's number, η the viscosity of the solvent, and r the radius of the particle. If the diffusion coefficient for a substance with molecules which are very large and roughly spherical is experimentally determined, the radius of the molecules may be approximately deduced from equation (10).

Equation (10) may be derived as follows. A particle moving at a moderate speed under the influence of an external driving force F in a viscous medium experiences a resistance force, fv, where v is the velocity of the particle and f is the "frictional force" or "frictional resistance." A steady condition is reached when the driving force and the resistance force acting on the particle are equal ($F = fv$) and the particle moves with a constant velocity

$$v = \frac{F}{f} \qquad (11)$$

For a spherical particle which is very large as compared with the molecules of the viscous liquid, the "frictional force" is defined by Stokes' formula

$$f = 6\pi\eta r \qquad (12)$$

and the velocity of the particle will be

$$v = \frac{F}{6\pi\eta r} \qquad (13)$$

The force acting on a single particle is equal to the force acting on one mole of the particles (called "total driving force" in Teorell's equation (2)), divided by Avogadro's number N

$$F = \frac{\text{Total driving force}}{N} \qquad (14)$$

Moreover, the velocity of particles multiplied by the molar concentration of the substance gives the flux of the substance in moles per unit area per unit time

$$\text{Flux} = vc \qquad (15)$$

(the amount of substance penetrating across a unit area per unit time will be contained in a cylinder with a base of unit area and a height equal to v).
On combining (13), (14), and (15) we obtain

$$\text{Flux} = \frac{1}{N\,6\pi\eta r}\, c\,\text{Total driving force} \qquad (16)$$

By comparison of (16) with (2) one can see that mobility of large and spherical particles can be expressed by

$$U = \frac{1}{N\,6\pi\eta r} \qquad (17)$$

If the diffusion coefficient D is defined as RTU (as in equation (8)), the formula (10), called sometimes the Einstein–Stokes equation, follows immediately.

It may be mentioned that concepts such as velocity of particles diffusing in a gradient of chemical potential are only useful fictions, for the process of diffusion is due to random thermal movements of particles and "...in a pure diffusion in a simple solution, no molecule has any finite average velocity in any preferred direction" (Hartley and Crank, 1949). Net flows by diffusion are due to differences in the numbers of particles diffusing in opposite directions and not to differences in their average velocities. The use of such fiction is justified by the previously discussed equivalence of the process of diffusion and of the process of migration, to which these concepts apply without theoretical objections.

Another limiting case of the relationship between the size of solute and solvent particles, i.e., when both are of the same size, has been theoretically solved by Eyring (1936) from the point of view of the theory of absolute reaction rates. According to this approach the diffusion of a particle is achieved by a series of successive jumps between holes in a liquid lattice. The distance between two successive energy minima separated by an energy barrier is λ. If the concentration changes in the direction of the x-axis only, the concentration at one minimum will be c and at the next minimum it will be $c + \lambda(\partial c/\partial x)$. Flux in the positive direction will be $\lambda k_1 c$, in the opposite direction $\lambda k_1[c + \lambda(\partial c/\partial x)]$ so that there will be a net flux of

$$\lambda^2 k_1 \frac{\partial c}{\partial x} = D \frac{\partial c}{\partial x} \tag{18}$$

in the negative direction. This net flux will be compensated by a corresponding flux of other molecules in the opposite direction and, according to the assumption, these will be of the same size. The theory is best suited for describing self-diffusion of labeled molecules of a liquid. Then the absolute rate for the transition k_1 may be derived from the formula for viscosity. An approximate formula for viscosity η follows from the theory of absolute reaction rates (Eyring, 1936):

$$\eta = \lambda_1 \frac{RT}{N} (\lambda^2 \lambda_2 \lambda_3 k_1)^{-1} \tag{19}$$

where λ_1 is the perpendicular distance between two neighboring layers of molecules sliding past each other, λ_2 is the distance between neighboring molecules in the direction of motion, and λ_3 the analogous distance in the direction normal to the other two.

From (18) it follows that

$$D = \lambda^2 k_1 \tag{20}$$

and on combining (19) and (20) we obtain

$$D = \frac{RT}{N} \frac{\lambda_1}{\lambda_2 \lambda_3 \eta} \tag{21}$$

The theory was very successfully applied to the diffusion of D_2O into H_2O (Eyring, 1936).

Mathematical consequences of Fick's first law, such as spatial distribution and temporary changes of concentrations due to pure diffusion, may be found as solutions of the partial differential equation known as Fick's second law. This law may be derived for the simple case of a unidirectional concentration gradient (in the direction of the x-axis) from Fick's first law as follows:

Let us consider two parallel planes, one at the distance x, the other at the distance $x + dx$. If the concentration gradient at the first one is $\partial c / \partial x$, the concentration gradient at the other will be

$$\frac{\partial c}{\partial x} + \frac{\partial(\partial c / \partial x)}{\partial x} \, dx = \frac{\partial c}{\partial x} + \frac{\partial^2 c}{\partial x^2} \, dx$$

Consequently, the flux across the first plane will be (eq. (8))

$$\frac{1}{A} \frac{dn}{dt} = -D \frac{\partial c}{\partial x}$$

whereas across the second it will be

$$\frac{1}{A} \frac{dn'}{dt} = -D \left(\frac{\partial c}{\partial x} + \frac{\partial^2 c}{\partial x^2} \, dx \right)$$

The rate of change of the number of moles in a small cylinder with base of unit area on both planes will then be

$$\frac{1}{A} \frac{dn - dn'}{dt} = D \frac{\partial^2 c}{\partial x^2} \, dx$$

Now, $A \, dx$ is the volume of the small cylinder and, therefore,

$$\frac{1}{A \, dx} \frac{dn - dn'}{dt}$$

is the rate of change of concentration $\partial c / \partial t$.

★
$$\frac{\partial c}{\partial t} = D \frac{\partial^2 c}{\partial x^2} \tag{22a}$$

is Fick's second law for unidirectional diffusion. A more general expression in Cartesian coordinates

$$\frac{\partial c}{\partial t} = D \left(\frac{\partial^2 c}{\partial x^2} + \frac{\partial^2 c}{\partial y^2} + \frac{\partial^2 c}{\partial z^2} \right) \tag{22b}$$

may be derived analogously. In other systems of coordinates Fick's second law (also called the *diffusion equation*) will take still other forms but for all of them the general formula

$$\frac{\partial c}{\partial t} = \Delta c = \nabla^2 c \qquad (22c)$$

can be used, where Δ or ∇^2 is the Laplacian operator. With appropriate initial and boundary conditions and provided that the diffusion coefficient D is considered as a constant the diffusion equation can be solved and functions like $c = f(x, y, z, t)$ can be found.

Solutions of the diffusion equation for a number of different cases may be found in Crank (1956) or in treatises on heat conduction, where analogous problems are encountered. Several important solutions of the diffusion equation were published by Hill (1928) and some of them are presented here in the way of illustration.

1. *Diffusion from or into a plane sheet.* A plane sheet (theoretically infinite, in practice so large that the surface of its edges is negligible in comparison with the surfaces of the two parallel planes) of thickness d is exposed at $t = 0$ on both sides to concentration c_o (of, for instance, a labeled substance, whereas originally it was equilibrated with the unlabeled one). The appropriate form of the diffusion equation is (22a); x is the distance from one of the planes. The solution is an infinite series

$$c = c_o\left[1 - \frac{4}{\pi}\left(e^{-D\pi^2 t/d^2}\sin\frac{\pi x}{d} + \frac{1}{3}e^{-9D\pi^2 t/d^2}\sin\frac{3\pi x}{d}\right.\right.$$
$$\left.\left. + \frac{1}{5}e^{-25D\pi^2 t/d^2}\sin\frac{5\pi x}{d} + \cdots\right)\right] \qquad (23)$$

where D is the diffusion coefficient and t time. The series converges rapidly; when $e^{-D\pi^2 t/d^2}$ is 0.6, the second term $e^{-9D\pi^2 t/d^2}$ is 0.01 and may be therefore neglected together with all the subsequent terms for any t greater than that yielding $e^{-9D\pi^2 t/d^2}$ less than, say, 0.6. When a fixed value is assigned to t, eq. (23) describes the spatial distribution of c at that time; if x is fixed, it describes the temporal variation at the point x.

Fractional equilibration, i.e., the ratio of the amount of labeled substance m_t present in the sheet at time t to the amount m_o present when full equilibrium has been reached may then be expressed as follows:

$$\bigstar \quad \frac{m_t}{m_o} = \frac{\int_0^d c\,dx}{c_o d} = 1 - \frac{8}{\pi^2}\left(e^{-D\pi^2 t/d^2} + \frac{1}{9}e^{-9D\pi^2 t/d^2} + \frac{1}{25}e^{-25D\pi^2 t/d^2} + \cdots\right)$$
$$(24a)$$

The opposite process (i.e., the disappearance of the unlabeled substance in the above example) will be described by

$$1 - \frac{m_t}{m_o} = \frac{8}{\pi^2} \left(e^{-D\pi^2 t/d^2} + \frac{1}{9} e^{-9D\pi^2 t/d^2} + \frac{1}{25} e^{-25D\pi^2 t/d^2} + \cdots \right) \quad (24b)$$

Again, for longer times all the terms except the first may be neglected and the process will be described with sufficient accuracy by a single exponential term.

The half-time $t_{0.5}$ of diffusion from or into a plane sheet may be expressed using the equation calculated by Dainty and House (1966a):

$$t_{0.5} = \frac{0.38 \, d^2}{D} \quad (25)$$

2. The formula describing *fractional equilibration of a cylinder* of radius r is

★ $$\frac{m_t}{m_o} = 1 - 4 \left(\frac{1}{\mu_1^2} e^{-\mu_1^2 Dt/r^2} + \frac{1}{\mu_2^2} e^{-\mu_2^2 Dt/r^2} + \cdots \right) \quad (26)$$

where μ's are the zeros of the Bessel function J_0 (the values of x for which the Bessel function $J_0(x)$ becomes zero). The first five of these roots have values: $\mu_1 = 2.4048$, $\mu_2 = 5.5201$, $\mu_3 = 8.6537$, $\mu_4 = 11.7915$, and $\mu_5 = 14.9309$.

3. Finally, *fractional equilibration of a sphere* of radius r is expressed by

★ $$\frac{m_t}{m_o} = 1 - \frac{6}{\pi^2} \left(e^{-D\pi^2 t/r^2} + \frac{1}{4} e^{-4D\pi^2 t/r^2} + \frac{1}{9} e^{-9D\pi^2 t/r^2} + \cdots \right) \quad (27)$$

Equations (24a), (26), and (27) may be used for the description of diffusion of various substances into spaces of appropriate shape exposed to constant concentrations of these substances, provided that these spaces are not surrounded by rate-limiting membranes. The presence of such membranes would actually simplify the mathematical description and the appropriate formulae will be derived in the section on membrane processes (p. 157). Equations for diffusion equilibration are thus suitable for describing diffusion into extracellular spaces or in situations where substances would permeate through membranes so rapidly that not the permeation across the cell membrane but the diffusion in the protoplasm will be rate-limiting. It should be also stressed that there are always unstirred layers present at both sides of the membrane and the greater the ease with which the substance permeates through the membrane, the greater the importance of diffusion across the unstirred layers in the overall process (Dainty, 1963; Ginzburg and Katchalsky, 1963; Dainty and House, 1966a). The role of unstirred layers will be discussed again in connection with membrane processes (p. 116).

The diffusion coefficients D most often encountered are of the order of 10^{-5} cm$^2 \cdot$sec^{-1} (e.g., 2.5×10^{-5} for the self-diffusion of water and 0.5×10^{-5} for sucrose). The diffusion coefficient for inulin with molecular weight of about 5100 daltons is 0.15×10^{-5} cm$^2 \cdot$sec^{-1} (Villegas, 1963). The self-diffusion coefficient for the sodium ion in 0.1 M NaCl at 25°C is 1.48×10^{-5} cm$^2 \cdot$sec^{-1} whereas in the corium of the frog skin it is reduced to about 0.3×10^{-5} cm$^2 \cdot$sec^{-1} (Winn et al., 1964) or 0.4×10^{-5} cm$^2 \cdot$sec^{-1} (Hoshiko et al., 1964).

Some of the complications brought about by mass flow will be considered. The first is the fact that diffusion coefficients are modified for cases where a net flow of substances takes place, i.e., where diffusing molecules of solute are being replaced not by analogous solute molecules as in self-diffusion but by molecules of the solvent and *vice versa*. Hartley and Crank (1949) showed that "in a binary solution, the net rate of transfer of either component is the result of a transfer by pure diffusion coupled with a transfer of that component due to a mass flow of the whole solution. The mass flow is due to the fact that, in general, the intrinsic rates of pure diffusion of the two components will differ and hence there is a tendency to set up a hydrostatic pressure in the solution which is relieved by a mass flow." If one examines the diffusion across a surface with no accompanying net transfer of volume,

$$D_s' \bar{V}_s \frac{\partial c_s}{\partial x} + D_w' \bar{V}_w \frac{\partial c_w}{\partial x} = 0 \tag{28}$$

applies, expressing the fact that the transfer of volume across a unit surface due to solute movement plus the transfer due to solvent movement equals zero. ($\bar{V}_s$ and $\bar{V}_w$ are the partial molal volumes of solute and of solvent, respectively.) In a binary solution, only molecules of solute and of solvent are present and the unit volume of solution may be expressed as

$$\bar{V}_s c_s + \bar{V}_w c_w = 1 \tag{29}$$

which, on differentiation with respect to x, gives

$$\bar{V}_s \frac{\partial c_s}{\partial x} + \bar{V}_w \frac{\partial c_w}{\partial x} = 0 \tag{30}$$

Comparing (*28*) with (*30*) we see that

$$D_s' = D_w'$$

i.e., the apparent diffusion coefficients D' (unlike the intrinsic ones, measured in the processes of self-diffusion) will be equal. The interdiffusion of two components may thus be expressed by a single diffusion coefficient depending on the conditions of the experiment and called the *mutual diffusion coefficient*. This situation obtains in experiments where changes of concentration of substances are determined by chemical analysis; when no net flows occur and when the movement of substances is followed by isotopic labeling, laws of pure diffusion and self-diffusion coefficients are relevant.

Finally, an example of interaction of diffusion with mass flow of known magnitude may be given. Let us consider a sheet of thickness d across which a volume flow per unit area of

$$\frac{1}{A} \frac{dV}{dt} = v$$

takes place (Fig. 2.1). This may occur in a layer of connective tissue across which an osmotic flow takes place (Natochin *et al.*, 1965). The sheet is exposed from one side (at $x = d$) to the solution of some substance of concentration c_o so that the substance diffuses into the sheet against the volume flow. The amount of substance diffusing per unit area to the left (in the negative direction of the x-axis) will, for any value of x, be equal

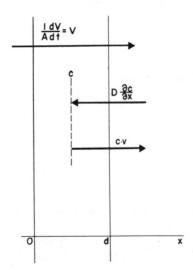

FIG. 2.1. Diffusion against a solvent flow.
For explanation see the text.

(by eq. (8)) to

$$\frac{dn}{dt} = D \frac{\partial c}{\partial x} \qquad (31)$$

According to eq. (1), the amount carried by the mass flow in the opposite direction is equal to

$$\frac{dn}{dt} = c \frac{dV}{dt} = cv \qquad (32)$$

A steady state will be reached when the two become equal

$$D \frac{\partial c}{\partial x} = cv \qquad (33)$$

Solution of the differential equation (33) for the boundary condition $c = c_o$ at $x = d$ gives the concentration profile of the substance:

$$c = c_o \, e^{v(x-d)/D} \qquad (34)$$

The ratio of the amount of the substance present inside the sheet to that which would be present in the absence of volume flow is then

$$r = \frac{\int_0^d c \, dx}{c_o d} = \frac{D}{vd} \, (1 - e^{-vd/D}) \qquad (35)$$

2.2. MIGRATION AND ELECTRODIFFUSION OF IONS

Another kind of transport becomes possible when the transported particles are electrically charged because the movement of such particles is subject to electrical forces. Biological fluids abound in charged particles, originating by dissociation of salts, acids, bases, and other molecules. It should be recalled that the existence of these ions freely wandering in solution does not depend on the presence of external electrical fields; their free mobility is due to the fact that water (or some other solvent of high dielectric constant) weakened the electrical forces previously holding the ions firmly in crystals or molecules. For this reason, however weak the external electrical field applied to a salt solution through electrodes may be, current will flow between them showing that mobile charged particles are present. The names of electrodes are defined according to the direction in which the current flows; the positively charged cations are attracted toward the cathode, the negative anions toward the anode. The current thus flows in the solution from the anode to the cathode, in the external circuit in the opposite direction. In a homogeneous field, i.e., a field of equal intensity at every point, only particles carrying a net charge are transported, whereas molecules having their positive and negative charges spatially separated but equal in magnitude (i.e., dipoles like amino acids at their isoelectric point) are merely aligned along the direction of such an uniform field. However, it should be remembered that in a nonuniform field also particles with no net charge can be transported; even if the separated charges are equal, the forces exerted on them may differ. In this section we shall describe first the movement of ions in a solution without concentration gradients, i.e., the so-called migration in an electrical field, to be followed by a consideration of transport due to the combined action of an electrical field and of concentration gradients, the electrodiffusion.

For describing the effects of an electrical field on the transport of ions, information on the electrical field intensity E at every point of the space in question is useful. The electrical field intensity at any point is the force which would act on a unit positive charge at this point and hence represents a vectorial quantity. Another description of an electrical field is obtained if at every point a scalar quantity, the electrical potential ψ, is given. The difference between the electrical potential of two points is the work performed when a unit charge is transferred between these points. According to this definition the electrical potential is known up to an arbitrary additive constant. The value of this constant is usually fixed by ascribing a zero value to the electrical potential either at an infinitely distant point (in theoretical calculations) or to earth (in practice). In most cases, however, it is only the difference of the electrical potential between two points that is important. The electrical field intensity is related to the electrical potential as follows: The electrical field intensity is the negative of the gradient of electrical potential:

$$E = -\operatorname{grad} \psi \tag{36}$$

If the change of the potential is considered in one direction only, say, in the direction of the x-axis, a derivative will take the place of the gradient:

$$E = -\frac{\partial \psi}{\partial x} \tag{37}$$

The charge carried by an ion is equal to ze. Here e is the elementary charge, the smallest amount of electricity obtainable. The negative charge of this size is carried by an electron, the positive charges e are the charges of protons and positrons. The value of e is 4.803×10^{-10} of absolute electrostatic units (esu) or 1.602×10^{-19} coulombs (C). It may be useful to remember that it is the ratio of the Faraday constant F (the charge of a gram equivalent of an ion) which is 96487 $C \cdot mol^{-1}$ to the Avogadro constant N which is approximately 6.02×10^{23} mol^{-1}; z is a small integer provided with a sign, indicating the number of elementary charges carried by the ion. The driving force acting on an ion of charge ze in a solution will be

$$F = zeE \tag{38}$$

where E is the electrical field intensity. The ion will be accelerated by this force but proportionally to its velocity the resistance force fv will increase. (Here, as on p. 31, f is the "frictional resistance" and v the velocity.) Very rapidly, the ion will attain a constant velocity of migration

$$v = \frac{zeE}{f} \tag{39}$$

If Stokes' formula is used (as on p. 31 and with the same limitations) to express the frictional resistance we have

$$v = \frac{zeE}{6\pi\eta r} \tag{40}$$

By multiplying the numerator as well as the denominator by Avogadro's number N, the Faraday constant F will appear in the numerator instead of the elementary charge e:

$$v = \frac{zFE}{N\,6\pi\eta r} \tag{41}$$

Remembering that the flux Φ of particles is equal to their velocity v multiplied by their concentration c (see p. 32) we obtain

$$\Phi = c\,\frac{1}{N\,6\pi\eta r}\,zFE \tag{42}$$

In the term $1/N6\pi\eta r$ we recognize immediately the mobility U from Teorell's equation (17) and finally, using relation $E = -\,\text{grad}\,\psi$, we may write

$$\Phi = -cUzF\,\text{grad}\,\psi \tag{43}$$

If changes of electrical potential in only one direction (say, in the direction of the x-axis) are considered, a derivative will take the place of the gradient:

$$\Phi = -cUzF\frac{\partial \psi}{\partial x} \tag{44}$$

Equations (43) or (44) relate the flux of an ion at any point to the change of electrical potential in the proximity of this point; if no other driving force is present they describe the situation completely. If, however, not only a gradient of electrical potential but also a gradient of chemical potential is present, a flux due to diffusion will be superimposed on the flux due to migration. The resulting process of electrodiffusion will be described by the algebraic sum of the expression for diffusion flow (5) with equation (43)

$$\Phi = -cU \, \text{grad}(\mu + zF\psi) = -cU \, \text{grad} \, \bar{\mu} \tag{45}$$

where $\bar{\mu}$ is the electrochemical potential $\bar{\mu} = \mu + zF\psi$. For a unidirectional case and dilute solutions (so that the approximation (4) $\mu = \mu_0 + RT \ln c$ is applicable) equations (6) and (44) will be summed:

$$\Phi = -cU\left(RT\frac{\partial \ln c}{\partial x} + zF\frac{\partial \psi}{\partial x}\right) \tag{46}$$

An important note concerning the mobility U from Teorell's equation should be inserted here. Mobility is defined as the velocity acquired due to a unit force. The natural units of the driving forces acting on a mole of a substance (i.e., the gradients of chemical potential, of the electrical potential multiplied by the charge of one mole of solute, or of their sum, the so-called electrochemical potential) are joule·mol^{-1}·cm^{-1}. The dimension of mobility U is therefore cm·sec^{-1}/joule·mol^{-1}·cm^{-1} = cm^2·sec^{-1} × joule^{-1}·mol. In electrochemistry, however, the gradient of the electrical potential in volts/cm is usually considered as the driving force and hence the dimension of mobility u thus defined is cm^2·sec^{-1}·volt^{-1}. Any one of these two mobilities may be easily calculated from the other, the relation between them being

$$u = zFU \tag{47}$$

Mobility U has been called the "diffusion mobility" and mobility u the "electrical mobility" by Spiegler and Wyllie (1956).

Mobility u was already seen in eq. (39) where $u = z/f$. Both mobilities will be used in the following derivations: u will be considered as a magnitude provided with a sign (+ for cations and − for anions), whereas U will always be a positive number.

Using the equality $d \ln y = dy/y$ (from differential calculus) eq. (46) may be written as

★
$$\Phi = -RTU\frac{\partial c}{\partial x} - zFcU\frac{\partial \psi}{\partial x} \qquad (48)$$

which may serve as a basic equation for further theoretical calculations. Equation (48) is a differential equation, describing the situation at a point and including values which are not accessible to direct experimental measurement—the derivatives of concentration and of electrical potential at the point. To relate quantities which are directly measurable, like the potential difference between two planes, the distance between which is finite (Fig. 2.2), to the differences of concentrations of the ion species present, as exist between these planes, or to derive an expression for the flux of a single ion species as a function of its concentration difference and of the potential difference between such planes, eq. (48) must be integrated. For carrying out the uniquely correct integration for a given case one would have to assume that the course of the change of electrical potential as well as the concentration profiles of the ions present across the region between the planes are known. This is ordinarily not the case and thus various simplifying assumptions must be made and equations obtained, which are amenable to experimental verification.

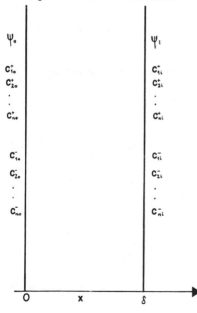

FIG. 2.2. Two planes at a finite distance between two solutions. Between the two planes there is either no mixing, only diffusion taking place, as corresponds to the assumptions of Planck (such a situation may obtain on using a porous plug of inert material), or else the solution is mixed, as corresponds to the assumptions of Henderson. ψ Electrical potential, c_j^+ concentration of cations, c_j^- concentration of anions. The distance between the two planes, called arbitrarily the "outer" o and "inner" i planes, is δ.

Planck's Solution

One of the oldest integrations of eq. (*48*) and perhaps the most rigorous one is due to Planck. In its derivation, electroneutrality is assumed to prevail throughout the whole region (microscopic electroneutrality), but no special attention is paid to concentration and potential profiles. However, even this solution has an approximative character, ionic concentrations rather than activities being used already in the basic equation, from which the derivation proceeds. The derivation of Planck's equation (Planck, 1890) which was published by MacInnes (1961) in more detail will be shown here, using the symbols employed in this book.

The amount of ions of a given species carried across a unit area at the coordinate x during time Δt is $\Phi \Delta t$, whereas at $x + \Delta x$ it is

$$\left(\Phi + \frac{\partial \Phi}{\partial x} \Delta x \right) \Delta t$$

Accumulation of ions between the two unit areas at x and $x + \Delta x$ during the time Δt is thus given by

$$- \frac{\partial \Phi}{\partial x} \Delta x \, \Delta t$$

and the change in ion concentration between x and $x + \Delta x$ is obtained by dividing accumulation by the corresponding volume (Δx cm^3 due to the unit cross section)

$$\Delta c = - \frac{\partial \Phi}{\partial x} \Delta t$$

Hence, in the limit the rate of the concentration change is given by

$$\frac{\partial c}{\partial t} = - \frac{\partial \Phi}{\partial x} \tag{49}$$

which is the equation of continuity, expressing the fact that no diffusing ions are lost; a spatial change in their flux appears as a temporal change in their concentration. If it is assumed that the mobilities U are independent of concentrations, expressions for flux of the type of eq. (*48*) may be inserted into eq. (*49*) and we can write

$$\frac{\partial c_j^+}{\partial t} = U_j^+ \left[RT \frac{\partial^2 c_j^+}{\partial x^2} + F \frac{\partial}{\partial x} \left(c_j^+ \frac{\partial \psi}{\partial x} \right) \right] \tag{50}$$

$$\frac{\partial c_j^-}{\partial t} = U_j^- \left[RT \frac{\partial^2 c_j^-}{\partial x^2} - F \frac{\partial}{\partial x} \left(c_j^- \frac{\partial \psi}{\partial x} \right) \right] \tag{51}$$

for the rate of change of a univalent cation concentration (c_j^+) and a univalent anion concentration (c_j^-), respectively. Here U_j^+ is the mobility of the j-th cation and U_j^- of the j-th anion.

In view of the prevailing electroneutrality

$$\sum_{j=1}^{n} c_j^+ = \sum_{j=1}^{n} c_j^- = c \tag{52}$$

is valid at any point, c being the "total concentration." Hence

$$\sum_{j=1}^{n} \frac{\partial c_j^+}{\partial t} - \sum_{j=1}^{n} \frac{\partial c_j^-}{\partial t} = 0 \tag{53}$$

By substituting (50) and (51) into (53) we obtain

$$\sum_{j=1}^{n} U_j^+ RT \frac{\partial^2 c_j^+}{\partial x^2} + \sum_{j=1}^{n} U_j^+ F \frac{\partial}{\partial x}\left(c_j^+ \cdot \frac{\partial \psi}{\partial x}\right) - \sum_{j=1}^{n} U_j^- RT \frac{\partial^2 c_j^-}{\partial x^2}$$

$$+ \sum_{j=1}^{n} U_j^- F \frac{\partial}{\partial x}\left(c_j^- \frac{\partial \psi}{\partial x}\right) = 0 \tag{54}$$

which, on integration, gives

$$\sum_{j=1}^{n} U_j^+ RT \frac{\partial c_j^+}{\partial x} + \sum_{j=1}^{n} U_j^+ Fc_j^+ \frac{\partial \psi}{\partial x} - \sum_{j=1}^{n} U_j^- RT \frac{\partial c_j^-}{\partial x}$$

$$+ \sum_{j=1}^{n} U_j^- Fc_j^- \frac{\partial \psi}{\partial x} = 0 \tag{55}$$

Setting

$$\sum_{j=1}^{n} U_j^+ c_j^+ = U \qquad \sum_{j=1}^{n} U_j^- c_j^- = V \tag{56}$$

eq. (55) may be solved for $\partial \psi / \partial x$

$$\frac{\partial \psi}{\partial x} = -\frac{RT}{F} \frac{\dfrac{\partial(U - V)}{\partial x}}{U + V} \tag{57}$$

This equation cannot be integrated at once, the dependence of the c's on x not being known. When, however, a steady state is approximated, the time derivatives in equations (50) and (51) become negligibly small, so that equations

$$RT \frac{\partial^2 c_j^+}{\partial x^2} + F \frac{\partial}{\partial x}\left(c_j^+ \frac{\partial \psi}{\partial x}\right) = 0 \tag{58a}$$

and

$$RT \frac{\partial^2 c_j^-}{\partial x^2} - F \frac{\partial}{\partial x} \left(c_j^- \frac{\partial \psi}{\partial x} \right) = 0 \tag{58b}$$

are valid from $x = 0$ to $x = \delta$. Their integration gives

$$RT \frac{\partial c_j^+}{\partial x} + F c_j^+ \frac{\partial \psi}{\partial x} = A_j \tag{59a}$$

and

$$RT \frac{\partial c_j^-}{\partial x} - F c_j^- \frac{\partial \psi}{\partial x} = B_j \tag{59b}$$

where A_j and B_j are integration constants.

Setting

$$\sum_{j=1}^{n} A_j = \mathbf{A} \qquad \sum_{j=1}^{n} B_j = \mathbf{B}$$

the sums of equations of type (59a) and (59b) can be written, respectively,

$$RT \frac{\partial c}{\partial x} + Fc \frac{\partial \psi}{\partial x} = \mathbf{A} \tag{60a}$$

and

$$RT \frac{\partial c}{\partial x} - Fc \frac{\partial \psi}{\partial x} = \mathbf{B} \tag{60b}$$

where c is the total concentration defined by eq. (52). Addition of (60a) and (60b) yields

$$2RT \frac{\partial c}{\partial x} = \mathbf{A} + \mathbf{B} \tag{61}$$

which integrates to

$$2RTc = (\mathbf{A} + \mathbf{B})x + \text{const}$$

i.e., the *total* concentration is a linear function of the coordinate x.

Since $c = c_o$ at $x = 0$ and $c = c_i$ at $x = \delta$ this function may be seen to be

$$c = \frac{c_i - c_o}{\delta} x + c_o \tag{63}$$

Subtracting eq. (60b) from eq. (60a)

$$2Fc \frac{\partial \psi}{\partial x} = \mathbf{A} - \mathbf{B} \tag{64}$$

and inserting (63) into (64), yields

$$\frac{\partial \psi}{\partial x} = \frac{(\mathbf{A} - \mathbf{B})\, \delta}{2F[(c_i - c_o)x + c_o\delta]} \tag{65}$$

Integrating from $x = 0$ to $x = \delta$ we obtain

$$\psi_i - \psi_o = \frac{(\mathbf{A} - \mathbf{B})\, \delta}{2F(c_i - c_o)} \ln \frac{c_i}{c_o} \tag{66}$$

If we define ξ as

$$\xi = \left(\frac{c_i}{c_o} \right)^{\frac{(\mathbf{A}-\mathbf{B})\delta}{2(c_i - c_o)RT}} \tag{67}$$

eq. (66) becomes

$$\psi_i - \psi_o = \frac{RT}{F} \ln \xi \tag{68}$$

Multiplication of equations of type ($59a$) and ($59b$) by the corresponding mobilities and summation yields

$$\sum_{j=1}^{n} U_j^{+} A_j = RT \frac{\partial \mathbf{U}}{\partial x} + F\mathbf{U} \frac{\partial \psi}{\partial x} \tag{69}$$

$$\sum_{j=1}^{n} U_j^{-} B_j = RT \frac{\partial \mathbf{V}}{\partial x} - F\mathbf{V} \frac{\partial \psi}{\partial x} \tag{70}$$

where $\mathbf{U}$ and $\mathbf{V}$ are quantities defined by equations (56). By inserting the expression for $\partial \psi / \partial x$ from eq. (57) into (69) and (70) it may be easily shown that their right-hand sides are equal and hence the same is true for their left-hand sides

$$\sum_{j=1}^{n} U_j^{+} A_j = \sum_{j=1}^{n} U_j^{-} B_j = \mathbf{C} \tag{71}$$

this being a constant sum of integration constants, multiplied by presumably constant mobilities.

Combination of (71) and (65) with (69) and (70), respectively, results in

$$\frac{\partial \mathbf{U}}{\partial x} + \mathbf{U} \frac{(\mathbf{A} - \mathbf{B})\, \delta}{2RT[(c_i - c_o)x + c_o\delta]} = \frac{\mathbf{C}}{RT} \tag{72}$$

$$\frac{\partial \mathbf{V}}{\partial x} - \mathbf{V} \frac{(\mathbf{A} - \mathbf{B})\, \delta}{2RT[(c_i - c_o)x + c_o\delta]} = \frac{\mathbf{C}}{RT} \tag{73}$$

These are first-order linear equations, i.e., they are of the form

$$\frac{dy}{dx} + f(x)\, y = g(x)$$

with the solution

$$y = e^{-\int f(x)dx} \left[\int e^{\int f(x)dx}\, g(x)\, dx + \text{const} \right]$$

Hence

$$\mathbf{U} = \frac{2\mathbf{C}[(c_i - c_o)x + c_o\delta]}{2(c_i - c_o)\, RT + (\mathbf{A} - \mathbf{B})\, \delta} + [(c_i - c_o)x + c_o\delta]^{-\frac{(\mathbf{A}-\mathbf{B})\delta}{2(c_i-c_o)RT}} \times \text{const.}$$

and

$$\mathbf{V} = \frac{2\mathbf{C}[(c_i - c_o)x + c_o\delta]}{2(c_i - c_o)\, RT + (\mathbf{A} - \mathbf{B})\, \delta} + [(c_i - c_o)x + c_o\delta]^{\frac{(\mathbf{A}-\mathbf{B})\delta}{2(c_i-c_o)RT}} \times \text{const.}$$

If $x = 0$, then $\mathbf{U} = \mathbf{U}_o$ and $\mathbf{V} = \mathbf{V}_o$ and if $x = \delta$, then $\mathbf{U} = \mathbf{U}_i$ and $\mathbf{V} = \mathbf{V}_i$, so that the last equations may be written in pairs for both boundaries, and the constants of the last integration may be eliminated from each pair. Using ξ as defined by eq. (67) we may write

$$\xi \mathbf{U}_i - \mathbf{U}_o = \frac{2\mathbf{C}\,\delta(\xi c_i - c_o)}{2(c_i - c_o)\, RT + (\mathbf{A} - \mathbf{B})\, \delta} \tag{74}$$

and

$$\mathbf{V}_i - \xi \mathbf{V}_o = \frac{2\mathbf{C}\,\delta(c_i - \xi c_o)}{2(c_i - c_o)\, RT - (\mathbf{A} - \mathbf{B})\, \delta} \tag{75}$$

By dividing (74) by (75) constant $\mathbf{C}$ is eliminated

$$\frac{\xi \mathbf{U}_i - \mathbf{U}_o}{\mathbf{V}_i - \xi \mathbf{V}_o} = \frac{2(c_i - c_o)\, RT - (\mathbf{A} - \mathbf{B})\, \delta}{2(c_i - c_o)\, RT + (\mathbf{A} - \mathbf{B})\, \delta} \frac{\xi c_i - c_o}{c_i - \xi c_o}$$

Finaly, substituting for $(\mathbf{A} - \mathbf{B})$ from eq. (67), we obtain

★
$$\frac{\xi \mathbf{U}_i - \mathbf{U}_o}{\mathbf{V}_i - \xi \mathbf{V}_o} = \frac{\ln \dfrac{c_i}{c_o} - \ln \xi}{\ln \dfrac{c_i}{c_o} + \ln \xi} \frac{\xi c_i - c_o}{c_i - \xi c_o} \tag{76}$$

This is Planck's equation and it contains ξ in a transcendent form. If ξ is found by trial and error or graphically for c_i, c_o, $\mathbf{U}_o$, $\mathbf{V}_o$, $\mathbf{U}_i$, and $\mathbf{V}_i$,

the potential difference between the two planes in Fig. 2.2. can be calculated
from eq. (68)

★
$$\psi_i - \psi_o = \frac{RT}{F} \ln \xi$$

The graphical method consists of plotting the left-hand side as well as the
right-hand side of eq. (76) for given values of c_i, c_o, U_o, U_i, V_o, and V_i
against different values of ξ. Two curves are thus obtained and their point
of intersection corresponds to ξ, thus solving the equation for the given
case (see MacInnes, 1961).

Henderson's Solution

Solution of Planck's equation is not easy and attempts to extend this
equation to cases where ions of various valencies are present would en-
counter mathematical difficulties. For this reason, various simplifying assumptions concerning the concentration profiles of individual ions or the
electrical potential profile were sought. Integrations carried out after such
assumptions yield not only explicit formulae for potential differences but
also useful equations relating the flux of an individual ion between two
points to the difference of its concentration and the difference of the electrical potential between these points.

One of these assumptions, which leads to Henderson's equation (Henderson, 1907, 1908; MacInnes, 1961) is that of the linear concentration
gradient of the individual ions. Henderson's equation is applicable if the
solution between the two planes in Fig. 2.2 is being efficiently mixed, for
then the concentrations at any point will approximate a linear mixture of
the solutions at the boundaries. Here a kinetic derivation of Henderson's
equation will be shown.

Because of the linear gradient the concentration of each ion between
$x = 0$ and $x = \delta$ is given by

$$c_j = c_{jo} + \frac{c_{ji} - c_{jo}}{\delta} x \tag{77}$$

and hence

$$\frac{\partial c_j}{\partial x} = \frac{c_{ji} - c_{jo}}{\delta} \tag{78}$$

Eq. (48) for the jth ion may thus be written

★
$$\Phi_j = -RTU_j \frac{c_{ji} - c_{jo}}{\delta} - z_j FU_j \left(c_{jo} + \frac{c_{ji} - c_{jo}}{\delta} x \right) \frac{\partial \psi}{\partial x} \tag{79}$$

If it is now assumed that no electrical current flows between the two planes in Fig. 2.2 we may set the sum of all currents, carried by the individual ions, equal to zero. Current is the flux in moles, multiplied by the charge of one mole; $u = zFU$ is the electrical mobility, discussed on p. 42.

$$\sum_{j=1}^{n} zF\Phi_j = -\frac{RT}{\delta}\left(\sum_{j=1}^{n} u_j c_{ji} - \sum_{j=1}^{n} u_j c_{jo}\right) - F\frac{\partial\psi}{\partial x}\sum_{j=1}^{n} z_j u_j c_{jo}$$
$$- F\frac{\partial\psi}{\partial x}\left(\sum_{j=1}^{n} z_j u_j c_{ji} - \sum_{j=1}^{n} z_j u_j c_{jo}\right)\frac{x}{\delta} \tag{80}$$

which may be written as

$$\partial\psi = -\frac{RT}{F}\frac{1}{\delta}\left(\sum_{j=1}^{n} u_j c_{ji} - \sum_{j=1}^{n} u_j c_{jo}\right)\frac{\partial x}{\displaystyle\sum_{j=1}^{n} z_j u_j c_{jo} + \frac{x}{\delta}\left(\sum_{j=1}^{n} z_j u_j c_{ji} - \sum_{j=1}^{n} z_j u_j c_{jo}\right)}$$

and integrated at once, giving

$$\bigstar \qquad \psi_i - \psi_o = -\frac{RT}{F}\frac{\displaystyle\sum_{j=1}^{n} u_j c_{ji} - \sum_{j=1}^{n} u_j c_{jo}}{\displaystyle\sum_{j=1}^{n} z_j u_j c_{ji} - \sum_{j=1}^{n} z_j u_j c_{jo}}\ln\frac{\displaystyle\sum_{j=1}^{n} z_j u_j c_{ji}}{\displaystyle\sum_{j=1}^{n} z_j u_j c_{jo}} \tag{81}$$

which is Henderson's equation. The slightly different form in which it is usually given (see, for instance, Spiegler and Wyllie, 1956) is due to the fact that here concentrations in mol per unit volume, rather than in gram equivalents per unit volume, are used.

An expression relating the flux of an individual ion to quantities which may be measured experimentally results from rewriting eq. (79) and leaving out the subscript j is

$$\partial\psi = \frac{-\Phi - RTU\dfrac{c_i - c_o}{\delta}}{zFU}\frac{\partial x}{c_o + \dfrac{c_i - c_o}{\delta}x} \tag{82}$$

In a steady state, when $\Phi = $ constant, integration from $x = 0$ to $x = \delta$ gives

$$\Phi = -\frac{RTU}{\delta}(c_i - c_o) - \frac{UzF}{\delta}\frac{c_i - c_o}{\ln c_i - \ln c_o}(\psi_i - \psi_o) \tag{83}$$

If $\psi_i = \psi_o$ eq. (83) reduces to

$$\Phi = -\frac{RTU}{\delta}(c_i - c_o) \tag{84}$$

which is the equation of diffusion in a linear gradient. If, on the other hand, $c_i = c_o$, the first term on the right-hand side is zero, whereas the second term is an undetermined quantity, $0/0$. The concentration c_i may however, be considered as a variable approaching c_o, and L'Hôpital's rule can be applied so that

$$\frac{d(c - c_o)}{d(\ln c - \ln c_o)} = \frac{dc}{d \ln c} = c$$

and eq. (83) reduces to

$$\Phi = - \frac{UzF}{\delta} c(\psi_i - \psi_o) \tag{85}$$

which is the equation of electrical migration in a homogeneous solution.

Goldman's Solution

Another simplifying assumption is due to Goldman (1943). It was introduced for describing electrodiffusion across membranes and the resulting formulae are widely used in membrane biophysics. It is assumed that the potential profile between the two planes in Fig. 2.2 is linear, or, which is the same thing, that the electrical field between them is constant. For this reason Goldman's equations are often called *constant-field equations*. The condition of a constant field may be written as

$$\frac{\partial \psi}{\partial x} = \frac{\psi_i - \psi_o}{\delta} \tag{86}$$

Under steady-state conditions, Φ will be a constant for each ion and concentration will be a function of the coordinate x only, so that ordinary derivatives rather than partial ones may be used. The basic equation (48) will have the form

$$\Phi = - RTU \frac{dc}{dx} - zFcU \frac{\psi_i - \psi_o}{\delta} \tag{87}$$

After separation of the variables the equation may be integrated from $x = 0$ to $x = \delta$

$$\int_0^\delta dx = -RTU \int_{\Phi + zFU \frac{\psi_i - \psi_o}{\delta} c_o}^{\Phi + zFU \frac{\psi_i - \psi_o}{\delta} c_i} \frac{dc}{\Phi + zFU \frac{\psi_i - \psi_o}{\delta} c}$$

giving

$$\delta = \frac{RT\delta}{zF(\psi_i - \psi_o)} \ln \frac{\Phi + zFU \dfrac{\psi_i - \psi_o}{\delta} c_i}{\Phi = zFU \dfrac{\psi_i - \psi_o}{\delta} c_o} \tag{88}$$

By solving (88) for Φ we have

$$\bigstar \qquad \Phi = zFU \frac{\psi_i - \psi_o}{\delta} \frac{c_i - c_o e^{-zF(\psi_i - \psi_o)/RT}}{e^{-zF(\psi_i - \psi_o)/RT} - 1} \tag{89}$$

which is Goldman's equation for flux of an individual ion species. It may be written in several equivalent forms which result from multiplying the numerator as well as the denominator on the right-hand side by

$$e^{zF(\psi_i - \psi_o)/RT}$$

and/or by -1. For $c_i = c_o = c$, eq. (89) reduces to eq. (85), whereas for $\psi_i \to \psi_o$ it may be shown, using L'Hôpital's rule, to reduce to (84). If only univalent ions are present, Goldman's equation for the potential difference $\psi_i - \psi_o$ may be derived easily. For the jth univalent cation, eq. (89) has the form

$$\Phi_j{}^+ = F \frac{\psi_i - \psi_o}{\delta} \frac{U_j{}^+ c_{ji}^+ - U_j{}^+ c_{jo}^+ e^{-F(\psi_i - \psi_o)/RT}}{e^{-F(\psi_i - \psi_o)/RT} - 1} \tag{90}$$

For the j-th univalent anion it may be written as

$$\Phi_j{}^- = -F \frac{\psi_i - \psi_o}{\delta} \frac{U_j{}^- c_{ji}^- - U_j{}^- c_{jo}^- e^{F(\psi_i - \psi_o)/RT}}{e^{F(\psi_i - \psi_o)/RT} - 1} \frac{-e^{-F(\psi_i - \psi_o)/RT}}{-e^{-F(\psi_i - \psi_o)/RT}}$$

$$= F \frac{\psi_i - \psi_o}{\delta} \frac{U_j{}^- c_{ji}^- e^{-F(\psi_i - \psi_o)/RT} - U_j{}^- c_{jo}^-}{e^{-F(\psi_i - \psi_o)/RT} - 1} \tag{91}$$

When no current flows between the two planes the sum of the fluxes of all the cations must be equal to the sum of the fluxes of all the anions, if all the ionic species are univalent:

$$\sum_{j=1}^{n} \Phi_j{}^+ = \sum_{j=1}^{n} \Phi_j{}^-$$

i.e.,

$$F \frac{\psi_i - \psi_o}{\delta} \frac{1}{e^{-F(\psi_i - \psi_o)/RT} - 1} \left(\sum_{j=1}^{n} U_j{}^+ c_{ji}^+ - \sum_{j=1}^{n} U_j{}^+ c_{jo}^+ e^{-F(\psi_i - \psi_o)/RT} \right)$$

$$= F \frac{\psi_i - \psi_o}{\delta} \frac{1}{e^{-F(\psi_i - \psi_o)/RT} - 1} \left(\sum_{j=1}^{n} U_j{}^- c_{ji}^- e^{-F(\psi_i - \psi_o)/RT} - \sum_{j=1}^{n} U_j{}^- c_{jo}^- \right)$$

Hence

$$\sum_{j=1}^{n} U_j^+ c_{ji}^+ + \sum_{j=1}^{n} U_j^- c_{jo}^- = e^{-F(\psi_i - \psi_o)/RT} \left(\sum_{j=1}^{n} U_j^+ c_{jo}^+ + \sum_{j=1}^{n} U_j^- c_{ji}^- \right)$$

and

★
$$\psi_i - \psi_o = \frac{RT}{F} \ln \frac{\sum_{j=1}^{n} U_j^+ c_{jo}^+ + \sum_{j=1}^{n} U_j^- c_{ji}^-}{\sum_{j=1}^{n} U_j^+ c_{ji}^+ + \sum_{j=1}^{n} U_j^- c_{jo}^-}$$
(92)

This is Goldman's equation for the potential difference.

3. PASSIVE MEMBRANE TRANSPORT OF NONELECTROLYTES

3.1. PERMEATION BY SIMPLE DIFFUSION

Transport of nonelectrolytes across cell membranes encompasses any penetration through these membranes by un-ionized substances, whatever the mechanism of the process may be. It will be seen in the following sections that for many, perhaps all, metabolically important substances such as sugars and amino acids efficient mechanisms operate in the cell membranes, transporting such molecules often in a highly specific manner and in most cases against their concentration gradients. However, many of the above substances—as well as a host of other, biochemically little-important, compounds—can enter the cell by paths obeying the laws of simple diffusion. A particular interest for the membrane physiologist is held by un-ionized molecules which seem to penetrate into cells by diffusion across thin lipid layers. This conclusion was reached at the end of the last century by Overton who found that such substances pass across the cell membranes the more rapidly the greater their partition coefficient between oil and water. The correlation between the permeability constant and the partition coefficient was further supported by studies of Collander (see Collander, 1949; Danielli, 1943, 1952; or Stein, 1967; for extensive discussion and references) who compared the rate of permeation of various nonelectrolytes like alcohols, amides, substituted ureas, etc. into the alga *Chara ceratophylla* with their olive oil:water partition coefficients. If the product of the permeability constant P and of the square root of the molecular weight $M^{1/2}$ is plotted against the partition coefficient an even better correlation is

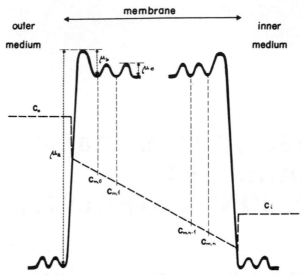

FIG. 3.1. Potential energy diagram of a thin lipid membrane according to Danielli (1943, 1952). The dashed line indicates the concentrations outside (c_o, c_i), as well as inside ($c_{m,0}, \ldots, c_{m,n}$) the membrane.

obtained. For larger molecules, a plot of $PM^{1/3}$ should yield still superior results (Stein, 1967).

Correlations between the rate of the transport process and the equilibrium distribution of the substance in question between oil and water (which is the meaning of the partition coefficient) suggest a new and important factor in the diffusion across thin lipid membranes not present in the diffusion in homogeneous media. This factor is the transition between two phases, occurring at each boundary of the lipid membrane. A rather simple and instructive theory of the diffusion of nonelectrolytes across thin lipid membranes, taking into consideration phenomena at the interfaces, is due to Danielli (1943, 1952).

As may be seen in Fig. 3.1, the theory envisages the membrane as a series of energy barriers, one at each interface between the membrane and the media and a number of them inside the membrane. Considering first the diffusion process inside the membrane, it is obvious that a molecule may diffuse only if it possesses the minimum kinetic energy μ_e necessary to break the hydrogen bonds and to overcome the Van der Waals and other forces restricting diffusion. The net flux of the diffusing substance between the zeroth and the first minimum will be

$$\Phi = e(c_{m,0} - c_{m,1}) \tag{1}$$

where e is a constant, the magnitude of which depends on μ_e. When steady state is reached the flux across each maximum is the same, so that

$$\Phi = e(c_{m,0} - c_{m,1}) = e(c_{m,1} - c_{m,2}) = \ldots = \frac{e}{n}(c_{m,0} - c_{m,n}) \qquad (2)$$

The concentration difference $c_{m,0} - c_{m,n}$ in the steady state, when the concentration profile is linear, is simply the n-multiple of the concentration difference between any two successive minima. In the steady state, moreover, the flux across the major barriers at the interfaces

$$\Phi = ac_o - bc_{m,0} = bc_{m,n} - ac_i = \frac{a}{2}(c_o - c_i) - \frac{b}{2}(c_{m,0} - c_{m,n}) \qquad (3)$$

is equal to the flux across any of the minor maxima, given by (2). (Here a and b are rate constants, the value of which depends on μ_a and μ_b, respectively.) Thus, on combining (2) and (3) we have

$$c_{m,0} - c_{m,n} = \frac{a}{b + 2e/n}(c_o - c_i) \qquad (4)$$

Substituting for $(c_{m,0} - c_{m,n})$ in (2) from (4), we may write

$$\Phi = \frac{e}{n} \frac{a}{b + 2e/n}(c_o - c_i) \qquad (5)$$

The ratio of flux Φ to the concentration difference across the membrane is defined as the permeability constant P,

$$\Phi = P(c_o - c_i) \qquad (6)$$

so that

★
$$P = \frac{ae}{nb + 2e} \qquad (7)$$

This is the fundamental equation of Danielli, defining the permeability of a homogeneous lipid membrane. According to Danielli, for many substances penetrating across the membranes rather slowly, μ_b is greater than μ_e and, therefore, b is much less than e. If the membrane is thin, n is small and nb may be neglected as compared with $2e$. Hence,

$$P = \frac{a}{2} \qquad (8)$$

and, for molecules passing across thin membranes very slowly, the transition from water to lipid is thus often rate-limiting. On the other hand, for rapidly penetrating molecules, $2e$ may be often neglected against nb, so that (7) simplifies to

$$P = \frac{a}{b}\,\frac{e}{n} \qquad\qquad (9)$$

Finally, Danielli shows that in both these limiting cases, the expression $PM^{1/2}$ is, to a first approximation, a linear function of the partition coefficient a/b.

Recently, the diffusion of nonelectrolytes across thin lipid membranes was discussed in detail by Stein (1967). Of his conclusions, let us mention that for each hydrogen bond which has to be broken when a molecule enters the membrane the transfer rate is lowered by 6- to 10-fold and that "each bare —CH_2— group in the permeant will increase the transfer rate by some twofold."

Another approach, alternative or supplementary to the above theory of diffusion across lipid membranes and aiming especially at the explanation of permeability of lipid-insoluble substances, assumes the presence of water-filled pores in the cell membrane. The problem of the pores in the cell membrane is a rather complicated one. In the first chapter of this book those aspects of the electron microphotographs and of models of the cell were already discussed which appear to be compatible with the pore hypothesis. In the section concerned with water transport another argument for the pore hypothesis will be encountered, based on differences between so-called diffusion and osmotic water permeabilities for a given membrane. Such differences will be seen to originate from the fact that friction per molecule of water during water flow across a membrane is greater if each single molecule has to travel separately, i.e., to diffuse, than when it takes place in osmotic bulk flow through the pores in the membrane. In the present section an attempt will be described, the aim of which was to determine the so-called equivalent pore radius, i.e., the radius of idealized water-filled cylindrical pores which, if present in the membrane, would invest it with some permeability properties experimentally observed. The method discussed here briefly was used by Goldstein and Solomon (1961; Solomon, 1961) to determine the equivalent pore radius in the human red cell membrane with the result of 4.2 Å. It is perhaps the most refined of the methods used for analogous determinations and we shall discuss it here for the emphasis it places on the permeation of a nonelectrolyte molecule through pores. Results of theoretical considerations and experimental measurements by a great number of workers were combined in the theory underlying this method and only some will be mentioned here.

Pappenheimer and co-workers (1951) considered two factors restricting the diffusion of spherical molecules through a cylindrical pore, the first of them being the steric hindrance to the entrance of the molecule into the pore, the second the viscous resistance inside the pore. The steric hindrance to entrance may be visualized as in Fig. 3.2. A molecule of radius a can penetrate only if the centre of the molecule falls within the area of a circle of radius $(r - a)$, so that the virtual area for penetration is $A = \pi(r - a)^2$ and its ratio to the actual area of the opening A_0, is

$$\frac{A}{A_0} = \left(1 - \frac{a}{r}\right)^2 \tag{10}$$

To express the second restricting factor, the frictional resistance within the pores, Pappenheimer and co-workers (1951) used an empirical equation, whereas somewhat later Renkin (1954) used another equation derived on theoretical grounds by Faxén (see Renkin, 1954, for references):

$$\frac{A}{A_0} = 1 - 2.104 \left(\frac{a}{r}\right) + 2.09 \left(\frac{a}{r}\right)^3 - 0.95 \left(\frac{a}{r}\right)^5 \tag{11}$$

The total restriction to diffusion due to the two factors was thus given by Renkin (1954) as

$$\frac{A'}{A_0} = \left(1 - \frac{a}{r}\right)^2 \left[1 - 2.104 \left(\frac{a}{r}\right) + 2.09 \left(\frac{a}{r}\right)^3 - 0.95 \left(\frac{a}{r}\right)^5\right] \tag{12}$$

When filtration across the pore proceeds (i.e., when the flow of solvent and of solute due to pressure takes place) the first of the two factors, expressing the steric hindrance at the entrance of the pore, must be modified. This is because the velocity of flow within the cylinder of radius $(r - a)$ (see Fig. 3.2) is greater than the mean velocity of flow through the entire pore. Renkin used the following expression for the steric hindrance at the

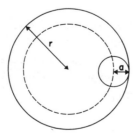

FIG. 3.2. Steric hindrance to the entrance of spherical molecules into cylindrical pores of radius r. (According to Pappenheimer *et al.*, 1951.)

entrance to the pore during ultrafiltration, derived by Ferry (see Renkin, 1954, for reference):

$$\frac{A_f}{A_0} = 2\left(1 - \frac{a}{r}\right)^2 - \left(1 - \frac{a}{r}\right)^4 \tag{13}$$

Renkin's equation for restricted filtration is then (cf. Solomon, 1960b)

$$\frac{A_f'}{A_0} = \left[2\left(1 - \frac{a}{r}\right)^2 - \left(1 - \frac{a}{r}\right)^4\right]\left[1 - 2.104\left(\frac{a}{r}\right) + 2.09\left(\frac{a}{r}\right)^3 - 0.95\left(\frac{a}{r}\right)^5\right] \tag{14}$$

Using Renkin's equation for filtration, Goldstein and Solomon (1961) expressed the ratio of the apparent area for filtration of solute (A_{sf}) to the apparent area for filtration of water (A_{wf})

$$\frac{A_{sf}}{A_{wf}} = \frac{[2(1-a/r)^2-(1-a/r)^4][1-2.104a/r+2.09(a/r)^3-0.95(a/r)^5]}{[2(1-a_w/r)^2-(1-a_w/r)^4][1-2.104a_w/r+2.09(a_w/r)^3-0.95(a_w/r)^5]} \tag{15}$$

To evaluate this ratio of apparent or restricted areas for filtration and thus to obtain equations relating the equivalent pore radius to the radius of the test molecule, Goldstein and Solomon (1961) made use of Staverman's coefficient σ, which will be discussed in some detail in section 4.2. in connection with the irreversible thermodynamics approach to permeability studies. It may do at this point to realize that osmotic pressure developed on the membrane by a substance to which the membrane is permeable is always smaller than that produced by a nonpermeating substance. If the solute molecules permeate across the membrane through the same channels as water, the movements of solute and water interact in the membrane and the observed osmotic pressure π_{obs} may be considerably lower than the theoretical osmotic pressure π_{theor}. The ratio of the two pressures is called σ

$$\sigma = \pi_{obs}/\pi_{theor} \tag{16}$$

and as long as van't Hoff's expression for osmotic pressure $\pi = RTc$ is valid, it is equal to the inverse ratio of the concentration of the given substance to the concentration of an impermeating substance developing the same osmotic pressure. It was stated by Durbin and co-workers (1956) and shown in detail by Durbin (1961) that

$$1 - \sigma = A_{sf}/A_{wf} \tag{17}$$

Goldstein and Solomon (1961) developed an elegant zero-time extrapola-tion method by which concentrations of various nonelectrolytes, poorly soluble in lipids, which were just required to prevent volume changes of human red cells, were determined. The σ coefficients were calculated from these concentrations for the individual substances. Since, from (15) and (17),

$$1 - \sigma = \frac{[2(1-a/r)^2-(1-a/r)^4][1-2.104a/r+2.09(a/r)^3-0.95(a/r)^5]}{[2(1-a_w/r)^2-(1-a_w/r)^4][1-2.104a_w/r+2.09(a_w/r)^3-0.95(a_w/r)^5]}$$

★   (18)

Goldstein and Solomon plotted $(1 - \sigma)$ as a function of the radius of the test molecule for various equivalent pore radii. A family of theoretical curves was thus obtained, with the equivalent pore radius as the parameter. The experimetally determined dependence of $(1 - \sigma)$ values on the radii of the test molecules (calculated as the mean radii from direct measurements of the dimensions of scale models of the molecules) was found to be best fitted by the curve for the pore radius of 4.2 Å and such is, therefore, the equivalent pore radius for the human red cell membrane. Analogous values were found for some other membranes (squid axon, kidney slices of *Nectu-rus*; see Solomon, 1961, for references). It is noteworthy that these values are in fair agreement with those assumed on the basis of a globular micelle model of the plasma membrane as discussed in chapter 1 of this book (p. 21) On the other hand, the approach using the value of $(1 - \sigma)$ for calculating the equivalent pore radius without appropriate corrections being made has been criticized (Dainty, 1963; Dainty and House, 1966b). The main point of the criticism is that, as will be seen later, $(1 - \sigma)$ is always different from zero for a permeating solute, even if there are no pores in the membrane (see section 4.2.).

3.2. MEDIATED DIFFUSION AND RELATED PHENOMENA

3.2.1. General Considerations

It has been shown in the foregoing section that the net flux of a sub-stance moving across a membrane by simple diffusion is proportional to the concentration difference along which the substance is moving (eq. 6). In other words, the amount of solute translocated per unit time across a membrane of unit area into a solute-free compartment should be directly proportional to the solute concentration at the starting side of the mem-brane. However, numerous observations made in different cells with different solutes indicated that the flux of a substance across the cell membrane does not increase linearly with concentration but rather tends toward a maximum value.

Moreover, it has been found repeatedly that closely related substances (D-glucose and L-glucose, or D-alanine and L-alanine) are transported at vastly different rates, contrary to what one would expect in simple diffusion.

It has been established by many investigators that the membrane passage of various substances was associated with a high temperature quotient ($Q_{10} = 2$–3) whereas simple diffusion would allow for a Q_{10} of about 1.03 in the generally used temperature range.

All these peculiarities indicated that the transport of the solutes in question (the first observations were made on monosaccharide transport across the placenta and in erythrocytes; Widdas, 1952, 1954) involves a chemical reaction, however transient it may be (LeFevre, 1948; Wilbrandt and Rosenberg, 1950). Two hypotheses offered themselves in this connection.

1. The solute is bound inside the cell to reactive sites of biopolymers, such as proteins and possibly polysaccharides.

2. The solute reacts transiently with a membrane component or with an array of membrane reactive sites and, either by a series of jumps or by translocation while bound to the membrane component, is transported inward.

The first-named hypothesis constitutes the basis of various sorption theories (*cf.* chapter 7) that might have limited validity and will be dealt with in a subsequent section of this book. Intracellular binding would account for the limitation of transport velocity at high solute concentrations and, if the sites were highly selective, might explain some (but not all) differences in the rates of uptake of closely related substances. The temperature quotient of the binding could also be rather high. However, apart from more sophisticated pieces of evidence, sorption theories cannot explain and even do not attempt to explain many observations on the uptake of nonelectrolytes, regarding its specificity, final level of distribution, and genetic control.

The second of the above hypotheses appears to be more likely and actually has been confirmed in many cell types and for a number of solutes transported so that at present hardly any doubt is entertained as to the cell membrane playing a specific role in the transport of substrates into cells.

In the following we shall derive the appropriate equations for such a mediated transport, involving a mobile membrane component, called simply a *carrier*, that does not proceed against a concentration gradient. In this sense, it can be termed passive equilibrating transport and it will be considered synonymous with *facilitated* or *mediated diffusion*.

Historically, the term facilitated diffusion has been used (e.g., Stein and Danielli, 1956) for describing transport up to diffusion equilibrium by different mechanisms, some of them not involving a mobile carrier (expanded lattice and polar pore hypotheses).

Several models, proceeding from the simpler to the more complex, will now be described. It will be generally assumed that (a) the transport substrate is not metabolized in the cell (for a treatment of transport of metabolic substrates see p. 244); (b) the carrier can move back and forth through the membrane but cannot leave it (although, particularly in growing cells, a more or less continuous supply of the carrier to the membrane must be assumed if only to replace older carrier molecules, all being subject to a constant turnover); thus, the total concentration of the carrier in the membrane is constant and the sum of carrier fluxes through the membrane (this being a vectorial quantity) is zero; (c) the solute (or substrate of the carrier system) can move across the membrane only when bound to the carrier (this restriction excludes various pump-and-leak systems to be dealt with later).

3.2.2. Model I

The restrictive assumptions of the present model are the following:

1. Only one substrate molecule is bound to each carrier molecule (for treatment of polyvalent carriers see p. 82).

2. The reaction between the carrier and its substrate proceeds at a much greater rate than the actual movement of the carrier and carrier-substrate complex so that an equilibrium is assumed to exist at both sides of the membrane with respect to substrate and carrier. (This simplification seems to be justified in most cell types. For a fuller treatment of other limiting processes, the steady-state kinetics, see p. 79.)

3. The mobilities of the free carrier and of the carrier–substrate complex are equal. (This is not necessarily the case in nature; the full treatment is found on p. 75.)

The transport model to be treated here is shown in the scheme

$$
\begin{array}{ccc}
\text{I} & \begin{array}{ccc}
\text{CS}_\text{I} & \xrightleftharpoons{\ D'\ } & \text{CS}_\text{II} \\
k_1 \big\|\, k_{-1} & & k_{-2} \big\|\, k_2 \\
\text{S}_\text{I} + \ \ \text{C}_\text{I} & \xrightleftharpoons{\ D'\ } & \text{C}_\text{II} \ \ + \text{S}_\text{II}
\end{array} & \text{II}
\end{array}
$$

The rate constants of the reactions at the membrane sides are designated

k_1, k_{-1}, k_2, and k_{-2}. Hence $k_{-1}/k_1 = K_{CS_I}$ and $k_{-2}/k_2 = K_{CS_{II}}$, the dissociation constants of the carrier–substrate complex. Since we are dealing here with an equilibrating transport, $K_{CS_I} = K_{CS_{II}}$. The dimension of K_{CS} is moles/liter (like concentration). D' is the rate constant of the movement of the carrier (or the carrier–substrate complex) in the membrane. It can be defined as DA/δ where D is the diffusion coefficient in $cm^2 \cdot sec^{-1}$, A the total area across which transport occurs in cm^2, and δ the thickness of the membrane in cm. Thus the dimensions of D' are $cm^3 \cdot sec^{-1}$.

In some cases, it is advantageous to disregard the dimensions of the membrane, these being generally constant in a given experiment. Then D' can be considered as a first-order rate constant, expressed in sec^{-1}. In such treatment, the quantity c_t to be described below also changes dimensions from $g \cdot cm^{-3}$ to g.

The designation with D' might be somewhat misleading as the carrier may move through the membrane not only by diffusion (this is actually rather unlikely) but perhaps by rotation or invagination of the membrane (*cf.* chapter 8). However, the carrier movement will always be expressed in the same units (either as volume flow or as rate constant) as the mathematical formalism and in all cases will be the same.

Let us now examine the flow of substrate from side I to side II of the membrane. Since equilibrium exists (k_1, etc. $\gg D'$) between S_I and C_I we can write $K_{CS} = s_I c_I / cs_I$ and $cs_I = s_I c_I / K_{CS}$. As the total carrier concentration at side I is given by $c_{tI} = c_I + cs_I = c_I(1 + s_I/K_{CS})$ we can derive that $cs_I = c_{tI} s_I / (K_{CS} + s_I)$. Here, as in all subsequent derivations, lowercase *italics* stand for concentrations of reactants, the designations and amounts of which are printed in roman type. Subscripts I and II refer to extracellular and intracellular compartments, respectively.

By analogy with elementary enzyme kinetics where the rate of the enzyme reactions is proportional to the concentration of the enzyme–substrate complex ES, the rate of transport here will be proportional to the concentration of CS. The proportionality factor, corresponding to k_2 of enzyme kinetics, will be D'. The quantity c_t represents the capacity component of the expression (in moles/liter or $g \cdot cm^{-3}$). Hence the rate of movement of substrate S from side I to side II is defined as

$$v_{S_I} = c_{tI} D' \frac{s_I}{s_I + K_{CS}} \quad \text{(in moles/sec or g. sec}^{-1}\text{)} \qquad (19a)$$

Symmetrically, the rate of transport from side II to side I is given by

$$v_{S_{II}} = -c_{tII} D' \frac{s_{II}}{s_{II} + K_{CS}} \qquad (19b)$$

It will be seen that the rates of movement from side I to side II are arbitrarily positive, opposite rates are negative.

The net rate of transport is then obtained by adding the two unidirectional fluxes (these can be viewed as independent of each other in a particular sense of the word as they are not affected by substrate concentration on the opposite side of the membrane). As in this case $c_t = c_{tI} = c_{tII}$ it can be written that

★ $$v_S = c_t D' \left(\frac{s_I}{s_I + K_{CS}} - \frac{s_{II}}{s_{II} + K_{CS}} \right) \tag{19c}$$

It should be observed at this point that the carrier concentration c_t refers to one membrane side only. More correctly, the total carrier concentration is $2c_t = c_{tI} + c_{tII}$, a relationship that becomes important when the carrier concentrations at the two membrane sides are not equal. Thus, the factor c_t of eq. (19c) is actually equal to $2c_t/2$. Being ignorant of the true concentration of the carrier, it is sometimes more expedient to speak about carrier amounts, these being in any given experiment related to the concentration by the same factor.

One of the obvious consequences of eq. (19c) is that for $v_S = 0$ (after sufficiently long reaction period), $s_I = s_{II}$, this being a characteristic of equilibrating transport.

It is readily seen that the equation can be rewritten as $v_S = D'(cs_I - cs_{II})$, this being identical with the rate equation for simple diffusion. However, the actual concentrations of CS in the membrane are not accessible to direct measurement so that recourse must be made to indirect approaches. Only under one type of conditions is the rate v_S truly identical with a diffusion movement—when $s \ll K_{CS}$, i.e., when the carrier is far from saturation. Equation (19c) then becomes

$$v_S = \frac{c_t D'}{K_{CS}} (s_I - s_{II}) \tag{20}$$

and the movement proceeds according to first-order reaction kinetics. Under these conditions, one may speak of the so-called D-kinetics (D for diffusion). The dependence of v_S on s_I is a straight line which is only slightly shifted by small changes of s_{II} (Fig. 3.3).

If, on the other hand, $s \gg K_{CS}$, eq. (19c), first converted to

$$v_S = c_t D' K_{CS} \frac{s_I - s_{II}}{(s_I + K_{CS})(s_{II} + K_{CS})}$$

becomes

$$v_S = c_t D' K_{CS}(1/s_{II} - 1/s_I) \tag{21}$$

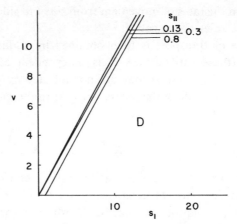

Fig. 3.3. Diagnostic plot of D-kinetics of transport (Wilbrandt, 1954). The carrier movement is rate-limiting, low concentrations of substrate are used.

In this form, it obeys what is called E-kinetics (E for enzyme). The most striking feature of transport at these high (saturation) concentrations is that the rate is markedly affected by s_{II}, particularly when s_{II} is small. The dependence of v_S on s_I is represented by a convex curve tending to a limit, the limiting value being greatly depressed by small increases of s_{II} (Fig. 3.4).

Let us now assume that two substrates, R and S, are present at the membrane interface and compete for the carrier. Then $c_t = c + cs + cr$

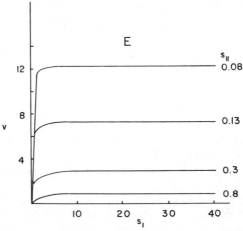

Fig. 3.4. Diagnostic plot of E-kinetics of transport (Wilbrandt, 1954). The carrier movement is rate-limiting, saturation concentrations of substrate are used.

$= c + s \cdot c/K_{CS} + r \cdot c/K_{CR} = c(1 + s/K_{CS} + r/K_{CR})$. Hence $cr = c_t r(1 + s/K_{CS} + r/K_{CR})/K_{CR}$ and the rate of transport of R (proportional to the amount of CR) is given by

$$v_{R(S)} = c_t D'\left(\frac{r_I}{r_I + K_{CR}(1 + s_I/K_{CS})} - \frac{r_{II}}{r_{II} + K_{CR}(1 + s_{II}/K_{CS})}\right) \quad (22a)$$

It will be seen that the terms of the equation are identical with those derived for competitive inhibition of an enzyme reaction since here the two substrates act as mutual competitors (the "reaction product" of the carrier reaction is an unchanged but translocated substrate).

For practical purposes the equation is usually written in terms of relative concentrations as follows:

$$v_R' = \frac{r_I'}{r_I' + s_I' + 1} - \frac{r_{II}'}{r_{II}' + s_{II}' + 1} \quad (22b)$$

The relative concentrations $s' = s/K_{CS}$ and $r' = r/K_{CR}$ are useful for two reasons: (1) They can be easily handled in the rate equations; (2) they supply immediate information on the degree of saturation of the enzyme (or carrier). Analogously, v' is sometimes used in place of $v/c_t D'$.

The above equation predicts several interesting phenomena, among them first-order kinetics for the movement of labeled substrate, counter-transport induced by the flow of a competing substrate from the other side of the membrane, and competitive acceleration. Let us deal with these phenomena one by one.

Movement of Labeled Substrate. If R is the labeled form of S, their carrier–substrate dissociation constants will be the same ($K_{CS} = K_{CR} = K$). Moreover, the isotopically labeled form will be present in a negligible concentration so that $s \gg r$. Equation (22b) can be rewritten as

$$v_R' = \frac{r_I'}{s_I' + 1} - \frac{r_{II}'}{s_{II}' + 1} \quad (23a)$$

If conditions are so chosen (by preincubating cells with substrate S) such that at the moment of adding R, $s_I = s_{II} = s$,

$$v_R' = (r_I - r_{II})/(s + K) \quad (23b)$$

The usefulness of this equation will be documented on p. 72 where the half-times of equilibration of a tracer are calculated from the integrated equation. This type of equation can be used for calculating total fluxes of substrate by simply multiplying the rate obtained by the reciprocal of specific activity $(s + r)/r$.

Countertransport (Rosenberg and Wilbrandt, 1957b). This type of behavior (the only instance in equilibrating transport when a substrate moves against its concentration gradient) is to be expected in all systems where a membrane component moves from one side of the membrane to the other (or is exposed alternately to the one and to the other membrane side). It can be demonstrated in two simple ways.

1. Let us equilibrate cells with substrate R so that $r_I = r_{II} = r$ and then add substrate S (either the same nonlabeled or a competing species) to the outside. At the moment of addition, $s_I = s$, $s_{II} = 0$. The movement of r will then be given by

$$v_R' = \frac{r'}{r' + s' + 1} - \frac{r'}{r' + 1} \qquad (24a)$$

Then, necessarily, $v_R < 0$, indicating that substrate R will move out of the cell against its concentration gradient (the process starts when $r_I = r_{II}$!). This movement, however, is not independent but is rather closely associated with the simultaneous movement of S which, immediately after addition, proceeds according to

$$v_S' = \frac{s'}{s' + r' + 1} \qquad (24b)$$

This movement persists until the condition is reached when

$$r_I/r_{II} = (s_I' + 1)/(s_{II}' + 1) \qquad (24c)$$

From there on, substrate R begins to move back into the cell until, in the final equilibrium, $r_I = r_{II}$ and $s_I = s_{II}$ (Fig. 3.5).

2. Let us incubate cells with substrate S, remove the medium, and resuspend the cells in a solution of a rather low concentration of R. R will move into the cell according to eq. (22b), this being a differential equation that must be solved in conjunction with that for the movement of S, the limits being 0 and r_{II} for r and s_{II} and 0 (if the volume of the medium is very large compared to that of the cells) for s. An analysis of such equations (either numerically according to a procedure like that of Runge–Kutta or using an analog computer) shows that the curve for influx of R should pass through a maximum (Fig. 3.6), the part AB representing in fact transport of R against its concentration gradient (Miller, 1965b).

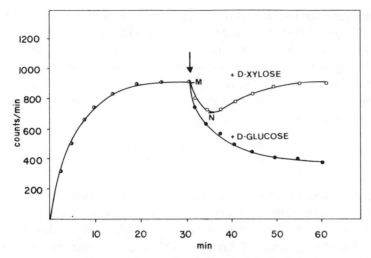

FIG. 3.5. Countertransport minimum of monosaccharides in baker's yeast. The cells were pre-incubated with labeled D-xylose and then either nonlabeled D-xylose or D-glucose were added. Between M and N, the labeled D-xylose moves out of the cell against its concentration gradient. After addition of glucose, the intracellular level of labeled D-xylose remains depressed because of metabolism of glucose inside the cells and hence a relatively lower competition with xylose transport outward than inward (*cf.* p. 244).

It is evident that the mechanism underlying the two types of counter-transport is the same—competition with the opposite flux by another transport substrate. Both types of countertransport demonstration require that s be rather high so that the carrier is near saturation. It is useful to have R as the labeled form of S as then $K_{CS} = K_{CR}$ and, moreover, the movement of R is then readily followed.

The existence of countertransport is an intrinsic feature of all mobile systems or, more precisely, will occur if the movement of the solute across

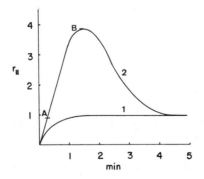

FIG. 3.6. Countertransport maximum in the transport of monosaccharides in human erythrocytes. Cells were preincubated in buffer (**1**) or in nonlabeled galactose (**2**), separated from the medium and resuspended in labeled galactose. The intracellular amount of label (r_{II}) is plotted against time. (Adapted from Miller, 1965 *b*.)

the membrane occurs, at least in part, as that of a component of the membrane (Ussing, 1949). Ussing derived the condition for simple (passive) diffusion (*cf.* p. 136) that the ratio of unidirectional fluxes (Φ_{in}/Φ_{ex}) of solute be equal to the ratio of activities (or concentrations in an approximation) of the solute at the two membrane sides (s_{in}/s_{ex}), taken with an opposite sign. It is readily seen that in a carrier transport (*cf.* eq. (19a) and (19b)) the flux ratio is given by

★
$$\frac{\Phi_{in}}{\Phi_{ex}} = -\frac{s_I(s_{II} + K_{CS})}{s_{II}(s_I + K_{CS})} \tag{25}$$

and hence Ussing's criterion is valid only for the case when $s_I = s_{II}$. However, it is approximated also when s_I, $s_{II} \gg K_{CS}$, i.e., at saturation concentrations, when an "exchange diffusion" takes place. If the flux resistance is defined as the reciprocal of flux ($F = 1/\Phi$) and the experimental arrangement described under (1) of the paragraph on countertransport is applied, the flux resistances for the movement of R will be

$$F_I = \frac{r' + s_I' + 1}{c_t D'} \tag{26a}$$

for the inward direction and

$$F_{II} = \frac{r' + s_{II}' + 1}{c_t D'} \tag{26b}$$

for the outward direction. The inequality of these resistances, the driving force proportional to r being the same at both sides of the membrane, will cause countertransport.

Countertransport is not expected to occur in a system where the substrate molecule moves from one fixed site in the membrane to another (adsorption-membrane hypothesis; LeFevre, 1948). There the flux of substrate R in one direction is given by

$$\Phi_{in} = a\left(n \frac{r_I'}{r_I' + s_I' + 1} n \frac{1}{r_{II}' + s_{II}' + 1}\right) \tag{27a}$$

and, similarly, in the opposite direction, by

$$\Phi_{ex} = a\left(n \frac{1}{r_I' + s_I' + 1} n \frac{r_{II}'}{r_{II}' + s_{II}' + 1}\right) \tag{27b}$$

where a is the rate-limiting factor (analogous to D') and n is the number of adsorption sites in a membrane layer; $r' = r/K_L$ where K_L is the equilibrium constant of an adsorption surface. Then

$$v_R = an^2 \left(\frac{r_I' - r_{II}'}{(r_I' + s_I' + 1)(r_{II}' + s_{II}' + 1)} \right) \tag{27c}$$

Here, then, $v_R > 0$ under all conditions, the substrate S having no effect on reversing the direction of flow of R. Intuitively, the situation is understood when we realize that the membrane sites in this model are exposed to both sides of the membrane at the same time so that the events taking place at one side of the membrane are "equilibrated" with those at the other membrane side, unlike when a carrier intervenes between the two sides. Correspondingly, Ussing's criterion for simple diffusion (or rather nonmoving membrane components) is met since here $\Phi_{in}/\Phi_{ex} = -r_I/r_{II}$.

Competitive Acceleration. Equation (22b) shows that under certain conditions, when a competitive inhibitor R is present at both sides of the membrane at equal concentrations, the actual rate of transport of substrate can be greater than in the absence of the inhibitor (Wilbrandt, 1961a). On comparing equations

$$v_S' = s_I'/(s_I' + 1) - s_{II}'/(s_{II}' + 1)$$

and

$$v_{S(R)}' = s_I'/(s_I' + r' + 1) - s_{II}'/(s_{II}' + r' + 1)$$

(simply by subtracting the first from the second one and introducing an inequality sign between the two rates) it will be seen that $v_{S(R)} > v_S$ if

$$s_I' s_{II}' > (r' + 1) \tag{28}$$

It should be observed that competitive acceleration never occurs with the initial rate of transport where always $v_{S(R)_0} < v_{S_0}$.

To be able to predict quantitatively the course of uptake of substrate according to the present model, eq. (19c) can be written in the differential form as

$$\frac{ds_{II}}{dt} = c_t D' \left(\frac{S_I/V_I}{S_I/V_I + K_{CS}} - \frac{S_{II}/V_{II}}{S_{II}/V_{II} + K_{CS}} \right) \tag{29a}$$

where S_I and S_{II} are the amounts of substrate at the two sides and V_I and V_{II} refer to the volumes in which the substrate is dissolved. The equation assumes that S_I does not change appreciably during transport since $V_I \gg V_{II}$. Hence $S_I/V_I = s_I$. When dealing with cells that are not osmotically responsive (like yeast and plant cells) or when using low concentrations of solute, one can assume the intracellular volume to be constant and hence S_{II}/V_{II} ($= s_{II}$) to vary in the same way as S_{II}.

For integration, the equation can be rewritten as follows:

$$K_{CS}c_tD' \, dt = \frac{(s_I + K_{CS})(s_{II} + K_{CS})}{s_I - s_{II}} \, ds_{II}$$

$$= (s_I + K_{CS}) \left(K_{CS} \, \frac{ds_{II}}{s_I - s_{II}} + \frac{s_{II} \, ds_{II}}{s_I - s_{II}} \right)$$

Integration between 0 and s_{II} for the time span from 0 to t yields

$$K_{CS}c_tD't = (s_I + K_{CS}) \left[(K_{CS} + s_I) \ln \frac{s_I}{s_I - s_{II}} - s_{II} \right] \qquad (29b)$$

This equation can be applied to an estimation of either K_{CS} or c_tD' directly if s_I and s_{II} can be estimated.

The equation simplifies greatly when the uptake of a labeled substrate is followed at equilibrium of the unlabeled form. Then the pertinent equation for the uptake of R (this being an isotopically labeled form of S) is

$$dr_{II}/dt = c_tD' \left(\frac{r_I - r_{II}}{s + K_{CS}} \right) \qquad (30a)$$

since S is present in great excess and at equal concentration at both sides of the membrane. Integration of this equation yields

$$c_tD't = (s + K_{CS}) \ln \frac{r_I}{r_I - r_{II}} \qquad (30b)$$

From this, the half-time of uptake (when $r_I = 2r_{II}$) is given by

★ $$t_{0.5} = 0.693(s + K_{CS})/c_tD' \qquad (31)$$

In cells which respond to the extracellular tonicity by changing their volume (like erythrocytes or microbial spheroplasts), eq. (29a) cannot be simplified and changes of V_{II} occurring with the penetration of substrate must be taken into account. Moreover, such cells must be incubated in approximately isotonic media so that another factor (external salt concentration) enters the equation. The integrated equation, called sometimes the swelling equation, is rather cumbersome to handle. If the concentrations are expressed in isotones and if the isotonic salt concentration in the medium is equal to 1, the equation is

$$K_{CS}D'c_t t = (s_I + K_{CS})\left[(1 - V_{II} - s_I V_{II})(1 + s_I + K_{CS})\right.$$
$$\left. + (K_{CS} + s_I + s_I^2 + s_I K_{CS}) \ln \frac{s_I}{(1 - V_{II})(1 + s_I)}\right] \quad (32)$$

where the intracellular volume V_{II} is related to the substrate amounts by $V_{II} = (1 + S_{II})/(1 + S_I/V_I)$.

However, here again, when the equilibration of a minute concentration of a labeled substrate is examined, eq. (30b) and (31) can be applied so that the desired constants can be estimated in a straightforward manner.

3.2.2.1. Determination of Transport Parameters

The expression for inward movement of substrate is identical with the common form of the Michaelis–Menten equation $v = Vs/(s + K_M)$ where V is the maximum rate of the reaction ($= c_t D'$) and K_M is the half-saturation Michaelis constant, shown to be identical with K_{CS} in the above simplified model. This enables us to apply the common techniques of determining K_M and V (see chapter 11). It is necessary only to determine the initial rates of influx (or efflux) and both K_M and V are readily obtained.

The reason for taking initial rates is that, as follows from the integrated equation (29b), the process is of a mixed order and, particularly near saturation concentrations, the time curve of uptake cannot be rectified by any simple mathematical procedure. The problem can be circumvented only by using tracer equilibration (cf. eq. (31)).

1. The half-time of tracer equilibration is determined at several concentrations of S using always the same concentration of R. This will yield a set of equations, from any two of which the unknown K_{CS} and $c_t D'$ can be computed. Alternatively, the values of $t_{0.5}$ can be plotted against s, the resulting straight line having a slope of $0.693/c_t D'$ and intersecting the ordinate at a point equal to $0.693 K_{CS}/c_t D'$ and the abscissa at a point equal to $-K_{CS}$.

2. The other technique that was found useful for calculating the transport parameters is based on estimating the rate of efflux from cells preloaded with substrate into a medium with relatively low substrate concentrations (Sen and Widdas, 1962). The appropriate integrated equation for the efflux will be

$$K_{CS}c_t D't = (s_I + K_{CS})\left[s_E - s_{II} + (s_I + K_{CS}) \ln \frac{s_E - s_I}{s_{II} - s_I}\right] \quad (33a)$$

where s_E is the equilibrium intracellular concentration at the beginning of the efflux, the other symbols being identical with those in eq. (29b). It is assumed here that the outside medium is so large that any contribution to its concentration by the substrate coming out of the cells can be neglected. It will be observed that the exit curve is linear for a great part of its course (i.e., when the carrier is fairly saturated on the inside interface of the membrane but unsaturated outside). This indicates that during the period corresponding to the linear part of the curve the logarithmic term of the above equation plays only an insignificant role (this will be true particularly if $(K_{CS} + s_I)$ is small relative to s_E). When the approximation is made that at about two-thirds of the efflux curve (measured on the y-axis) the logarithm of the fraction is equal to unity, the equation at that point will be

$$K_{CS}c_t D't = (s_I + K_{CS})(K_{CS} + s_I - s_{II} + s_E) \qquad (33b)$$

Extrapolating the linear part to the level where $s_I = s_{II}$ we obtain

$$t = (s_E + K_{CS})(s_I + K_{CS})/K_{CS}c_t D' \qquad (34)$$

The time thus measured is then a linear function of s_I, all the remaining quantities being constant in a given experiment. When $s_I = 0$ the time obtained is t_0 and the concentration at which $t = 2t_0$ is equal to K_{CS}, which is (in this particular model) both the dissociation constant of the carrier-substrate complex and the half-saturation constant.

The reader will have noticed that $c_t D'$ has been treated in the foregoing analysis of transport parameters simply as V of enzyme kinetics. Unfortunately, at the present stage of research only little more can be attempted. The amount of carrier in the membrane has been assessed in several cases with an accuracy of one or two orders of magnitude (cf. p. 191) and hence the rate constant D' cannot be calculated with any higher accuracy. The rate constant itself, being a rather vaguely defined quantity, ranging from a modified diffusion coefficient to a constant involving the probability of adsorption on lipid–water interfaces, is not amenable to direct estimation.

The model described on the preceding pages was used to demonstrate various characteristics of carrier transport, bearing in mind the fact that it cannot explain all the subtle phenomena observed. Still, some types of transport (e.g., that of monosaccharides in rabbit erythrocytes) are explained by it quite adequately. In the subsequent models, some of the simplifications will be omitted.

3.2.3. Model II

The second mobile-carrier model differs from the first one in that equal rates of movement of the free carrier and of the carrier–substrate complex are not assumed, only restrictions 1 and 2 on p. 63 being valid (e.g., Kotyk, 1966). Schematically, then,

$$\text{I} \qquad \begin{array}{c} \text{CS}_\text{I} \xrightleftharpoons{D'_\text{CS}} \text{CS}_\text{II} \\ k_1 \updownarrow k_{-1} \qquad k_{-2} \updownarrow k_2 \\ \text{S}_\text{I} + \quad \text{C}_\text{I} \xrightleftharpoons[D_\text{C'}]{} \text{C}_\text{II} \quad + \text{S}_\text{II} \end{array} \qquad \text{II}$$

In deriving the equations for the transport rate of S it is useful to start from the premise that the sum of the carrier fluxes in the membrane is zero. Since $c = c_t K_{CS}/(s + K_{CS})$ and $cs = c_t s/(s + K_{CS})$ (cf. p. 64) the rate of movement of the free carrier is defined by

$$\Phi_\text{C} = D_\text{C}'\left(c_{tI}\frac{1}{s_I' + 1} - c_{tII}\frac{1}{s_{II}' + 1}\right) \tag{35a}$$

and that of the carrier–substrate complex by

$$\Phi_\text{CS} = D_\text{CS}'\left(c_{tI}\frac{s_I'}{s_I' + 1} - c_{tII}\frac{s_{II}'}{s_{II}' + 1}\right) \tag{35b}$$

For $\Phi_\text{C} = -\Phi_\text{CS}$ it must hold that $c_{tI} \neq c_{tII}$. Referring to the note on p. 65 we can write $2c_t = c_{tI} + c_{tII}$. Let $\alpha = 2c_t/c_{tI} = 1 + c_{tII}/c_{tI}$. On adding equations (35a) and (35b) we obtain

$$\frac{D_\text{C}' + D_\text{CS}' s_I'}{s_I' + 1} = (\alpha - 1)\frac{D_\text{C}' + D_\text{CS}' s_{II}'}{s_{II}' + 1} \quad \text{or} \quad x_1/y_1 = (\alpha - 1)x_2/y_2$$

From this

$$c_{tI} = 2c_t x_2 y_1/(x_1 y_2 + x_2 y_1) \quad \text{and} \quad c_{tII} = 2c_t x_1 y_2/(x_1 y_2 + x_2 y_1)$$

The rate of transport of S which is identical with eq. (35b) is then defined by

$$\bigstar \qquad v_S = 2D_\text{C}' D_\text{CS}' c_t \frac{s_I' - s_{II}'}{2(D_\text{C}' + D_\text{CS}' s_I' s_{II}') + (D_\text{C}' + D_\text{CS}')(s_I' + s_{II}')} \tag{36a}$$

Here again, as in Model I, for $v_S = 0$ the concentrations on both sides of the membrane must be equal. It is an important aspect of the equation that, unlike in Model I, the unidirectional fluxes of S are not independent of the substrate concentration at the other side of the membrane. This produces a different type of flux interdependence, superimposed on that caused by carrier mediation.

The *initial rate* of uptake is defined by

$$v_{S_0} = 2D_C'D_{CS}'c_t \frac{s_I'}{2D_C' + (D_C' + D_{CS}')s_I'} \tag{36b}$$

This expression can be written as $v_{S_0} = Vs_I/(s_I + K_T)$ which is formally identical with the Michaelis–Menten type (and hence the Model I equation for transport rate), with

$$V = 2D_C'D_{CS}'c_t/(D_C' + D_{CS}') \tag{37a}$$

and

$$K_T = 2D_C'K_{CS}'/(D_C' + D_{CS}') \tag{37b}$$

From this comparison it follows that cells in which transport proceeds according to Model II will also display simple kinetics where the initial transport rate of single substrates will be involved. This is because the truly initial rate is mediated by the carrier available at side I and this (in the absence of any substrate) will be in the same concentration as on side II. When $D_C' = D_{CS}'$ in the above expressions for V and K_T, V becomes $D'c_t$ and K_T becomes K_{CS} as in the simple Model I.

Model II begins to differ from Model I more strikingly if actual fluxes and their interactions are estimated. This can be done again by means of tracer labeling and, as before, equations derived for the movement of two competing substrates are best suited to this purpose.

In the presence of substrates R and S the movement of R will be described by

$$v_{R(S)} = 2D_{CR}'c_t \frac{(D_{CS}'s_{II}' + D_{CR}'r_{II}' + D_C')r_I' - (D_{CS}'s_I' + D_{CR}'r_I' + D_C')r_{II}'}{(D_{CS}'s_I' + D_{CR}'r_I' + D_C')(1 + s_{II}' + r_{II}') + (D_{CS}'s_{II}' + D_{CR}'r_{II}' + D_C')(1 + s_I' + r_I')} \tag{38a}$$

If R is the labeled form of S, $K_{CR} = K_{CS}$ and $D_{CR}' = D_{CS}'$ and, moreover, generally $r \ll s$. In that case, eq. (38a) becomes

$$v_{R(S)} = 2D_{CR}'c_t \frac{(D_{CR}'s_{II}' + D_C')r_I' - (D_{CR}'s_I' + D_C')r_{II}'}{(D_C' + D_{CR}'s_I')(1 + s_{II}') + (D_C' + D_{CR}'s_{II}')(1 + s_I')} \tag{38b}$$

This equation already predicts what has been termed the "preloading effect," consisting in an increased rate of uptake of a substrate brought about by the presence of substrate on the other side of the membrane. To illustrate this phenomenon, let us incubate cells with substrate S, then replace the external solution with one containing only a small amount of R (r_I) so that $s_I = 0$ and $s_{II} = s$. For estimating the initial rate of its uptake (when $r_{II} = 0$) the pertinent equation is

$$v_{R(S)_0} = 2D'_{CR}c_t \frac{(D'_C + D'_{CR}s'_{II})r'_I}{D'_C(2 + s'_{II}) + D'_{CR}s'_{II}} \qquad (39a)$$

Setting $D'_{CR}/D'_C = \varrho$, we can write

$$v_{R(S)_0} = 2D'_{CR}c_t \frac{r'_I(1 + \varrho s'_{II})}{2 + s'_{II}(1 + \varrho)} \qquad (39b)$$

It can be seen that for $\varrho > 1$, increasing concentrations of S will bring about an increase of the initial rate of uptake of R. Thus, for example, for $\varrho = 3$, it follows that $v' = (1 + 3s'_{II})r' : (2 + 4s'_{II})$ which, for $s'_{II} = 0$, gives a value of $r'/2$, while for $s'_{II} = 100$ a value of about $3r'/4$ is obtained. Such an acceleration is not predictable in Model I where, under similar experimental conditions, $v' = r'$ so that s does not enter into the picture.

The net (not initial) flow of tracer (estimating the half-time of uptake) is under such conditions (when $s_I \neq 0$ and $s_{II} \neq 0$) slowed down by the presence of S both in Model I and Model II since the pertinent equations contain s only in the denominator. However, the accompanying unidirectional flux of S, determined from the specific radioactivity of the tracer flux, is not altered by preloading in Model I where it will be defined by

$$\Phi_{in} = v_R(s_I + r_I)/r_I = c_t D' \frac{r_I(s_I + r_I)}{(s_I + K_{CS})r_I} \approx c_t D' s_I/(s_I + K_{CS}) \qquad (40a)$$

but in Model II, where it is defined by

$$\Phi_{in} = v_R(s_I + r_I)/r_I$$

$$\cong 2D'_{CR}c_t \frac{(D'_C + D'_{CR}s'_{II})s'_I}{(D'_C + D'_{CR}s'_I)(1 + s'_{II}) + (D'_C + D'_{CR}s'_{II})(1 + s'_I)} \qquad (40b)$$

it will be greater than when no preloading has taken place ($s_{II} = 0$) for similar reasons as in eq. (39a).

Countertransport. As in Model I, induced countertransport will occur under suitable conditions. When cells are preincubated with R until $r_I = r_{II} = r$ and then S is added, at the moment of addition (when $s_{II} = 0$)

eq. (*38a*) will become

$$v_{R(S)} = 2D'_{CR}c_t \frac{-D'_{CR}r's_I'}{(D_C' + D'_{CS}s_I' + D'_{CR}r')(1+r') + (D_C' + D'_{CR}r')(1+s_I'+r')} \quad (41a)$$

and hence $v_{R(S)} < 0$. The turning point of countertransport can be computed from eq. (*38a*) to occur when

$$\frac{D_C' + D'_{CS}s_{II}' + D'_{CR}r_{II}'}{D_C' + D'_{CS}s_I' + D'_{CR}r_I'} = r_{II}/r_I \quad (41b)$$

Competitive Acceleration. Here again, Models I and II differ only in the role of the different mobilities of the free and of the loaded carrier and competitive acceleration is predictable under certain conditions. However, the range in which this phenomenon can be expected to occur is defined by such a complex expression that it is of no practical use.

3.2.3.1. Estimation of Transport Parameters

The constants V and K_T (the half-saturation concentration of the initial rate of uptake) can be obtained from a Lineweaver–Burk-type plot as in Model I. The dissociation constant K_{CS} can be computed from the half-time of tracer equilibration as there the different mobilities D_C' and D'_{CS} cancel out and the constant in eq. (*23b*) is truly K_{CS}.

What is interesting to establish is the ratio of the rates of movement $D'_{CS}/D_C' = \varrho$. This can be done in at least two ways.

1. By comparing the K_T and K_{CS} using eq. (*37b*), from which it follows that

★ $$\varrho = 2K_{CS}/K_T - 1 \quad (42)$$

2. By comparing the initial rate of exit of substrate from preloaded cells into a substrate-free medium with that into an equilibrium concentration of substrate. The exit of R into a substrate-free medium is characterized by the initial rate

$$v_{R(0)} = -2c_t \frac{D_C'D'_{CS}r_{II}}{2D_C'K_{CS} + s_{II}(D_C' + D'_{CS})} \quad (43a)$$

while that into an equilibrium concentration of substrate by

$$v_{R(\infty)} = -c_t \frac{D'_{CS}r_{II}}{K_{CS} + s_{II}} \quad \text{(here } s_{II} = s_I = s) \quad (43b)$$

From a comparison of these equations one obtains

$$\varrho = \frac{2K_{CS}(1 - x) + s(2 - x)}{xs} \tag{44}$$

where $x = v_{R_{(0)}}/v_{R_{(=)}}$. For very high s this expression reduces to $\varrho = 2/x - 1$ where $x = K_T/K_{CS}$ by analogy with eq. (42). It is of interest that the above equations for ϱ set a limit on the ratio of K_{CS}/K_T which cannot be less than 0.5 if positive values for the rates of movement are to be obtained.

The preloading effect and the acceleration of unidirectional fluxes of substrate by saturation of the carrier constitute what has been sometimes called the *exchange diffusion* mechanism. The term exchange diffusion has been applied rather indiscriminately to a number of transport phenomena: (1) To the movement of a substrate on saturated carrier bringing about the phenomenon of countertransport. (2) To the movement of a carrier that transports different species of substrate in opposite directions. (3) To the movement of a carrier that binds more readily a substrate molecule by "exchange" for one that is already bound, this involving less activation energy than when a free carrier binds a substrate molecule. (4) To the movement of a carrier that is accelerated by its binding a substrate molecule (the present definition). This list of interpretations warrants a cautious approach to the use of the term under all circumstances. It is perhaps best applicable to the situation where a carrier moves back and forth while fully saturated (always carries a substrate molecule with it) but it is advisable to define any of the above mechanisms in more explicit terms.

Apart from uphill transport mechanisms that will be dealt with later, Model II can explain most sugar transports in microorganisms, particularly that of monosaccharides in various yeast species, as well as in some erythrocytes.

3.2.4. Model III

In this treatment we shall omit another of the restrictions enumerated on p. 63, that the movement of the carrier across the membrane be limiting (hence only restriction 1 on p. 63 is now valid). The model thus does not describe an equilibrium between carrier and substrate at the two membrane interfaces but rather a steady state which is defined by two fundamental conditions, one being common to the preceding models,

$$2c_t = c_I + cs_I + c_{II} + cs_{II} \tag{45}$$

the other being new, expressing the fact that the flow of substrate S through the individual steps of the model must be equal to the overall transport across the system:

$$v_S = s_I c_I k_1 - c_I s k_{-1} = D'_{CS}(cs_I - cs_{II})$$
$$= cs_{II} k_{-2} - s_{II} c_{II} k_2 = D_C'(c_{II} - c_I) \qquad (46)$$

It will be seen that in the derivation we assume the rates of movement of the free and of the loaded carrier to be different from each other but equal in both directions for each of the species. Moreover, we assume (this being again a transport leading up to a diffusion equilibrium) that the dissociation constants of the carrier–substrate complex are equal at both sides of the membrane ($k_{-1}/k_1 = k_{-2}/k_2 = K_{CS}$). The rate of uptake will be proportional to the net amount of CS crossing the membrane per unit time but this amount is not a simple function of s and of the dissociation constant of CS but rather a complex function involving the rates of formation and dissociation of the CS complex. On eliminating c_I, c_{II}, cs_I, and cs_{II} from eqs. (45) and (46), we obtain for the flux inward

$$\Phi_{in} = c_t \frac{s_I D'_{CS} f}{1 + s_I' + \dfrac{(s_I - s_{II})(k_{-1}D'_{CS} - k_{-1}D_C' - 2D_C'D'_{CS})f}{2k_{-1}D'_{CS}(D_C' + s_{II}f)}} \qquad (47a)$$

where

$$f = \frac{k_1 k_2 K_{CS}}{D'_{CS}(k_{-2} + k_{-1}) + k_{-1}k_{-2}}$$

Similarly, the outward flux will be defined by an analogous formula

$$\Phi_{ex} = c_t \frac{s_{II} D'_{CS} f}{1 + s_{II}' + \dfrac{(s_{II} - s_I)(k_{-2}D'_{CS} - k_{-2}D_C' - 2D_C'D'_{CS})f}{2k_{-2}D'_{CS}(D_C' + s_I f)}} \qquad (47b)$$

The transport rate is then $v_S = \Phi_{in} - \Phi_{ex}$. For the initial rate of uptake, when $s_{II} = 0$,

$$v_{S_0} = c_t D'_{CS} f \frac{s_I}{1 + \dfrac{s_I[2D_C'D'_{CS}k_{-1} + fK_{CS}(k_{-1}D'_{CS} - k_{-1}D_C' - 2D_C'D'_{CS})]}{2k_{-1}D'_{CS}D_C'K_{CS}}} \qquad (48)$$

It will be seen that this relationship resembles formally the well-known Michaelis–Menten formula where both V and K_M are replaced by very complex constants involving the various rate constants. Hence, even in the present model, the fundamental transport parameters, the maximum rate

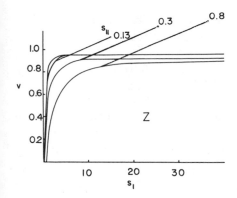

FIG. 3.7. Diagnostic plot of Z-kinetics of transport (Wilbrandt, 1954). The surface reaction is rate-limiting, the carrier is generally saturated.

and the half-saturation constant, are preserved and can be determined (Regen and Morgan, 1964).

The present model, provided that the rates of the substrate–carrier reactions are similar in magnitude to those of the actual carrier movement, demonstrates even countertransport and competitive acceleration. Only if the rate of movement of the carrier is very much greater than the rates of the interface reactions is countertransport no longer apparent since c_I and c_{II} can then equilibrate very rapidly and the situation begins to resemble the adsorption-membrane mechanism where there is no mediator between the two membrane sides.

Thus, apart from this extreme case, how can one distinguish between the mechanisms of, say, Model I and Model II, on the one hand, and that of Model III, on the other? This is most easily done by plotting the dependence of v on s_I which is not linear, the slope of the curve decreasing with increasing s_{II} in the low concentration range (Fig. 3.7). This is the case designated by Rosenberg and Wilbrandt as Z-kinetics.

A novel approach to the distinction between a "surface-equilibrium" and a "steady-state" model was presented by Blumenthal and Katchalsky (1969) who compared the relaxation times of the diffusional process (τ_{diff}) and that of the chemical reaction of association and dissociation (τ_{chem}). For the former, Einstein derived $2D\tau_{\mathrm{diff}} = (\varDelta x)^2$ where D is the diffusion coefficient and $\varDelta x$ the path travelled by the carrier–substrate complex. For the latter, let us assume that $\delta s = 0$ and hence $\delta c = -\delta cs$ where δs, δc, and δcs are the deviations of local concentrations s, c, and cs from the mean equilibrium concentrations—$\bar{s}$, $\bar{c}$, and $\bar{cs}$—in the membrane. The rate of the reaction is then given by the perturbation of δcs, thus

$$\frac{d(\delta cs)}{dt} = -(k_1\bar{s} + k_{-1})\,\delta cs \qquad (49)$$

Setting $k_1\bar{s} = 1/\tau_{chem}$ and using the simplified equation for flow of S (see Blumenthal and Katchalsky, 1969) one can derive that

$$\Phi_{S(\text{steady-state})} = \Phi_{S(\text{equilibrium})} \left[1 + \frac{\tau_{chem}}{\tau_{diff}} \right]^{-1} \qquad (50)$$

Thus, if the relaxation time of the chemical reaction is much shorter than that of the transmembrane movement, the two fluxes are equal. This was shown to be the case for a model (nontransport) system of binding of nicotinamide-adenine dinucleotide to D-glyceraldehyde 3-phosphate associated with a conformational change of the enzyme. Hence, if a conformational change ("allosteric transition") is involved in the transmembrane movement of the carrier, one is supported in the assumption of equilibrium kinetics of mediated transport.

It should be mentioned at this point that a model in which the substrate can be bound to the carrier and/or released from it only through the mediation of an enzyme has also been described (Rosenberg and Wilbrandt, 1955). The kinetics of such systems are even more complex since there not only the limitation of the rate by the carrier movement but also limitations at one of the enzyme–substrate reactions can bring about kinetics resembling the type designated as E (see p. 66). It seems that enzymes are involved in many transports (even equilibrating ones). However, the cases are mostly such that the rate of carrier movement is rate-limiting for the whole process and mostly D-kinetics will then be expected for low s and E-kinetics for high values of s.

3.2.5. Model IV

Whereas in all the previous models only one substrate molecule was assumed to be bound to the carrier the present model drops this limitation but, on the other hand, assumes the movement of the carrier to be rate-limiting so that only restriction 2 on p. 63 is now valid. If both the possibilities of rate limitation were taken into account the equation would become prohibitively complicated (Britton, 1966).

For the demonstration of the specific features of a polyvalent carrier a simpler model will be more expedient (Wilbrandt and Kotyk, 1964).

Let us assume that the carrier has two sites to which substrate can be bound so that it exists as C, CS, and CSS. The dissociation constants of the complexes are then K_{CS} ($= c \cdot s / cs$) and K_{CSS} ($= cs \cdot s / css$). Even in the rather unusual case that more than two substrate molecules should be bound to one carrier molecule, the salient features of the kinetics discussed here will be present.

The carrier fluxes across the membrane will then be

$$\Phi_C = D_C' \left(c_{tI} \frac{1}{1 + s_I' + s_I's_I''} - c_{tII} \frac{1}{1 + s_{II}' + s_{II}'s_{II}''} \right) \quad (51a)$$

$$\Phi_{CS} = D_{CS}' \left(c_{tI} \frac{s_I'}{1 + s_I' + s_I's_I''} - c_{tII} \frac{s_{II}'}{1 + s_{II}' + s_{II}'s_{II}''} \right) \quad (51b)$$

$$\Phi_{CSS} = D_{CSS}' \left(c_{tI} \frac{s_I's_I''}{1 + s_I' + s_I's_I''} - c_{tII} \frac{s_{II}'s_{II}''}{1 + s_{II}' + s_{II}'s_{II}''} \right) \quad (51c)$$

where $s' = s/K_{CS}$ and $s'' = s/K_{CSS}$. Just as in the treatment of Model II, for the sum of the carrier fluxes to be zero, necessarily $c_{tI} \neq c_{tII}$.

The transport of substrate S will be defined by a sum of equations (51b) and (51c) so that

★
$$v_S = 2c_t \frac{x_2 y_1 - x_1 y_2}{x_2 w_1 + x_1 w_2} \quad (52)$$

where

$$x = D_C' + D_{CS}'s' + D_{CSS}'s's''; \quad y = D_{CS}' + 2D_{CSS}'s's''; \quad w = 1 + s' + s's''$$

The unidirectional fluxes (as in Model II) will depend on both substrate concentrations s_I and s_{II}. The initial rate of uptake is given by

$$v_{S_0} = 2c_t \frac{D_C'y_1}{D_C'w_1 + x_1} = 2c_t \frac{D_C'(D_{CS}'s_I' + 2D_{CSS}'s_I's_I'')}{D_C'(s_I' + s_I's_I'' + 2) + D_{CS}'s_I' + D_{CSS}'s_I's_I''} \quad (53a)$$

and, if the assumption is made that $D_C' = D_{CSS}' \gg D_{CS}'$

$$v_{S_0} = 2c_t \frac{2D_C'D_{CSS}'s_I's_I''}{D_C'(s_I' + s_I's_I'' + 2) + D_{CSS}'s_I's_I''} \quad (53b)$$

An important feature of this rate is that at low substrate concentrations it will increase with the square of the concentration while at high concentrations it will approach a saturation value (Fig. 3.8). The first phenomenon can be intuitively grasped by realizing that the functional carrier–substrate complexes are only those with two substrate molecules attached. The amount of any such divalent complex (such as CSS) will be proportional to $c \cdot s \cdot s$ by the mass action law. The saturation kinetics at high concentrations are readily understood since here, again, a limited number of carrier sites only can be occupied even at extreme concentrations.

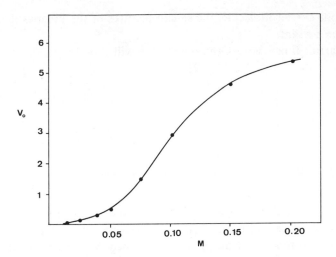

FIG. 3.8. Initial rate of uptake of D-fructose by human erythrocytes. The signoid character of the curve indicates a plurivalent carrier.

Several important features of this model emerge from the consideration of transport of two substrates, R and S. Here the rate of transport of S in the presence of R will be given by the same type of equation as (52) where, however,

$$x = s'(D'_{CS} + D_{CSS}s'' + D_{CSR}r''_s) + r'(D'_{CR} + D'_{CRR}r'' + D'_{CRS}s''_r) + D'_{C} \quad (54a)$$

$$y = s'(D'_{CS} + 2D'_{CSS}s'' + D'_{CSR}r''_s) + D'_{CRS}s''_r r' \quad (54b)$$

and

$$w = 1 + s'(1 + s'' + r''_s) + r'(1 + r'' + s''_r) \quad (54c)$$

Here $r''_s = csr/c \cdot s'$ and $s''_r = crs/c \cdot r'$.

Equations (54a,b,c) were derived by simply adding four more carrier fluxes to the set on p. 83—those for CR, CRR, CRS, and CSR—and deriving the zero net flux condition. The rate of uptake of S is then given by

$$v_S = \Phi_{CS} + 2\Phi_{CSS} + \Phi_{CSR} + \Phi_{CRS}$$

If we let cells incubate with different concentrations of S until $s_I = s_{II}$ and then add always the same minute amount of R (this being the labeled form of S), we can plot the rate of uptake of R against s. If a double-logarithmic plot is used the simple relationship can be derived that for low s (but always greater than r) $\log v_R/\log s = +1$, while for high s, $\log v_R/\log s$

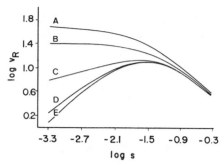

FIG. 3.9. Theoretical curves of the relationship between the logarithm of uptake rate of labeled substrate R and the logarithm of the same nonlabeled substrate S equilibrated with cells before adding R, assuming the existence of carrier species C, CS, CSS, CR, CRR, CRS, and CSR, the transmembrane rate constants being such that $D_C' = D_{CSS}' = 1$ and $K_{CSS}/K_{CS} = 4$ (on statistical grounds) for different values of D_{CS}' (A 1, B 0.5, C 0.1, D 0.01, E 0). (Adapted from Wilbrandt and Kotyk, 1964.)

$= -1$ (Fig. 3.9). This contrasts with the behavior of a monovalent carrier, where for low s the slope of the curve is 0. (High concentrations of s remain equal to -1.) The maximum observed in the figure is not to be expected in any type of monovalent carrier and serves thus as a defining criterion for polyvalent carriers. Actually, however, when $D_{CS}' = D_{CSS}'$, other conditions being identical, no such maximum is to be expected. It occurs again, even if the rates of movement of the complexes are equal, for the condition that $K_{CSS} \ll K_{CS}$.

The second important feature of the polyvalent transport system can be derived from eq. (28) and has been termed tentatively *cotransport*. This phenomenon is observed under conditions similar to those of countertransport except that the concentrations of substrate must be low.

For the sake of clarity, let us assume that S is chemically identical to R and that hence

$$K_{CS} = K_{CR} = K \qquad K_{CSS} = K_{CRR} = K_{CSR} = K_{CRS} = K'$$
$$D_C' = D_{CSS}' = D_{CRR}' = D_{CRS}' = D_{CSR}' = D \qquad D_{CS}' = D_{CR}' = 0$$
$$s'' = s_r'' \qquad r'' = r_s''$$

Equations (54a,b,c) then become

$$x = D[1 + (s'' + r'')(s' + r')] \qquad y = D[s'(2s'' + r'') + s''r']$$
$$w = 1 + (s' + r')(1 + s'' + r'')$$

Substituting these in eq. (52) we are interested primarily in the sign of the

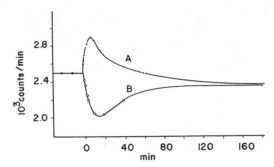

FIG. 3.10. Cotransport (A) and countertransport (B) of D-arabinose by erythrocytes of some human subjects. At time zero, a low (A) or a high (B) concentration of nonlabeled sugar was added to a suspension preincubated with a minute concentration of labeled D-arabinose. (According to Wilbrandt and Kotyk, 1964.)

numerator, since the denominator will be always positive. If the cells have been pre-incubated with S so that $s_I = s_{II} = s$ and at the moment of adding R, $r_{II} = 0$,

$$v_s \text{ (numerator)} = \varkappa[s(KK' + s^2)(s + r) - s^2(s^2 + r^2 + 2rs + KK')] \quad (55)$$

where $\varkappa$ is a constant. Hence, upon addition of a substrate R to the equilibrium of a competing substrate S, S will transiently move against its concentration gradient. At low concentrations the linear terms in (55) will predominate and the movement will be inward (cotransport). At higher s, the movement of S will be outward (countertransport) (Fig. 3.10).

In the way of example, let us consider two cases, one where $s = 10^{-3}\ M$, $r = 10^{-3}\ M$, the other where $s = 10^{-1}\ M$ and $r = 1\ M$. Let $K = 10^{-3}\ M$ and $K' = 10^{-2}\ M$. Then in the first case the numerator above will be $(2.2 - 1.4)10^{-11}$ and hence positive, in the second case it will be $(0.11 - 1.2)10^{-3}$ and hence negative.

Competitive acceleration is also predictable in this model although defined by a very complex expression.

The type of polyvalent transport described above seems to fit some cases of monosaccharide transport in human erythrocytes and possibly that of amino acids in specialized cells.

3.2.6. Inhibition of Mediated Diffusion

Many of the transport systems exhibiting features of facilitated diffusion have been found to be inhibited in various ways (for the individual cases the last section of this book should be consulted). Although inhibitions of cell metabolism usually do not affect these transports (2,4-dinitrophenol, sodium azide, iodoacetate, etc.) other compounds, such as SH-group poisons or agents interacting with protein groups, are effective. The mechanism of inhibition may be rather varied.

In addition to the competition for the carrier site displayed by many related substrates which has been treated in ample detail on the preceding pages (S *vs.* R), there may be competitive inhibition simulated by compounds which are not transported into the cell but are bound to the membrane carrier immobilizing it (e.g., glucose under anaerobic conditions in the yeast *Rhodotorula gracilis*). The pertinent equation is then identical with eq. (22a), written as

$$v = V \frac{s}{s + K_s(1 + i/K_i)} \tag{56}$$

There may be cases of noncompetitive inhibition where the inhibitor is bound to the membrane near the substrate binding site, preventing the access of substrate to the carrier but itself being unaffected in its binding by the presence of substrate (e.g., corticosteroids in the sugar transport by human erythrocytes). In this case, the transport of S will be defined (in the simplest case of Model I) by

$$v = V \frac{s}{(s + K_s)(1 + i/K_i)} \tag{57}$$

where i is the inhibitor concentration and K_i the enzyme-inhibitor dissociation constant.

There may be cases of inhibition which are only partly competitive or partly noncompetitive. In the first case, it is assumed that the inhibitor affects the affinity of the transport system but combines with a different site than the proper substrate, the complexes CS and CIS breaking down at the same rate. The appropriate rate equation is then

$$v = V \frac{s[1 + (i/K_i)(K_s/K_s')]}{s[1 + (i/K_i)(K_s/K_s')] + K_s(1 + i/K_i)} \tag{58}$$

where K_s' is the equilibrium constant of the CIS complex.

In the second case, the complexes CS and CIS break down at different rates (or are transported at different rates) and the pertinent equation for the movement of S is

$$v = V \frac{s[1 + (i/K_i)(k'/k)]}{(s + K_s)(1 + i/K_i)} \tag{59}$$

where k and k' are the rate constants for the breakdown of CS and CIS, respectively, or, in the transport-limited process, the rate constants of movement across the membrane, i.e., D'_{CS} and D'_{CIS}, respectively.

The inhibitors discussed above are formally identical with those of enzyme kinetics and, in transport studies, may be applied directly to the movement of the carrier, this being (in the equilibrium models) the limiting step of the overall reaction (just as the breakdown of the ES complex is in equilibrium enzyme kinetics). However, carrier transport involves very probably catalysis by enzymes, either on one or on both sides of the membrane and various inhibitors can be visualized to interact either with the carrier or with the enzyme. Then, of course, the equilibrium treatment is no more valid and equations of Model III described above must be used. Assuming that $D_C' = D'_{CS}$, it is useful to rewrite the expression for rate v in terms of transport resistance $F = (s_I - s_{II})/v$. Then the expression is composed of additive terms representing the enzyme reaction on one side of the membrane, that on the other side of the membrane, and the transport across the membrane, respectively. Thus:

★
$$F = K_{CS}[(1 + s_I')(1 + s_{II}'')(1 + K_{EC}/c_t)/e_I k_1$$
$$+ (1 + s_I'')(1 + s_{II}'')(1 + K_{EC}/c_t)/e_{II} k_2$$
$$+ (1 + s_I'')(1 + s_{II}'')/c_t D'] \qquad (60)$$

where $s' = s/K_{ES}$, $s'' = s/K_{CS}$, k_1 and k_2 are the rate constants of the reaction

$$EC + S \underset{k_2}{\overset{k_1}{\rightleftharpoons}} ECS$$

(assumed to be equal on both sides of the membrane), and $K_{ES} = e \cdot s/es$ $= ec \cdot s/ecs$, $K_{EC} = e \cdot c/ec = es \cdot c/ecs$, and $K_{CS} = c \cdot s/cs$.

The inhibitor terms are simply added wherever appropriate. Thus, an inhibitor reacting with the enzyme on side I competitively with substrate will appear as $i' (= i/K_{I_{ES}})$ in the first parenthesis: $(1 + s_I' + i')$. An inhibitor reacting with the enzyme at side I competitively with carrier appears as $i' (= i/K_{I_{EC}})$ in the third parenthesis: $(1 + K_{EC}[1 + i']/c_t)$. An inhibitor reacting with the enzyme noncompetitively will appear as $i' (= i/K_{I_{En}})$ in $(1 + i')e_I k_1$. Similarly, an inhibitor reacting with the carrier C and not penetrating into the membrane will result in an extension of the last parenthesis but one as $i' (= ci/c)$. If the inhibitor is a penetrating one, the last term is extended by $i'[1 - s_I' s_{II}''/(1 + i')]$ where $i' = i/K_{I_{CS}}$.

The transport resistance increment ΔF obtained on application of the inhibitor is a useful quantity which can be plotted against s_{II} for various s_I

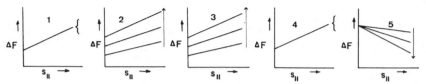

FIG. 3.11. Dependence of ΔF (the transport resistance increment) on substrate concentrations inside (s_{II}) and outside the cells (different straight lines for different concentrations) for various inhibitor types. The arrows in plot **2**, **3**, and **5** show the direction of increasing s_I, the braces in plots 1 and 4 indicate merging of lines at different values of s_I. Inhibitor affecting the reaction of: **1** enzyme with substrate, competitively; **2** enzyme with carrier, competitively; **3** enzyme with carrier, noncompetitively; **4** carrier with substrate (the inhibitor does not cross the membrane), **5** carrier with substrate (the inhibitor is itself transported).

resulting in families of curves as shown in Fig. 3.11, permitting a diagnosis of the inhibitor type (Rosenberg and Wilbrandt, 1962).

The inhibition by phloretin and its analogues of sugar transport in human erythrocytes was thus shown to be of the type I_{ES} or I_{CS} or, even better, of the type $I_{EC \cdot S}$, suggesting that phloretin interferes with substrate binding not only on the carrier but also on the enzyme.

4. PERMEATION OF IONS AND WATER

4.1. PERMEATION OF IONS AND MEMBRANE POTENTIALS

4.1.1. Introduction

For a discussion of the permeation of ions and of membrane potentials, the theoretical considerations concerning electrodiffusion of ions in homogeneous media (section 2.2.) are of utmost importance. However, the phenomena occurring on membranes—and especially on cell membranes—are more complicated than those in a homogeneous medium. The principal sources of these complications are associated with two factors.

1. The permeation across membranes, unlike the electrodiffusion in homogeneous media, involves transitions between different phases, occurring on the two surfaces of the membrane.

2. The flows of ions across membranes may be subject to various interactions. Direct coupling between the flow of an ion and cell metabolism, known as active transport, is the most important interaction taking place in cell membranes.

It appears that no rigorous derivation of formulae for ion fluxes and membrane potentials which would take into account these two factors is possible at present. Actually, most often such factors are neglected in the derivations of theoretical formulae altogether. The resistances encountered in phase transitions and the surface potentials are assumed to be negligibly small; the mechanisms by which the ionic pumps bring about membrane potentials are associated only with their ability to produce concentration gradients of actively transported ions and the custom is to designate as active simply those parts of ionic fluxes which cannot be explained by simple physicochemical forces. Still, the formulae obtained in this way do describe satisfactorily the observed phenomena in a number of cases. The description of the resting and the action potential in nerve by Hodgkin and Katz (1949) using a simple form of the Goldman equation is perhaps the best known example. This may suggest that the complicating factors are sometimes of little importance. In other cases, however, the situation is different: The presence of electrogenic pumps cannot be overlooked and equations of electrodiffusion are not applicable so that qualitative interpretations based on a consideration of simple electrical analogues may be the most that can be achieved at the moment.

Before a survey of some of the current ideas in this field is attempted, a short discussion of the relation between the membrane potential and the equilibrium distribution of ions across the membrane may be useful. This relation is valid independently of the complicating factors mentioned above and is often helpful in the analysis of membrane phenomena involving ions.

4.1.2. Equilibrium of Ions across Membranes

Let us consider a membrane separating two solutions which contain the same salt dissociating into two ionic species at different concentrations. Let the membrane be permeable to only one of the two ions present in the system, the ion of the opposite sign being, say, too large to penetrate through the narrow pores in the membrane or being expelled from the pores by membrane charges of the same sign situated there. (The way in which ions of one sign are prevented from crossing the membrane is immaterial; even charges fixed to a framework of macromolecules and confined thus to some region in solution may play the role of our nonpermeating ions, the boundary of the macromolecular region being considered as a "membrane.") In such a situation, ions of one sign cannot possibly cross the membrane, whereas ions of the opposite sign may be exchanged across the membrane quite freely. But it is important that no net transfer of an analytically estimable amount of the permeating ion can take place across the membrane, although there is a concentration gradient favoring this process. The reason for this is that the transfer of these ions involves separation of charges and hence generation of membrane potential (strictly speaking a difference of the electrical potential across the membrane which is called the membrane potential). The membrane potential, equivalent to the difference in the chemical potential by which the escape tendency of the permeating ion can be expressed, prevents any further transfer. Ions of both signs are thus in equilibrium but their equilibria are of fundamentally different kinds. The equilibrium of the nonpermeating ion is due to passive resistances preventing any movement across the membrane and represents thus a kind of static equilibrium. The equilibrium of the permeating ion, on the other hand, is due to the active tendencies of the system, which balance each other, and is thus a dynamic equilibrium. The condition which is to be satisfied in equilibria of this kind is that the electrochemical potential of the ion is the same everywhere in the system. The electrochemical potential, $\bar{\mu}_j$, of an ion j, may be expressed by the formula

$$\bar{\mu}_j = \mu_{0j} + RT \ln(fc_j) + z_j F\psi \tag{1}$$

where $f \cdot c_j$ is the activity of the ion (in dilute solutions, the activity coefficient f approaches 1 and may be neglected), ψ is the electrical potential, and other symbols have their usual meaning. In aqueous solutions of similar composition and at the same temperature the value of μ_0 (expressing that part of the partial molal free energy of the ion which is related neither to its concentration nor to the electrical potential) is likely to be the same. Hence from the equality of the electrochemical potential on one side of the membrane (in the "outer" solution, subscript o) with that on the other side (the "inner" solution, subscript i) it follows that

$$RT \ln(f_o c_o) + zF\psi_o = RT \ln(f_i c_i) + zF\psi_i \tag{2}$$

or

★ $$E_m = \psi_i - \psi_o = \frac{RT}{zF} \ln \frac{f_o c_o}{f_i c_i} \tag{3}$$

If the ionic strengths of the solutions on the two sides of the membrane are similar, so that $f_o \simeq f_i$ (or, as the first approximation, if the values of the activity coefficients cannot be evaluated)

$$E_m = \frac{RT}{zF} \ln \frac{c_o}{c_i} = \frac{RT}{zF} 2.303 \log \frac{c_o}{c_i} \tag{4}$$

Values of RT/F 2.303 at different temperatures are presented in Table 4.1.

It may be remembered that for showing that the gradient of chemical potential is the driving force of diffusion (section 2.1) the validity of eq. (4) was assumed. In order to avoid circular reasoning let us consider eq. (4) as an experimentally established fact.

Table 4.1. Value of $S = 2.303\ RT/F$ at Different Temperatures

C°	S	C°	S
0	54.2	24	58.9
5	55.2	25	59.1
10	56.2	26	59.3
15	57.2	27	59.5
20	58.1	28	59.7
21	58.3	29	59.9
22	58.5	30	60.1
23	58.7	37	61.5

A simple example of the above equilibrium is given in Fig. 4.1, where a membrane separates two solutions of potassium chloride, their concentrations being at a ratio of 10 : 1. If the membrane is ideally cation-permeable (e.g., made of a cation-exchange resin), an equilibrium is readily established when (at 20°C) the more concentrated solution is 58 mV more negative than the more dilute solution ($z = +1$ for K^+ and $\log 10 = 1$). Only a minute amount of potassium ions has been displaced across the membrane as just corresponds to providing the membrane with the separated charges so as to attain the appropriate membrane equilibrium potential. The deviations from electroneutrality are exceedingly small, out of the reach of analytical techniques and may be safely neglected when conditions of electroneutrality for the two solutions are written.

If the membrane in Fig. 4.1 were ideally anion-permeable, the sign of the membrane potential would be reversed ($z = -1$ for Cl^-).

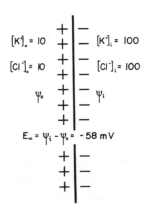

FIG. 4.1. Equilibrium of potassium cations across an ideally cation-permeable membrane. For explanation see the text.

Equilibria of bivalent ions correspond to equilibrium potentials one-half those of monovalent ions for the same concentration ratio, z in eq. (4) being ± 2.

It may be seen that the equilibria described so far are not complete; the possibility of water movement due to different osmotic pressure at the two sides of the membrane was ignored. To prevent such movement of water, hydrostatic pressure must be applied to the more concentrated solution or else the osmotic pressure of the dilute solution must be increased (e.g., by adding a nonpermeating nonelectrolyte to it); only then is a complete thermodynamic equilibrium attained.

It may be of interest to recall here how equilibrium membrane potentials like the one in Fig. 4.1 are measured. With electrodes reversible to the ion in equilibrium the potential difference measured between the two solutions would be zero (in the present case with potassium-selective electrodes; cf. chapter 12). With reference electrodes connected with the two solutions by means of salt bridges, the correct value of -58 mV would be approximately measured. With electrodes reversible to chloride ions (e.g., silver–silver chloride electrodes) placed directly into the two solutions a difference of 116 mV would be found. The concentration effects of ions on the electrode potentials of reversible electrodes are thus seen to be subtracted or added to the membrane potential in the circuit.

The equilibria considered above are special cases of Gibbs–Donnan equilibria. This name is usually associated with a more general situation where another permeating ion of the same sign is present in addition to the nonpermeating ion. The greater the amount of the salt containing permeating ions of both signs, the lower the ratio of ion distribution across the membrane and hence the lower the membrane potential. This consequence of the presence of permeable salts is often referred to as the suppression of Donnan effect. Equilibrium concentrations of the diffusible ions under the conditions of these more general equilibria may be calculated from equations like eq. (4) and from conditions of electroneutrality. As an example, let the situation shown in Fig. 4.2 be considered. The inner space shown in Fig. 4.2 is surrounded by a rigid membrane, so that its volume

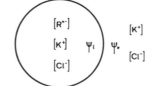

FIG. 4.2. An example of Gibbs–Donnan equilibrium. For explanation see the text.

is fixed and the concentration of a nonpermeating polyvalent anion $[R^{n-}]$ is, therefore, constant. For the sake of simplicity, the outer space is so large that the concentration of potassium cations and of chloride anions in it cannot appreciably change, being equal to a constant s:

$$[K^+]_o = [Cl^-]_o = s \qquad (5)$$

The membrane is permeable to both potassium cations and chloride anions; whatever their initial concentration, in equilibrium they must satisfy the equilibrium conditions of eq. (4), any possible effects of activity coefficients being neglected:

$$\bigstar \qquad E_m = \frac{RT}{F} \ln \frac{[K^+]_o}{[K^+]_i} = -\frac{RT}{F} \ln \frac{[Cl^-]_o}{[Cl^-]_i} = \frac{RT}{F} \ln \frac{[Cl^-]_i}{[Cl^-]_o} \qquad (6)$$

and hence

$$\frac{[K^+]_o}{[K^+]_i} = \frac{[Cl^-]_i}{[Cl^-]_o} = r \qquad (7)$$

where r is the so-called Donnan ratio.

Furthermore, the condition of electroneutrality for the inside ions must be satisfied (for the outer ions it is already implied in eq. (5)),

$$n[R^{n-}] + [Cl^-]_i = [K^+]_i \qquad (8)$$

By expressing $[Cl^-]_i$ from eq. (7) and using equations (5) and (8) we have

$$[Cl^-]_i = \frac{[K^+]_o[Cl^-]_o}{[K^+]_i} = \frac{s^2}{n[R^{n-}] + [Cl^-]_i}$$

or

$$[Cl^-]_i^2 + n[R^{n-}][Cl^-]_i - s^2 = 0$$

and hence

$$[Cl^-]_i = -\frac{n[R^{n-}]}{2} + \sqrt{\left(\frac{n[R^{n-}]}{2}\right)^2 + s^2} \qquad (9)$$

Similarly, for $[K^+]_i$

$$[K^+]_i = \frac{n[R^{n-}]}{2} + \sqrt{\left(\frac{n[R^{n-}]}{2}\right)^2 + s^2} \qquad (10)$$

From equilibrium concentrations given by equations (9) and (10) the Donnan ratio can be calculated and the membrane potential expressed. Both the potassium and chloride ions now being in a thermodynamic equilibrium, potassium- as well as chloride-reversible electrodes will measure no potential difference in the system. Reference electrodes connected with the solutions by means of salt bridges will show approximately the theoretical value.

Equations (*3*) or (*4*) are useful in analyzing ion distribution across biological membranes. If a time-invariant distribution of a permeating ion prevails across a membrane without these equations being satisfied, it may be suspected that the steady distribution is due to a continual expenditure of metabolic energy, i.e., to the active transport of the ion in question.

4.1.3. Nonequilibrium Membrane Potentials and Ion Fluxes

The electrical potential difference across the membrane proper can be satisfactorily expressed for a neutral membrane by Planck's equation (*2.76*), in which concentrations of ions on the membrane side of the boundary between membrane and solution are used, rather than their concentrations in the solutions. The derivation of this equation for a membrane was carried out by Polissar (1954) who also showed how, under the condition of equilibrium at the membrane surfaces, the concentrations of ions in solution can be used in Planck's equation and the total transmembrane potential (including two interfacial potential differences) used for calculations. A similar procedure will be described later in this section when discussing Goldman's equation.

Planck's equation is very complicated, does not give an explicit expression for the membrane potential, and it would be difficult to test its validity experimentally. For this reason, one resorts to approximate solutions of the electrodiffusion problem which give more manageable results. The approximations involved are probably less serious than the often obligatory disregard for interfacial phenomena and electrogenicity of ionic pumps.

In section 2.2, two simplifying assumptions were encountered, leading to solutions of electrodiffusion which, being rigorous under the specified conditions, could be used as approximations elsewhere. The first of them is the assumption of linear concentration gradients, which leads to Henderson's formula for the liquid junction potential (eq. (*2.81*)) and to the related formula for the flux of an individual ion (eq. (*2.83*)). These equations have a general validity for mixed boundaries, and, as pointed out by Polissar (1954), they are hence less suitable for describing observations in membranes in which the local concentration is a function not only of the position but also of the electrostatic pattern. In two special cases, the potential difference predicted by Henderson's formula is the same as that given by Planck's formula, irrespective of mixing and, as shown by Teorell (1953), not only in free solution but also in a "fixed charge" membrane (see later). The two cases are the following: (1) Only one uni-univalent electrolyte diffuses in the system. (2) The total concentrations in the two solutions are equal.

The equations used almost exclusively at present for describing passive transport of ions and membrane potentials across plasma membranes are Goldman's equations, called also constant-field equations. Equations of the form of eq. (2.89) are used to describe the passive fluxes of individual ionic species and an expression of the form of eq. (2.92) is used for defining the membrane potential. Generally only three ions, K^+, Na^+, and Cl^-, are taken into consideration. To be able to use their concentrations in solution, $[K^+]$, $[Na^+]$, and $[Cl^-]$, in place of their concentrations in the membrane, c_{K^+}, c_{Na^+}, and c_{Cl^-}, which are not accessible to direct measurement, Hodgkin and Katz (1949) made use of partition (or disrtibution) coefficients k_{K^+}, k_{Na^+}, and k_{Cl^-}

$$c_{K^+} = k_{K^+}[K^+] \qquad c_{Na^+} = k_{Na^+}[Na^+] \qquad c_{Cl^-} = k_{Cl^-}[Cl^-] \qquad (11)$$

an approximation which will be discussed later. In this way, the passive net flux of, for example, potassium ions across the membrane may be expressed by using eq. (2.89) as

$$\Phi_{K^+} = FU_{K^+}k_{K^+}\frac{E_m}{\delta}\frac{[K^+]_i - [K^+]_o\,e^{-FE_m/RT}}{e^{-FE_m/RT}} \qquad (12)$$

z being $+1$ for K^+; δ is the thickness of the membrane. The Goldman–Hodgkin–Katz formula for the membrane potential becomes (from eq. (2.92))

$$E_m = \frac{RT}{F}\ln\frac{U_{K^+}k_{K^+}[K^+]_o + U_{Na^+}k_{Na^+}[Na^+]_o + U_{Cl^-}k_{Cl^-}[Cl^-]_i}{U_{K^+}k_{K^+}[K^+]_i + U_{Na^+}k_{Na^+}[Na^+]_i + U_{Cl^-}k_{Cl^-}[Cl^-]_o} \qquad (13)$$

On introducing the permeability constant for ion j

$$P_j = \frac{RTU_jk_j}{\delta} \qquad (14)$$

equation (12) becomes

★ $$\Phi_{K^+} = P_{K^+}\frac{FE_m/RT}{e^{-FE_m/RT} - 1}([K^+]_i - [K^+]_o\,e^{-FE_m/RT}) \qquad (15)$$

Analogously, for sodium ions it may be written as

$$\bigstar \qquad \Phi_{Na^+} = P_{Na^+} \frac{FE_m/RT}{e^{-FE_m/RT} - 1} ([Na^+]_i - [Na^+]_o e^{-FE_m/RT}) \qquad (16)$$

and for chloride anions (for which $z = -1$ in eq. (2.95)) as

$$\bigstar \qquad \Phi_{Cl^-} = P_{Cl^-} \frac{FE_m/RT}{e^{-FE_m/RT} - 1} ([Cl^-]_i e^{-FE_m/RT} - [Cl^-]_o) \qquad (17)$$

Using definition (14) the equation for the membrane potential (13) becomes

$$\bigstar \qquad E_m = \frac{RT}{F} \ln \frac{P_{K^+}[K^+]_o + P_{Na^+}[Na^+]_o + P_{Cl^-}[Cl^-]_i}{P_{K^+}[K^+]_i + P_{Na^+}[Na^+]_i + P_{Cl^-}[Cl^-]_o} \qquad (18)$$

Equation (18) would be also obtained from the condition that in the absence of an external circuit the sum of the fluxes of the cations across the membrane (eq. (15) + eq. (16)) is equal to the flux of the anions (eq. (17)). In some cases eq. (18) may be simplified still further. For example, if chloride ions are in a thermodynamic equilibrium across the membrane, eq. (4) will relate their distribution to the membrane potential (possible differences in the activity coefficients being neglected). Moreover, an equation analogous to eq. (18) will be derived from the condition that, if there is no current flowing across the membrane, the sum of the passive potassium net flux (eq. (15)) and of the passive sodium net flux (eq. (16)) is zero, the passive net flux of chloride anions (eq. (17)) vanishing when chloride anions are in equilibrium. Hence

$$E_m = \frac{RT}{F} \ln \frac{P_{K^+}[K^+]_o + P_{Na^+}[Na^+]_o}{P_{K^+}[K^+]_i + P_{Na^+}[Na^+]_i} = \frac{RT}{F} \ln \frac{[Cl^-]_i}{[Cl^-]_o} \qquad (19)$$

is valid when chloride anions are in thermodynamic equilibrium.

Goldman's equations in the above form neglect both the possible complications brought about by the interfacial phenomena and the possible effects of electrogenic pumps on the generation of the membrane potential. Let us deal first with the interfacial phenomena.

We may imagine the total transmembrane potential E to be composed of three separate potential differences: E_o corresponding to the first transition region, E_m corresponding to the membrane proper, and E_i, corresponding to the other transition region. Thus

$$E = E_o + E_m + E_i \qquad (20)$$

Neglecting of the difference between E and E_m, implicit in the equations given above, is thus equivalent to the assumption that $E_o = -E_i$ or, in a special case, $E_o = E_i = 0$.

The surface potentials E_o and E_i were considered explicitly in the fixed-charge theory of Teorell, Mayer, and Sievers, as reviewed by Teorell (1953). Equilibrium is assumed to prevail here at the boundaries of the membrane (in other words, the membrane is assumed to be thick so that the main resistance to the ion fluxes is localized in the membrane proper and equilibria on the surfaces have enough time to be practically established) so that the surface potentials may be expressed from the Donnan ratios for the ions and these ratios, in their turn, from the charge density in the membrane, using equations such as (9) and (10). The fixed-charge theory, however, may be better applicable to artificial membranes than to cell membranes. Thus Solomon (1960) calculated that the absurd value of over 100 mol/liter for the density of fixed charges would be required to account for the permeability properties of the red cell membrane from the point of view of the fixed-charge theory. Solomon concluded that a combination of fixed-charge repulsion and of steric hindrance in narrow pores, as suggested by Sollner (1955), is a more likely explanation. For this reason, a more general approach to the surface potentials undertaken by Polissar (1954) will be briefly reported here.

If equilibria cannot be assumed to prevail on the membrane surfaces, terms taking into account the resistance of transition regions can be introduced into equations expressing the fluxes of the individual ion species (Polissar, 1954). But even if the main resistance encountered by the ions is localized inside the membrane proper so that equilibria are closely approximated at the membrane surfaces, rather important modifications of the flux and potential equations would still have to be carried out if the two surface potentials E_o and E_i are known. The character of these modifications may be illustrated by Polissar's (1954) development of the expression for the membrane potential which holds, in analogy with eq. (18), for a constant field within the membrane but takes into account the nonzero values of the surface potentials E_o and E_i and assumes equilibria at the membrane surfaces.

As pointed out by Polissar (1954), the partition (or distribution) coefficient in the usual sense vould give complete information on the equilibrium distribution of an ion between the solution and the membrane only if there were no potential difference between the two, or if the ion did not carry a charge. This being not the case, the partition coefficient must be multiplied by an exponential factor, containing the appropriate potential difference in the exponent. Thus in our terminology:

$$
\begin{aligned}
c_{K_o^+} &= k_{K^+}[K^+]_o\, e^{-FE_o/RT} \\
c_{K_i^+} &= k_{K^+}[K^+]_i\, e^{FE_i/RT} \\
c_{Na_o^+} &= k_{Na^+}[Na^+]_o\, e^{-FE_o/RT} \\
c_{Na_i^+} &= k_{Na^+}[Na^+]_i\, e^{FE_i/RT} \\
c_{Cl_o^-} &= k_{Cl^-}[Cl^-]_o\, e^{FE_o/RT} \\
c_{Cl_i^-} &= k_{Cl^-}[Cl^-]_i\, e^{-FE_i/RT}
\end{aligned}
\tag{21}
$$

Equations (21) rather than equations (11) should be introduced into eq. (2.98) and then, with the definition of the permeability constant (14), eq. (22) instead of eq. (18) is obtained:

$$
E_m = \frac{RT}{F}\, \ln \frac{P_{K^+}[K^+]_o e^{-FE_o/RT} + P_{Na^+}[Na^+]_o e^{-FE_o/RT} + P_{Cl^-}[Cl^-]_i e^{-FE_i/RT}}{P_{K^+}[K^+]_i e^{FE_i/RT} + P_{Na^+}[Na^+]_i e^{FE_i/RT} + P_{Cl^-}[Cl^-]_o e^{FE_o/RT}}
\tag{22}
$$

Equation (22) is valid for the constant field in the membrane and equilibria at the membrane surfaces. Recalling that the total transmembrane potential $E = E_o + E_m + E_i$ (eq. (20)) and since, obviously,

$$
E_o + E_i = \frac{RT}{F}\, \ln(e^{FE_o/RT} \cdot e^{FE_i/RT})
\tag{23}
$$

$$
E = \frac{RT}{F}\, \ln \frac{P_{K^+}[K^+]_o + P_{Na^+}[Na^+]_o + P_{Cl^-}[Cl^-]_i\, e^{F(E_o - E_i)/RT}}{P_{K^+}[K^+]_i + P_{Na^+}[Na^+]_i + P_{Cl^-}[Cl^-]_o\, e^{F(E_o - E_i)/RT}}
\tag{24}
$$

The surface potentials E_o and E_i are not accessible to direct measurement owing not only to the minute dimensions of the cell membrane, but primarily due to the fact pointed out by Gibbs (1899) and Guggenheim (1967) (see chapter 12) that potential differences between two different phases are excluded from physical measurement in principle. Consequently, their values cannot be determined by using partition coefficients of the ions, since their estimation would require measurements in which a zero potential difference between two different media would exist or "ions" carrying no charge would be used. However, by assuming various possible values of the partition coefficients Polissar (1954) was able to give an estimate of the quantitative significance of the two surface potentials. His procedure, applicable to neutral membranes, was the following.

Electroneutrality of ion concentrations inside the membrane is postulated. For the outer boundary of the membrane

$$
c_{K_o^+} + c_{Na_o^+} = c_{Cl_o^-}
\tag{25}
$$

By substituting the ion concentrations in the adjacent solutions from equations (21) we obtain

$$k_{K^+}[K^+]_o e^{-FE_o/RT} + k_{Na^+}[Na^+]_o e^{-FE_o/RT} = k_{Cl^-}[Cl^-]_o e^{FE_o/RT} \qquad (26)$$

and, by solving for E_o,

$$E_o = \frac{RT}{F} \ln \left(\frac{k_{Na^+}[Na^+]_o + k_{K^+}[K^+]_o}{k_{Cl^-}[Cl^-]_o} \right)^{1/2} \qquad (27)$$

An analogous equation may be written for the inner membrane potential. It may be seen that only the ratios of distribution coefficients are of importance in determining the surface potentials. Using the experimental data of Hodgkin and Katz (1949) obtained on the squid axon and assuming various ratios of the partition coefficients, Polissar (1954) reached the conclusion that: "...surface potentials are, in general, of the same order as the observed transmembrane potential, and that the value of E_m is apt to differ from the observed value of E by a large amount."

It may be perhaps concluded that surface potentials are likely to be of importance and that if a satisfactory fit for the membrane potential is not obtained using eq. (18) under various conditions, a variation of the surface potentials in eq. (24) may be responsible, rather than variation of permeability constants in eq. (18), reflecting structural changes in the membrane.

Another factor capable of influencing seriously the membrane potentials is the presence of an electrogenic pump, i.e., the active transport of ions not rigidly coupled with the active transport of the same number of ions of the same sign in the opposite direction or of the opposite sign in the same direction. The presence of such electrogenic pumps in the cell membrane makes itself apparent by generation of membrane potentials which are under specified conditions higher than the equilibrium potential of any of the ion species present. Electrogenic sodium pumps were demonstrated by careful measurements in the frog muscle (Kernan, 1962; Keynes and Rybová, 1963; Cross et al., 1965), in the epithelial cells of the toad bladder (Essig et al., 1963; Frazier and Leaf, 1963) as well as in rat proximal tubules (Marsh and Solomon, 1964).

The compatibility of Goldman–Hodgkin–Katz equation (18) with the presence of an electrogenic pump in the frog muscle was analyzed by Geduldig (1968) who concluded that: "...in frog muscle almost any measurable electrogenicity should indicate an imbalance of diffusion currents which are too great to justify rigorous use of the constant-field equation," whereas if "...pumps are only slightly electrogenic (in frog muscle, at any rate), the net diffusion current will not substantially affect the constant-field equation." By "measurable electrogenicity," a potential difference of about 3 mV was meant by the author, corresponding to current density of about 1 $\mu A \cdot cm^{-2}$ (approximately 10^{-11} $mol \cdot cm^{-2} \cdot sec^{-1}$). Fluxes much lower than that through the electrogenic pump may thus be tolerated, the constant-field equation still retaining its validity.

The difficulties encountered when attempting to account for the contribution of electrogenic pumps to the membrane potential, even under the constant-field assumption, are substantial. Frumento's (1965) treatment of the situation in which at $t = 0$ the pumps in a muscle fiber, previously inactivated (e.g., due to cold) begin to work, may well illustrate this point. Frumento's basic equation relates the fluxes of individual ions to the rate of the membrane potential change; any imbalance in ionic fluxes charges the membrane capacity C_m

$$\Phi_{Na^+,p} + \Phi_{K^+,p} + \Phi_{Na^+} + \Phi_{K^+} + \Phi_{Cl^-} = \frac{C_m \, dE_m}{F \, dt} \tag{28}$$

Here $\Phi_{Na^+,p}$ and $\Phi_{K^+,p}$ are fluxes through the pumps, Φ_{Na^+}, Φ_{K^+}, and Φ_{Cl^-} are passive fluxes which can be expressed by equations (15), (16), and (17). However, the intracellular concentrations in these equations are not constant but can be expressed by integrals

$$([Na^+]_i)_t = ([Na^+]_i)_0 + \frac{A}{V} \int_0^t \Phi_{Na^+} \, dt + \frac{A}{V} \int_0^t \Phi_{Na^+,p} \, dt$$

$$([K^+]_i)_t = ([K^+]_i)_0 + \frac{A}{V} \int_0^t \Phi_{K^+} \, dt + \frac{A}{V} \int_0^t \Phi_{K^+,p} \, dt \tag{29}$$

$$([Cl^-]_i)_t = ([Cl^-]_i)_0 + \frac{A}{V} \int_0^t \Phi_{Cl^-} \, dt$$

where A/V is the surface volume ratio for the fibre. Moreover, it is assumed that the pumped fluxes of sodium and potassium are linearly related

$$\Phi_{K^+,p} = r\Phi_{Na^+,p} \tag{30}$$

Sodium and potassium being pumped in opposite directions, r is a negative number; if the system behaves as the sodium electrogenic pump, it is a negative fraction. Combining equations (15), (16), (17), (29), (28), and (30), an integro-differential system is obtained

$$\Phi = P_{Na^+} \frac{FE_m/RT}{e^{-FE_m/RT} - 1} \times$$

$$\left\{ ([Na^+]_i)_0 + \frac{A}{V} \int_0^t \Phi_{Na^+} \, dt + \frac{A}{V} \int_0^t \Phi_{Na^+,p} \, dt - [Na^+]_o \, e^{-FE_m/RT} \right\} \quad (31)$$

$$\Phi = P_{K^+} \frac{FE_m/RT}{e^{-FE_m/RT} - 1} \times$$

$$\left\{ ([K^+]_i)_0 + \frac{A}{V} \int_0^t \Phi_{K^+} \, dt + \frac{rA}{V} \int_0^t \Phi_{Na^+,p} \, dt - [K^+] \, e^{-FE_m/RT} \right\} \quad (32)$$

$$\Phi = P_{Cl^-} \frac{FE_m/RT}{e^{-FE_m/RT} - 1} \times$$

$$\left\{ \left[([Cl^-]_i)_0 + \frac{A}{V} \int_0^t \Phi_{Cl^-} \, dt \right] e^{-FE_m/RT} - [Cl^-]_o \right\} \quad (33)$$

$$(1 + r)\Phi_{Na^+,p} + \Phi_{Na^+} + \Phi_{K^+} - \Phi_{Cl^-} - \frac{C_m}{F} \frac{dE_m}{dt} = 0 \quad (34)$$

The initial concentrations and permeabilities being given, Frumento's system of equations contains six unknowns: Φ_{Na^+}, Φ_{K^+}, Φ_{Cl^-}, $\Phi_{Na^+,p}$, r, and V. Two of these are thus to be assumed or determined experimentally. For instance, assuming that $\Phi_{Na^+,p}$ is represented by a sigmoid function and that one potassium ion enters for three sodium ions expelled (so that $r = -1/3$), Frumento (1965) was able to calculate the time change of the membrane potential E_m. To do so, further simplifying assumptions were necessary, namely that for slow changes of E_m the last term in eq. (34) may be neglected, whereas for rapid changes of E_m the intracellular concentrations may be considered constant for a short time and fluxes during that period may be calculated from equations (15), (16), and (17) instead of (31), (32), and (33). Still, the calculations are not simple and may be seen to contain a number of approximations—among others it is assumed that r is a constant, that the rate of the active ion transport by the electrogenic pump is independent of the membrane potentials, etc.

A different approach to the description of ionic fluxes and membrane potentials, perhaps less rigorous but often more convenient, is provided by the use of "electrical analogues" or "equivalent circuits." One of them, probably the most popular, is shown in Fig. 4.3. In the hands of Hodgkin and Huxley it proved its usefulness in explaining the excitability phenomena in the nerve membrane (see Hodgkin, 1958, for references). Its merits and implications were discussed by Dainty (1960) and by Finkelstein and Mauro (1963). In this analogue or model the current due to the jth species I_j related

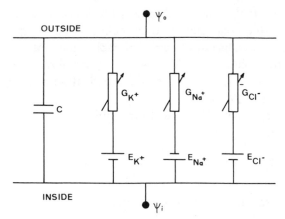

Fig. 4.3. Electrical analogue (or "model" or "equivalent circuit") of the cell membrane. C is the capacitance of the membrane, the other symbols being explained in the text.

to the flux of this ion Φ_j by $I_j = z_j F \Phi_j$ is given by

$$I_j = G_j(E_j - E_m) \tag{35}$$

where E_j is the equilibrium potential for the ion species in question

$$E_{K^+} = \frac{RT}{F} \ln \frac{[K^+]_o}{[K^+]_i} \qquad E_{Na^+} = \frac{RT}{F} \ln \frac{[Na^+]_o}{[Na^+]_i} \qquad E_{Cl^-} = \frac{RT}{F} \ln \frac{[Cl^-]_i}{[Cl^-]_o}$$

E_m is the membrane potential

$$E_m = \psi_i - \psi_o$$

and G_j is the conductance of the membrane for the jth ion. When the total membrane current is zero, the membrane potential is given by

$$\bigstar \qquad E_m = \frac{G_{K^+}E_{K^+} + G_{Na^+}E_{Na^+} + G_{Cl^-}E_{Cl^-}}{G_{K^+} + G_{Na^+} + G_{Cl^-}} \tag{36}$$

or, if the chloride ion happens to be in equilibrium, $E_{Cl^-} = E_m$

$$E_m = \frac{G_{K^+}E_{K^+} + G_{Na^+}E_{Na^+}}{G_{K^+} + G_{Na^+}} \tag{37}$$

and, moreover, if $G_{K^+} \gg G_{Na^+}$, then $E_m \simeq E_{K^+}$, whereas if $G_{Na^+} \gg G_{K^+}$, then $E_m \simeq E_{Na^+}$.

It might seem that if the membrane is not labile so that its structure does not change, the conductances G_j remain constant. However, this is rarely the case, the conductances being probably functions of the ion concentrations in the solutions as well as of the membrane potential. The physical meaning of G_j is made clear by the derivation of eq. (35) by Finkelstein and Mauro (1963).

The differential equation for the electrodiffusion of the ion (eq. (2.46)) may be written in the form

$$\Phi_j = -c_j U_j z_j F\left(\frac{RT}{z_j F}\frac{d\ln c_j}{dx} + \frac{d\psi}{dx}\right) \tag{38}$$

The current due to the jth ion is, therefore,

$$I_j = -c_j U_j z_j^2 F^2\left(\frac{RT}{z_j F}\frac{d\ln c_j}{dx} + \frac{d\psi}{dx}\right) \tag{39}$$

For the steady state in which I_j would be independent of x, the integration from the otuside (o) to the inside (i) membrane face gives

$$I_j \int_0^i \frac{dx}{c_j U_j z_j^2 F^2} = -\left[\frac{RT}{z_j F}\ln\frac{(c_j)_i}{(c_j)_o} + \psi_i - \psi_o\right]$$

$$= \frac{RT}{z_j F}\ln\frac{(c_j)_o}{(c_j)_i} - \dot{E}_m = E_j - E_m \tag{40}$$

On comparing eqs. (40) and (35) it may be seen that

$$G_j = \frac{1}{\displaystyle\int_0^i \frac{dx}{c_j U_j z_j^2 F^2}} \tag{41}$$

so that G_j is the integral conductance of the membrane proper toward the jth ion (the resistances of the transition regions being neglected).

The way in which the electrogenic pumps may be introduced into the electrical analogue of the cell membrane is suggested by the equivalent circuit of the isolated frog skin with the same solutions on both sides (Fig. 4.4) which was shown by Ussing and Zerahn (1951) to account well for the electrical properties of this membrane. In Fig. 4.4 $E_{Na\ active}$ is the electromotive force of the sodium pump, G_{Na} the conductance of the skin toward the actively transported sodium ions, and G_Σ the conductance toward all permeating ions. Equilibrium potentials of all ions are zero, the solutions of the same composition being applied to both sides of the skin. It appears that, except for the last condition, the cell membrane in which

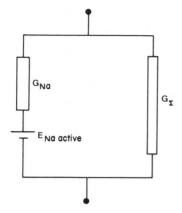

FIG. 4.4. Equivalent circuit for the frog skin suggested by Ussing and Zerahn (1951). Explanation in the text.

an electrogenic sodium pump is operative has much in common with a complex membrane like frog skin. Combining the features of the analogues in Fig. 4.3 and in Fig. 4.4, the nonzero equilibrium potentials can be accounted for in the analogue in Fig. 4.5. The circuit is almost identical with that used by Frumento (1965) when deriving eq. (28). In analogy to the discussion of eq. (37) we may assume that the membrane potential will approximate the electromotive force corresponding to the highest conductance. Thus when G_{K^+} is much higher than the other individual conductances —and this may intuitively correspond to the situation where concentrations of potassium ions at the two membrane sides are high, since then also the concentration of these ions in the membrane will be high—the membrane potential will be practically equal to the potassium equilibrium potential.

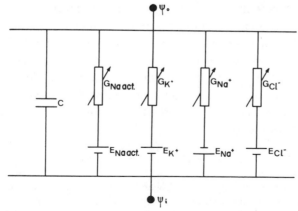

FIG. 4.5. An electrical analogue of the cell membrane with an electrogenic sodium pump. Explanation in the text.

This situation has been often reported in, for istance, muscle fibers. If, on the other hand, the concentration of the potassium ions is low on both sides of the membrane, the intracellular potassium in the fiber being exchanged for sodium by a leaching procedure in an ice-cold saline and when, subsequently at a higher temperature, a massive extrusion of sodium by the pump takes place (Kernan, 1962; Keynes and Rybová, 1963) the membrane potential will be nearer the electromotive force of the sodium pump. It may be the case that quantitative considerations of this analogue would finally lead to satisfactory models of the cell membrane containing electrogenic pumps.

4.2. TRANSPORT OF WATER

The phenomenon of water transport across an ideal semipermeable membrane will be discussed in this section, together with the differences between the so-called diffusional and osmotic permeability coefficients, their bearing on the problem of pores in membranes and the complications brought about by the presence of unstirred layers at the surface of the membrane. Subsequently, transport of water across membranes which are leaky for solutes will be considered and some principles of the thermodynamics of irreversible processes will be used to describe the interactions between water and solute flows in such membranes.

4.2.1. Ideal Semipermeable Membrane

An ideal semipermeable membrane is permeable to the solvent, e.g., to water, and impermeable to the solute. Theoretical considerations of water transport across such a membrane are much simpler than in the case of a leaky membrane, where interactions between the water and solute flows may take place and where the volume flow is due not only to the volume flow of water but also to the volume flow of solute. Since there are no interactions * in our present model, the water flow across an ideal semipermeable membrane may be expressed by Teorell's simple equation (2.11) which will have the form

$$\Phi_w = -U_w c_w \operatorname{grad} \mu_w \qquad (42)$$

* No interactions between *water flow* and *solute flow*; later it will be seen that there is another way of looking at an ideal semipermeable membrane, *viz.*, as a place of pronounced interaction between *volume flow* and *exchange flow*, but it need not concern us here.

or, with the observable movement across the membrane proceeding in the direction of the x-axis only,

$$\Phi_w = -U_w c_w \frac{d\mu_w}{dx} \tag{43}$$

Here Φ_w is the water flux—the amount of water penetrating per unit time per unit area normal to the direction of the transport—U_w is the mobility of water in the membrane, and $d\mu_w/dx$ is the driving force, the derivative of the chemical potential of water. For a thin membrane, the derivative may be replaced by the ratio of finite differences

$$\Phi_w = -U_w \bar{c}_w \frac{\Delta\mu_w}{\Delta x} \tag{44}$$

or

$$\Phi_w = -\frac{U_w \bar{c}_w}{x} (\mu_{wi} - \mu_{wo}) \tag{45}$$

where Δx is the thickness of the membrane, $\bar{c}_w$ is an average concentration of water in the membrane, and μ_{wi} and μ_{wo} are the chemical potentials of water on the inner and outer membrane surface, respectively. (See Fig. 4.6.)

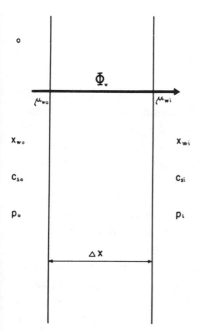

FIG. 4.6. Flux of water across a membrane. For explanation see the text.

If water equilibrium on the surfaces of the membrane is assumed, chemical potentials of water on the surfaces are equal to those in the adjacent solutions and expressions for potential in the adjacent solutions may be used in eq. (45). It will be seen later that if the permeation of water across the membrane is rapid, there will be concentration gradients in the unstirred layers at the membrane and it is the composition of the solution in the proximity of the membrane which must be assessed and manipulated, rather than that of the bulk of the solution. If the permeation is slow, the values for the bulk of the solutions can be used directly.

Chemical potential (see section 2.1) of water may be written, under the assumption that the solution is an ideal one, as

$$\mu_w = \mu_{w0} + RT \ln x_w + \bar{V}_w p \tag{46}$$

where x_w is the mole fraction of water ($= c_w/(c_w + c_s)$), $\bar{V}_w$ is the partial molal volume of water, approximately equal to $18 \text{ cm}^3 \cdot \text{mol}^{-1}$, and p is hydrostatic pressure. The term $\bar{V}_w p$, which can be often neglected in the expression for the chemical potential of a solute, is not negligible here; if the bulk of water in solution is considered rather than, say, its small labeled part, x_w is nearly unity and the term containing its logarithm is therefore small and comparable with $\bar{V}_w p$ even for small hydrostatic pressures. Inserting (46) into (44) we obtain

$$\Phi_w = - \frac{U_w \bar{c}_w}{\Delta x} (RT \, \Delta \ln x_w + \bar{V}_w \, \Delta p) \tag{47}$$

It would not be wrong to use the equation in this form but it is usually not done. The reason for this is that for experimental solutions exact concentrations of solutes rather than exact values of x_w are known. For this reason, simplified expressions for the chemical potential are inserted into equations for water flow, in which the mole fraction of water is expressed approximately by the known concentrations. The approximations used are, of course, different, according to whether the movement of water irrespective of its labeling (i.e., of water forming the bulk of the solutions) or the movement of its minute labeled fraction is considered. At first the equations concerning the former case will be derived and discussed.

We may begin by considering the conditions of water equilibrium across an ideal semipermeable membrane. The difference in the value of x_w at the two sides of the membrane will be due to different concentrations of impermeable solute at the two sides. In equilibrium, Φ_w will vanish and the expression in parentheses in eq. (47) is hence zero. Then

$$\bar{V}_w \, \Delta p = -RT \, \Delta \ln x_w \qquad (48a)$$

or

$$\Delta p = -\frac{RT}{\bar{V}_w} \, \Delta \ln x_w \qquad (48b)$$

The expression on the left-hand side is the difference in hydrostatic pressure, the expression on the right-hand side is the difference in the so-called osmotic pressure π

$$\pi = -\frac{RT}{\bar{V}_w} \ln x_w \qquad (49)$$

Since $x_w = c_w/(c_w + c_s)$ the expression for osmotic pressure may be re-written as

$$\pi = -\frac{RT}{\bar{V}_w} \ln \frac{c_w}{c_w + c_s} = \frac{RT}{\bar{V}_w} \ln \frac{c_w + c_s}{c_w} = \frac{RT}{\bar{V}_w} \ln \left(1 + \frac{c_s}{c_w}\right) \approx \frac{RT}{\bar{V}_w} \frac{c_s}{c_w} \qquad (50)$$

using the fact that $-\ln y = \ln(1/y)$ and $\ln(1 + x) \approx x$, as follows from Taylor's theorem for small values of x. Finally, $\bar{V}_w c_w \approx 1$, since in dilute solutions the unit volume to which concentrations are referred is almost entirely made up by the volume of water, this being the product of its partial molal volume and its concentration. If $\bar{V}_w$ is expressed in $cm^3 \cdot mol^{-1}$ the units of concentration must be $mol \cdot cm^{-3}$ or, if concentrations are expressed in $mol \cdot liter^{-1}$, $\bar{V}_w$ in $liter \cdot mol^{-1}$ must be used, as otherwise a numerical coefficient would have to be introduced at this stage.

We can then write

★ $$\pi = RT \, c_s \qquad (51)$$

which is van't Hoff's well-known formula for osmotic pressure, representing a very good approximation for dilute solutions. Equation (48b) then may be written as

$$\Delta p = RT \, \Delta c_s \qquad (52)$$

where, it should be stressed again, c_s is the concentration of impermeant solutes, for only ideal semipermeable membranes are discussed for the moment. Comparing eq. (48a) with eq. (52) we can see that the above development provided us with the approximation

$$-RT \, \Delta \ln x_w = \bar{V}_w RT \, \Delta c_s \qquad (53)$$

which can be inserted into eq. (47), giving

$$\Phi_w = -\frac{U_w \bar{c}_w \bar{V}_w}{\varDelta x}(\varDelta p - RT\,\varDelta c_s) \tag{54}$$

If the water content of the membrane were very high, the term $\bar{c}_w \bar{V}_w$ on the right-hand side of eq. (54) would approach unity and could be omitted; otherwise, it must remain in the expression as a numerical coefficient smaller than 1. For further convenience we may write

$$\bar{V}_w = \frac{1}{\tilde{c}_w} \tag{55}$$

instead of $\bar{V}_w$ in eq. (54), $\tilde{c}_w$ being considered simply as a quantity defined by eq. (55). In view of the fact that the solutions at both sides of the membrane are dilute we may handle $\tilde{c}_w$ as being equal to the concentration of water at either side of the membrane, or to an average of the two, without incurring a serious error. We can then write

$$\Phi_w = -\frac{U_w \bar{c}_w}{\varDelta x \tilde{c}_w}(\varDelta p - RT\,\varDelta c_s) \tag{56a}$$

or

$$\Phi_w = -\frac{U_w}{\varDelta x}k_w(\varDelta p - RT\,\varDelta c_s) \tag{56b}$$

where k_w is the distribution coefficient of water between the membrane and the solutions

$$k_w = \frac{\bar{c}_w}{\tilde{c}_w} \tag{57}$$

In osmotic experiments, the volume flow $J_V = \Phi_w \bar{V}_w$ is expressed in volume units per unit area per unit time rather than Φ_w in moles per unit area per unit time. The volume flow across an ideal semipermeable membrane is then

$$J_V = -\frac{U_w}{\varDelta x}k_w \bar{V}_w(\varDelta p - RT\,\varDelta c_s) \tag{58}$$

or

★
$$J_V = L_p(\varDelta p - RT\,\varDelta c_s) \tag{59}$$

where

$$L_p = -\frac{U_w}{\varDelta x}k_w \bar{V}_w \tag{60}$$

is called the hydraulic conductivity of the membrane.

Some consequences of eq. (59) seem to have been not quite clearly visualized in the past. Equation (59) implies that the flux of water is related by the same coefficient to the difference in hydrostatic pressure as it is to the difference in osmotic pressure, whatever the membrane structure may be (as long as the membrane is ideally semipermeable, of course). Whereas with membranes permitting only diffusional flow of water during which each water molecule travels separately it was considered natural that hydrostatic and osmotic pressure would produce the same kind of water movement, it was rather difficult to envisage the osmotic pressure producing the same hydraulic flow as that caused by the hydrostatic pressure in a membrane containing continuous water-filled pores. For the history and a penetrating physicochemical analysis of the problem the paper by Dainty (1963) should be consulted. Mauro (1960) demonstrated that a drop of hydrostatic pressure must somehow develop at the opening of the pores due to difference in osmotic pressure. Such hydrostatic pressure is equivalent to the hydrostatic pressure applied to the whole of the solution except that it does not push against the membrane as a whole, with the possibility of breaking it, this being the reason why much higher pressures can be achieved in practice in osmotic experiments than in hydrostatic experiments. The mechanism by which the drop in the hydrostatic pressure may be produced at the pore opening has been considered by Dainty and Meares and described by Dainty (1963). The authors approached the problem from the point of view of the kinetic theory of liquids. No solute can enter the pores in an ideal semipermeable membrane, so that if two adjacent layers at the pore openings are considered, one of them will be a layer of pure water, the other a layer of solution. Jumps of water molecules from one layer into the other leave vacancies behind. The water concentration being higher in the pure water, more jumps proceed from this layer and more vacancies are formed there per unit time than in the layer of the solution. The greater number of vacancies leads to a decrease in the density of water which can be related to the corresponding decrease in pressure using the experimental value of the compressibility of water. A result of the correct order of magnitude was obtained by Dainty and Meares by the above development (Dainty, 1963). It may be seen that the described mechanism is similar to that invoked by Hartley and Crank (1949) to explain the origin of the mass flow accompanying the diffusion of any solute in free solutions (see section 2.1).

Let us now consider another type of experiment in which water permeability is measured, the experiment in which permeation of labeled water under the condition of osmotic equilibrium is followed and related to the concentrations of labeled water at the two sides of the membrane. To simplify the situation still further we will assume that not only the difference in osmotic pressure is balanced by the difference in hydrostatic pressure, but that both the differences are actually zero. Equation (37) may be then rewritten to describe the flux of labeled water as follows:

$$\Phi_w{}^* = -\frac{U_w \bar{c}_w{}^*}{\varDelta x^*} RT \, \varDelta \ln x_w{}^* \tag{61}$$

where quantities marked with an asterisk correspond to labeled water. Usually, however, another equation is used to interpret the experimental data, relating, as mentioned above, the flux of labeled water to its concentration difference across the membrane:

$$\Phi_w{}^* = P_d \, \varDelta c_w{}^* \tag{62}$$

Here P_d is the permeability constant of water as determined in a tracer experiment, the subscript d being used to stress that, unlike in osmotic experiments where mass flow may be involved, the mechanism of water permeability is here certainly diffusional. P_d is called the diffusional permeability of the membrane.

Derivation of eq. (62) from eq. (61) involves a number of approximations. The expression for $\varDelta \ln x_w{}^*$ may be simplified as follows

$$\varDelta \ln x_w{}^* = \varDelta \ln \frac{c_w{}^*}{c_w{}^* + c_w} \approx \varDelta \ln \frac{c_w{}^*}{c_w}$$

$$= \varDelta(\ln c_w{}^* - \ln c_w) \approx \varDelta \ln c_w{}^* = \frac{\varDelta c_w{}^*}{\tilde{c}_w{}^*} \tag{63}$$

applying, in succession, relations $c_w{}^* \ll c_w$ (the concentration of labeled water is much less than the concentration of unlabeled water), $\varDelta \ln c_w \approx 0$ (the concentration of unlabeled water is practically the same at the two sides of the membrane), and $\varDelta \ln y = \varDelta y/y$ from differential calculus. Inserting eq. (63) into eq. (61) we have

$$\Phi_w{}^* = -\frac{RTU_w}{x} \frac{\bar{c}_w{}^*}{\tilde{c}_w{}^*} \varDelta c_w \tag{64}$$

where $\bar{c}_w{}^*$ is the average of the concentration of labeled water in the membrane, whereas $\tilde{c}_w{}^*$, as defined by eq. (63), is a kind of average of the concentrations of labeled water in the two media. The ratio $\bar{c}_w{}^*/\tilde{c}_w{}^*$ may thus be considered as being approximately equal to the distribution coefficient of water between the membrane and the solution k_w defined by eq. (57). Hence eq. (64) may be written as

$$\Phi_w{}^* = - \frac{RTU_w}{x} k_w \, \Delta c_w{}^* \tag{65}$$

On comparing eq. (65) with eq. (62) it may be seen that

$$P_d = - \frac{RTU_w}{x} k_w \tag{66}$$

and a comparison of (66) with eq. (60) shows that, if U_w is the same in both equations

★ $$L_p \frac{RT}{\bar{V}_w} = P_d \tag{67}$$

$L_p(RT/\bar{V}_w)$ is usually called the osmotic permeability coefficient of the membrane P_{os}. But U_w, the mobility of water in eq. (66), corresponds to diffusional mobility as it describes the movement of tracer in an experiment in which osmotic equilibrium prevails. Hence if eq. (67) is valid, it may be assumed that also U_w in eq. (60) corresponds to a diffusional movement. Since L_d is obtained in an osmotic experiment, such a result means that the membrane permits diffusional movement only, that there are no water-filled pores across the membrane for which eq. (67) is valid. On the other hand, the finding that

$$L_p \frac{RT}{\bar{V}_w} > P_d \tag{68}$$

is considered as evidence for the existence of pores in the membrane, in which mass flow may take place and thus a higher mobility of water may be found than would correspond to a diffusional process.

The use of inequality (68) as evidence for the presence of pores in membranes has been subjected to an important criticism by Dainty (1963). As shown by Dainty, inequality (68) is likely to hold for perhaps all biological membranes whether they contain pores or not, simply due to the fact that the present values of P_d for water (and, for that matter, P for all rapidly permeating solutes) are greatly underestimated, the reason for this being the negligence of the effects of unstirred layers adjacent to the membrane. Due to the presence of such unstirred layers (also called Nernst diffusion layers) the difference of the concentrations of the diffusing substance between the two solutions immediately at the two surfaces of the membrane may be much less than the difference between the bulks of the two solutions. If the difference of the bulk concentrations rather than the difference corresponding to the layers adjacent to the membrane is used in eq. (62), the value of P_d comes out clearly less than it should.

Unstirred layers are regions of slow laminar flow parallel to the membrane. According to Dainty (1963) the thickness of unstirred layers varies between 20 and 500 μ, depending on stirring that can be achieved in biological experiments. More rigorously, such would be the thickness of layers with linear concentration gradients, producing the same effects as those experimentally established with actual unstirred layers. Substances can be transported across the unstirred layers solely by the process of diffusion and each of the two layers adjoining the membrane thus acts itself as a membrane with the permeability constant $RTU/\delta = D/\delta$, where U is the mobility of the substance diffusing in a solution, D its diffusion constant (see section 2.1), and δ the thickness of the layer. The measured permeability constant thus corresponds to the series combination of the membrane and of the two unstirred layers

$$\frac{1}{P_{\text{measured}}} = \frac{1}{P_{\text{true}}} + \frac{\delta_1}{D_1} + \frac{\delta_2}{D_2} \tag{69}$$

where subscripts 1 and 2 correspond to the two unstirred layers. Hence

★ $$P_{\text{true}} = \frac{1}{\dfrac{1}{P_{\text{measured}}} - \dfrac{\delta_1}{D_1} - \dfrac{\delta_2}{D_2}} \tag{70}$$

If the constants of diffusion are known (for self-diffusion of labeled water, D is about 2.5×10^{-5} cm$^2 \cdot$sec^{-1}; see section 2.1) and the thickness of the unstirred layers is estimated (500 μ for systems with practically no stirring to 20 μ for solutions violently agitated) the range of the probable values of the true permeability constant may be derived.

Uncorrected permeability constants higher than 10^{-4} cm$\cdot$sec^{-1} are likely to be considerably affected by the presence of unstirred layers for the "permeability" of the unstirred layers may be of the same order of magnitude. Very high permeability constants cannot be determined with desirable accuracy, for the rate of permeation is then controlled by the diffusion across the adjacent unstirred layers and the correction for the unstirred layer effect becomes too great.

Unlike the diffusional permeability constant P_d, the hydraulic conductivity L_p of the membrane is only very little affected by the presence of unstirred layers (Dainty, 1963). Hence, if inequality (68) is valid for a membrane, possible effects of unstirred layers should be considered first and only when these are completely excluded, the inequality may be used as a criterion for the presence of mass flow (and hence presumably of water-filled pores) in the membrane.

Ginzburg and Katchalsky (1963) studied the effects of unstirred layers with artificial membranes. Dainty and House (1966a) determined the thickness of the unstirred layers in isolated frog skin from the time course of the electrical potential transients brought about by abrupt changes in ionic composition of bathing media. When stirring at 120, 300, and 500 rpm, the thickness of the outer unstirred layer was found to lie within the ranges of 40–60 μ, 30–50 μ, and 30–40 μ, respectively. The corresponding ranges for the inner unstirred layer (including, doubtless, the corium of the skin) were 150–230 μ, 120–200 μ, and 100–170 μ. Actually, examining the evidence for membrane pores in frog skin, Dainty and House (1966b) found an even greater increase of P_d with stirring rate than would be predicted from the above measurement and concluded that application of inequality (68) to frog skin cannot at present serve as definitive evidence for water-filled pores in this tissue.

An important attempt at interpreting the water permeability coefficients obtained in *osmotic* experiments on single cells is due to Dick (1964, 1966). Observing that these permeability coefficients may be correlated with the surface–volume ratio (higher permeability coefficients are found with large cells than with small ones) Dick concluded that slow diffusion of water in the cell interior is of importance in the process of osmotic swelling of cells. To explain the slowness of the diffusion inside the cell Dick invoked the notion of the mutual diffusion coefficient due to Hartley and Crank (1949) as discussed in section 2.1 of this book. It may be remembered that when interdiffusion of two components occurs the rate of this process is characterized by a single diffusion coefficient, the value of which depends on experimental conditions. Interpreting the osmotic swelling of the cells as interdiffusion of water and of proteins and nucleic acids Dick concluded that the diffusion coefficient for water inside the cell may approximate the values for self-diffusion coefficients of these macromolecules and hence be very low. From the assumption that the true water permeability of cell membranes of various cells is not likely to vary very widely—the view supported by the finding of Robertson that membranes of various cells have a similar structure (see chapter 1 of this book)—Dick suggests that permeability coefficients from 3×10^{-4} to 7×10^{-3} cm·sec^{-1} and intracellular diffusion coefficients from 8×10^{-10} to 2×10^{-8} cm²·sec^{-1} may correspond to the physical reality in the course of osmotic experiments with single cells. All available data for osmotic permeability coefficients of various cells, apart from the data obtained with a few cells isolated by dissections and having higher permeabilities, can be explained by the above combination of permeability and diffusion coefficients (Dick, 1964, 1966).

It should be perhaps stressed that the extremely low values of the intracellular water diffusion coefficients are likely to correspond only to the situations where, as in osmotic swelling or shrinking experiments, macromolecules are actually displaced; in experiments with labeled water, carried out in osmotic equilibrium, or in experiments in which a *steady-state* osmotic flow *across* cells or cellular layers is measured, the much higher self-diffusion coefficient of water appears to be relevant.

4.2.2. Membrane Permeable to the Solute

The theoretical considerations presented in this section so far were greatly simplified by the assumption that the membranes discussed were ideally semipermeable, i.e., the only osmotic phenomena considered were those produced by nonpermeant solutes. With permeant solutes the situation is more complicated primarily due to the fact that the flow of water and the flow of solute may be coupled together in the membrane. Interactions of this kind are aptly described by the phenomenological theory of irreversible processes, called the thermodynamics of irreversible processes or thermodynamics of the steady state. There are many excellent treatments of this subject, of which the short monograph by Denbigh (1958), *The Thermodynamics of the Steady State*, and that by Katchalsky and Curran (1965), *Nonequilibrium Thermodynamics in Biophysics*, are most suitable for a biologist interested in cell membrane transport. Some elements of this theory required further in this book will be briefly stated here.

In the various special cases of Teorell's equation (2.8) encountered so far it was assumed that the flux of a substance is directly proportional to the appropriate driving force which may be called its conjugated force. It was repeatedly postulated that the flux of a substance is linearly dependent on the gradient of the partial molal free energy of this substance and the equation describing this situation was written and then simplified, using various approximations, so that relations which could be verified by experiments were obtained. The above equations were of the form

$$J_1 = L_{11}X_1 \tag{71}$$

where J_1 was the flux of the substance, X_1 its conjugated force, and L_{11}

the appropriate coefficient, which is at least approximately constant in the most useful equations.

To describe the interactions between two different fluxes (or, indeed, among any number of them), which occur so often in nature, the thermodynamics of irreversible processes makes use of the more general assumption that each flux depends linearly on all driving forces operating in the system. Instead of eq. (71) more involved relations are used in the form

$$J_1 = L_{11}X_1 + L_{12}X_2 + \ldots + L_{1n}X_n$$
$$J_2 = L_{21}X_1 + L_{22}X_2 + \ldots + L_{2n}X_n$$
$$\cdot \; \cdot \; \cdot \; \cdot \; \cdot \; \cdot \; \cdot \; \cdot \; \cdot \; \cdot \; \cdot \; \cdot \; \cdot \; \cdot$$
$$J_n = L_{n1}X_1 + L_{n2}X_2 + \ldots + L_{nn}X_n$$

or, more concisely,

★
$$J_i = \sum_{j=1}^{n} L_{ij} X_j \tag{72}$$

Here n is the number of fluxes (and of their conjugated forces) in the system, L_{ii} are the classical coefficients expressing the dependence of fluxes on their conjugated forces and L_{ij} (where $i \neq j$) are the cross coefficients, expressing the dependence of fluxes on nonconjugated forces. Equations (72) are called phenomenological relations or thermodynamic equations of motion. If the fluxes and forces are chosen in such a way that the sum of the products of the fluxes and their conjugated forces is equal to the rate of entropy production in the system (or, in isothermal systems, to the dissipation function which is the rate of entropy production multiplied by the absolute temperature) Onsager showed that

★
$$L_{ij} = L_{ji} \tag{73}$$

The derivation of this important law is based on two assumptions. The first of them is the principle of microscopic reversibility, stating that in equilibrium each molecular process is balanced by the reverse of that process; the second assumption is that the rate of decay of statistical fluctuations may be described by the same phenomenological relations as macroscopic fluxes. The derivation may be found in the monographs by Denbigh and by Katchalsky and Curran cited above.

Thermodynamics of the irreversible processes does not guarantee that the linear relations (72) present a satisfactory description for all systems; it is to be found experimentally whether they represent at least a reasonable approximation. But if equations (72) with properly chosen fluxes and forces are valid, then relation (73) must be valid.

Phenomenological relations describing in a convenient form the fluxes of water and of other nonelectrolytes across membranes and the mutual interactions of these flows were developed by Kedem and Katchalsky (1958) and by Katchalsky (1961). Their development will be briefly described here.

For the sake of simplicity, only two fluxes will be assumed to proceed in the system, the flux of water and the flux of an uncharged solute. The flux of water is the number of moles of water transported through unit area per unit time and it is equal to the concentration of water (in the membrane expressed per unit volume of the membrane) times the velocity of water movement:

$$\Phi_w = \frac{dn_w}{dt} = c_w v_w \tag{74}$$

Analogously, the solute flux is equal to

$$\Phi_s = \frac{dn_s}{dt} = c_s v_s \tag{75}$$

The appropriate driving forces per mole are the negatives of the chemical potential gradients (or, in the unidirectional case, of the derivatives)

$$X_w = - \frac{d\mu_w}{dx} \tag{76}$$

$$X_s = - \frac{d\mu_s}{dx} \tag{77}$$

for then the sum of the products of the fluxes with the conjugated forces gives the dissipation function per unit area

$$T \frac{d_i S}{dt} = \Phi_w \left(- \frac{d\mu_w}{dx} \right) + \Phi_s \left(- \frac{d\mu_s}{dx} \right) \tag{78}$$

where $d_i S/dt$ is the rate of entropy production in the system and T is the absolute temperature.

Phenomenological relations expressing the two fluxes at a given point in the membrane thus may be written, using equations (76) and (77)

$$\Phi_w = L_{ww} \left(- \frac{d\mu_w}{dx} \right) + L_{ws} \left(- \frac{d\mu_s}{dx} \right) \tag{79}$$

$$\Phi_s = L_{ss} \left(- \frac{d\mu_s}{dx} \right) + L_{sw} \left(- \frac{d\mu_w}{dx} \right) \tag{80}$$

Neither the derivatives of chemical potentials at a point nor the fluxes at a point are amenable to a direct experimental measurement. To obtain the phenomenological relations in terms of measurable quantities, equations (79) and (80) must be integrated across the thickness of the membrane. Katchalsky (1961) performed the integration using the method of Kirkwood (1954) in which the plausible assumption is made that the fluxes reach steady values which are independent of x and equal to the measurable fluxes across the membranes and reached the conclusion that equations (79) and (80) may be transformed into

$$\Phi_w = L'_{ww} \Delta \mu_w + L'_{ws} \Delta \mu_s \tag{81}$$

$$\Phi_s = L'_{ss} \Delta \mu_s + L'_{sw} \Delta \mu_w \tag{82}$$

where the $\Delta\mu$'s are differences of the chemical potentials between the two surfaces of the membrane. Equality of the cross coefficients is again valid, so that $L'_{ws} = L'_{sw}$.

Equilibrium is usually assumed to obtain at the membrane surfaces, so that the chemical potentials at the membrane surfaces are considered to be equal to those in the adjacent solutions or, if the effects of unstirred layers may be neglected, to those in the bulk of the two solutions. If an ideal behavior of the solutions is assumed, eq. (46) will correspond to the chemical potential of water and an analogous expression will hold for the chemical potential of the uncharged solute (in case of an ion, an additional term, the partial molal free energy due to the charge of the ion, would have to be introduced). The $\Delta\mu$'s will then be expressed as

$$\Delta \mu_w = \bar{V}_w \Delta p + RT \Delta \ln x_w \tag{83}$$

$$\Delta \mu_s = \bar{V}_s \Delta p + RT \Delta \ln x_s \tag{84}$$

If it is, moreover, assumed that the solutions are dilute, approximations analogous to those carried out from eq. (50) to (53) may be used in eq. (83) and approximation (63) may be applied to the dilute solute. Equations (83) and (84) may then be written in the form

$$\Delta \mu_w = \bar{V}_w \Delta p - \frac{RT \Delta c_s}{\tilde{c}_w} = \bar{V}_w \Delta p - \frac{\Delta \pi}{\tilde{c}_w} \tag{85}$$

$$\Delta \mu_s = \bar{V}_s \Delta p - \frac{RT \Delta c_s}{\tilde{c}_s} = \bar{V}_s \Delta p + \frac{\Delta \pi}{\tilde{c}_s} \tag{86}$$

where $\Delta\pi$ is the difference of the osmotic pressure across the membrane.

Using these approximations, Kedem and Katchalsky (1958) and Katchalsky (1961) rewrote the dissipation function per unit area (eq. (78)) as

$$T\frac{d_iS}{dt} = \Phi_w\,\Delta\mu_w + \Phi_s\,\Delta\mu_s = \Phi_w\left(\bar{V}\Delta p - \frac{\Delta\pi}{\tilde{c}_w}\right) + \Phi_s\left(\bar{V}_s\,\Delta p - \frac{\Delta\pi}{\tilde{c}_s}\right)$$

$$= (\Phi_w\bar{V}_w + \Phi_s\bar{V}_s)\,\Delta p + \left(\frac{\Phi_s}{\tilde{c}_s} - \frac{\Phi_w}{\tilde{c}_w}\right)\Delta\pi = J_V\,\Delta p + J_D\,\Delta\pi \quad (87)$$

Two new fluxes which will be found useful later were thus properly defined by equation (87), where

$$\bigstar \qquad\qquad J_V = \Phi_w\bar{V}_w + \Phi_s\bar{V}_s \qquad\qquad (88)$$

is the total volume flow, conjugated with the driving force Δp. A special case of the volume flow was already encountered when ideal semipermeable membranes were discussed; then it was, of course, equal to the volume flow of water. The other flux, conjugated with $\Delta\pi$,

$$\bigstar \qquad\qquad J_D = \frac{\Phi_s}{\tilde{c}_s} - \frac{\Phi_w}{\tilde{c}_w} \qquad\qquad (89)$$

is called exchange flow and corresponds to the velocity of the solute relative to water in the membrane. It may be seen that both the new fluxes have the dimension of velocity.

Phenomenological relations for the total volume flow and for the exchange flow are then of the form

$$\bigstar \qquad\qquad J_V = L_p\,\Delta p + L_{pD}\,\Delta\pi \qquad\qquad (90)$$

$$\bigstar \qquad\qquad J_D = L_D\,\Delta\pi + L_{pD}\,\Delta p \qquad\qquad (91)$$

since Onsager's reciprocal relation $L_{pD} = L_{Dp}$ is again valid.

The meaning of the coefficients L_p, L_D, and L_{pD} becomes clearer from further development of equations (90) and (91) undertaken by Kedem and Katchalsky (1958). Some of the results will be shown here.

In a very coarse nonselective membrane (like a diaphragm made of sintered glass) the volume flow brought about by a difference in osmotic pressure, in the absence of a difference in hydrostatic pressure, is practically zero

$$J_V = L_{pD}\,\Delta\pi = 0 \qquad\qquad (92)$$

If, on the other hand, the difference in the concentrations of the solute is zero, no ultrafiltration is produced by the hydrostatic pressure difference on the nonselective membrane, i.e., the exchange flux is zero:

$$J_D = L_{pD}\, \Delta p = 0 \tag{93}$$

From equations (92) and (93) it may be seen that L_{pD} is zero for a nonselective membrane which does not discriminate between the solute and the solvent. In such membranes, therefore, the volume flow and the exchange flow are determined solely by their conjugated forces. The selectivity of the membrane depends obviously on both the membrane structure and the nature of the solute; if, for example, labeled water is considered as a solute and ordinary water as the solvent, all membranes will behave as practically nonselective.

As soon as the membrane is able to discriminate in some degree between the solute and the solvent, the phenomena of osmotic flow and of ultrafiltration become important; in other words, the cross coefficient L_{pD} is different from zero. The limiting case is the already discussed ideal semipermeable membrane. The flux of solute Φ_s across ideal semipermeable membrane is equal to zero. Hence, from eq. (88) and eq. (89), equations

$$J_V = \Phi_w \bar{V}_w \tag{94}$$

and

$$J_D = -\Phi_w/\tilde{c}_w \tag{95}$$

are valid for an ideal semipermeable membrane. However, since in dilute solutions $\bar{V}_w = 1/\tilde{c}_w$ (eq. (55)),

$$J_D = -J_V \tag{96}$$

in an ideal semipermeable membrane, where both flows are due only to the water flux. On combining eq. (96) with equation (90) and (91), we may write

$$(L_p + L_{pD})\, \Delta p + (L_D + L_{pD})\, \Delta \pi = 0 \tag{97}$$

which is satisfied for arbitrary differences in hydrostatic and osmotic pressure if and only if

$$L_p = -L_{pD} = L_D \tag{98}$$

The behavior of an ideal semipermeable membrane is thus completely characterized by a single coefficient, the hydraulic coefficient L_p, already encountered in eq. (59).

The essential and novel contribution of the theory of Kedem and Katchalsky (1958), however, lies in the description of the behavior of membranes occupying an intermediate position between coarse and ideal semipermeable membranes. For interpreting experimental results obtained with such membranes, it is convenient to modify equations (90) and (91) still further.

The volume flow J_V and the flow of solute Φ_s (rather than the exchange flow J_D) are measurable experimentally. Since in dilute solutions the volume fraction of the solute is very small, $c_s \bar{V}_s \ll 1$, it may be seen from equations (88) and (89) that

$$\Phi_s = (J_V + J_D)\tilde{c}_s \qquad (99)$$

Φ_s is often measured under the condition of zero volume flow, where (from eq. (90))

$$\Delta p = - \frac{L_{pD}}{L_p} \Delta \pi \qquad (100)$$

By substituting from eq. (100) and eq. (91) in eq. (99) for $J_V = 0$ we obtain

$$\Phi_s = \frac{L_p L_D - L_{pD}^2}{L_p} \tilde{c}_s \Delta \pi \qquad (101)$$

and, remembering that $\pi = RTc_s$ (eq. (51)) so that $\Delta \pi = RT \Delta c_s$

$$\Phi_s = \frac{L_p L_D - L_{pD}^2}{L_p} \tilde{c}_s RT \Delta c_s \qquad (102)$$

for $J_V = 0$. The flux of solute is seen to be proportional to the concentration difference of the solute across the membrane. New coefficients are conveniently introduced at this stage. Kedem and Katchalsky (1958) defined the coefficient ω as

$$\omega = \frac{L_p L_D - L_{pD}^2}{L_p} \tilde{c}_s \qquad (103)$$

Equation (97) may then be written

$$\Phi_s = RT\omega \Delta c_s \qquad (104)$$

for $J_V = 0$, i.e., for measurements carried out at constant volume. $RT\omega$ is then seen to be the permeability constant of an uncharged solute (as used in eq. (62) or eq. (3.6) measured under the condition of zero volume flow which is the simplest way of obtaining the data.

The other of the two coefficients used by Kedem and Katchalsky (1958) is the reflection coefficient σ introduced by Staverman and encountered already in section 3.1 (eq. (3.16)). σ is defined as

★
$$\sigma = -\frac{L_{pD}}{L_p} \qquad (105)$$

and with this substitution, eq. (100), valid for conditions of zero volume flow, becomes

$$\Delta p = \sigma RT \, \Delta c_s \qquad (106)$$

Thus in a membrane which is leaky for the solute the volume flow ceases at Δp ("observed osmotic pressure") which is less than the difference of osmotic pressure calculated from van't Hoff's formula, $\Delta \pi = RT \, \Delta c_s$ ("theoretical osmotic pressure"), since σ is always less than 1; only in the limiting case of the ideal semipermeable membrane $\sigma = 1$ since then $L_p = -L_{pD}$ (eq. (98)). The origin of eq. (3.16), $\sigma = \pi_{obs}/\pi_{theor}$, is thus clear.

Phenomenological equations may now be written in terms of the new coefficients. By substituting from eq. (105) in eq. (90) we obtain a new expression for the total volume flow

★
$$J_V = L_p(\Delta p - \sigma RT \, \Delta c_s) \qquad (107)$$

Derivation of the expression for the solute flow is rather more involved. With the definition of σ (eq. (105)) inserted into eq. (103) we have

$$\omega = (L_D - L_p \sigma^2)\tilde{c}_s \qquad (108)$$

Then, by combining eq. (99) with equations (107), (91) where $\Delta \pi = RT \, \Delta c_s$, (105), and (108), Kedem and Katchalsky (1958) obtained the expression for flux of solute

$$\Phi_s = c_s L_p(1 - \sigma) \, \Delta p + [\omega - \tilde{c}_s L_p(1 - \sigma)\sigma] \, RT \, \Delta c_s \qquad (109)$$

which, finally, together with eq. (107), gives

★
$$\Phi_s = \omega RT \, \Delta c_s - \tilde{c}_s(1 - \sigma) J_V \qquad (110)$$

Equations (107) and (110) describe completely the transport processes taking place across a leaky membrane with a dilute solution of a single uncharged solute at both sides, taking into account the interactions between the two flows. Thus, for a given solvent–solute combination, the values of L_p, ω, and σ characterize the properties of the membrane fully.

Equation (*107*) is most important in the present treatment of transport of water across membranes. However, even eq. (*110*) may occasionally prove useful for describing the flux of labeled water under the condition of a nonzero volume flow; σ will then be zero, since hardly any membrane can discriminate between ordinary and labeled water.

For a nonpermeant solute $\sigma = 1$, since $\sigma = -L_{pD}/L_p$ (eq. (*105*)) and $L_p = -L_{pD}$ for a membrane behaving toward a given solute as ideally semipermeable (eq. (*98*)). Equation (*107*) may be generalized to describe with sufficient accuracy the situation when a nonpermeant solute at concentration c_i is present in the system (Katchalsky, 1961)

★
$$J_V = L_p(\Delta p - RT\,\Delta c_i - \sigma RT\,\Delta c_s) \qquad (111)$$

Equation (*111*) may be also considered as a generalization of eq. (*59*) for the case when a permeant solute is added to one or to both sides of the membrane.

The rest of this section will concern the reflection coefficient σ and the problem of water movement linked to active transport of solutes, where the consideration of the reflection coefficient is relevant.

For nonelectrolytes, the value of σ ranges from 0 to 1. As was shown before, σ is zero when the membrane does not discriminate between the solute and the solvent molecules. In practice, this is the case when the membrane is rather coarse and the molecules of the solute and of water do not differ much in size, as otherwise they could be discriminated even in homogeneous solution (see the treatment of interdiffusion by Hartley and Crank, 1949, as discussed in section 2.1). An almost ideal case is that where the solute is just labeled water. For a nonpermeant solute, σ is 1, as follows from a comparison of the condition for zero volume flow across an ideal semipermeable membrane (eq. (52)) and across a leaky membrane (eq. (*106*)). For permeating solutes, σ is *always* less than 1. The greater the frictional interaction between the solute flow down its concentration gradient and the osmotic flow of water (due to the same gradient) in the opposite direction, the smaller the volume flow due to osmotic pressure and, hence, the smaller the reflection coefficient σ. But even if there is absolutely no frictional interaction between solute and solvent in the membrane, as is the case when the two permeate by two distinct pathways (e.g., water only through pores and solute only through the lipid portion of a lipid membrane containing pores), $\sigma < 1$, since the volume flow produced under these conditions is less than the volume flow due to the same concentration of nonpermeant solute, the decrement being equal to the volume flow due to the permeating solute itself. This is the basis of Dainty's criticism (Dainty, 1963; see section 3.1) of drawing conclusions on the equivalent pore radius under the assumption that σ is less than 1 solely due to the frictional interaction in the pores. Before evaluating the relative contributions of the two factors in lowering the value of σ, a few words about the measurement of σ might be in order.

Assuming the existence of the concentration gradient of a single solute across a membrane and taking the volume flow across the membrane to be zero, it follows from eq. (106) (valid for zero volume flow) that

$$\sigma = \frac{\Delta p}{RT \, \Delta c_s} \tag{112}$$

Hence σ is the ratio of hydrostatic pressure at which no volume flow occurs due to a concentration difference Δc_s of the solute for which σ is measured, or to the theoretical pressure calculated from van't Hoff's formula. Instead of hydrostatic pressure, osmotic pressure due to a nonpermeant solute $RT \, \Delta c_i$ can be used, the two being equivalent; see eq. (52). Then σ is simply

$$\sigma = \frac{\Delta c_i}{\Delta c_s} \tag{113}$$

where Δc_i and Δc_s are the concentration differences of a nonpermeant and of the tested solute, respectively, which produce equivalent effects in volume flow.

At zero total volume flow, the volume flow due to the solute plus the volume flow due to water is zero. Let us now assume that there are no frictional interactions between the solute and water fluxes in the membrane, the two permeating through discrete pathways. With no interactions with solute flow in the membrane the volume flow *due to water* is given by the classical formula of eq. (59)

$$\Phi_w \bar{V}_w = L_p(\Delta p - RT \, \Delta c_s) \tag{114}$$

The volume flow due to solute may be expressed using eq. (104) to which the phenomenological relation (110) reduces for zero total volume flow:

$$\Phi_s \bar{V}_s = \bar{V}_s \omega RT \, \Delta c_s \tag{115}$$

The sum of the two fluxes being equal to zero,

$$\frac{\Delta p}{RT \, \Delta c_s} = 1 - \frac{\bar{V}_s \omega}{L_p} \tag{116}$$

Comparison with eq. (*112*) yields

$$\bigstar \qquad\qquad \sigma = 1 - \frac{\bar{V}_s \omega}{L_p} \qquad\qquad (117)$$

for a membrane with no interactions between solvent and solute, the two migrating through distinct pathways. For a membrane where friction between solute and solvent is appreciable, σ will be still smaller, so that

$$\bigstar \qquad\qquad 0 < \sigma < 1 - \frac{\bar{V}_s \omega}{L_p} \qquad\qquad (118)$$

for a permeating solute, characterized by the partial molal volume $\bar{V}_s$ and the permeability constant $P = RT\omega$, the hydraulic permeability of the membrane being L_p. If inequality (*118*) rather than eq. (*117*) is valid for a given membrane, an interaction between solute and water in the membrane is indicated, implying a common pathway (most simply pores permeable for both substances).

If solute and water do interact in the membrane, an explicit expression for σ may be obtained if phenomenological relations are derived in terms of frictional and distribution coefficients. This was done by Dainty and Ginzburg (1963) for a lipid membrane containing water-filled pores. Their expression for σ is

$$\sigma = 1 - \frac{\omega \bar{V}_s}{L_p} - \frac{K_s^c f_{sw}^c}{f_{sw}^c + f_{sm}^c} \qquad\qquad (119)$$

where K_s^c is the distribution coefficient of the solute between solution and the pores in the membrane and f_{sw}^c and f_{sm}^c are frictional coefficients describing the frictional forces between solute and water and solute and the membrane, respectively. (Frictional coefficients multiplied by the velocity of solute molecules relative to water or to the membrane give the forces per mole of solute.)

Phenomenological relations may be extended to describe transport of water related to the transport of electrolytes. It has been shown that the reflection coefficient σ for an electrolyte may become negative in a charged membrane which leads to the phenomenon of *negative osmosis* (Kedem and Katchalsky, 1961; Katchalsky, 1961). Electro-osmosis and related electrokinetic phenomena have been treated from the point of view of the thermodynamics of irreversible processes by Dainty (1963).

Finally, a few words about the water transport linked to active transport of solute may be of interest. Flow of water has been demonstrated to take place across many biological membranes where no macroscopically measurable gradient of osmotic pressure between the two media bathing the membrane could be detected and still the flow had an osmotic origin. *Endogenous, intrinsic,* or *micro-osmosis* are the terms propounded by Smyth (1964) to designate this phenomenon, where osmotic forces are generated by the cells themselves, to distinguish it from *exogenous, extrinsic,* or *macro-osmosis* due to the osmotic pressure difference between the two bathing media. Again, the thermodynamics of irreversible processes is of great assistance in explaining endogenous osmosis. Let us turn our attention to Fig. 4.7, where a series arrangement of two membranes may be seen, each characterized by its reflection coefficient σ for a certain solute and by its hydraulic conductivity L_p. Let it be assumed that $\sigma_2 < \sigma_1$, whereas $L_{p2} > L_{p1}$. Let the osmotic pressure in the compartments A and C be the same, whereas in the compartment B between the two membranes the osmotic pressure is constantly somewhat higher owing to an active transport of the above solute into this space. Since $\sigma_1 > \sigma_2$ the osmotic flow of water will occur predominantly from compartment A to compartment B and only in a smaller degree from C to B. A hydrostatic pressure is built up in compartment B by this osmotic inflow and, since $L_{p2} > L_{p1}$, it results in a hydraulic flow of water predominantly from compartment B to C. Net flow of water is thus seen to take place from compartment A to C without being assisted (it may be even opposed) by the difference in osmotic pressure between C and B. The above model was suggested and its feasibility in experiments with artificial membranes proved by Curran and co-workers

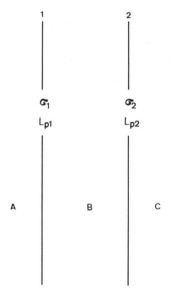

FIG. 4.7. Two-membrane system as a model for solute-linked water transport. For explanation see the text.

(Curran, 1960; Curran and McIntosh, 1962; Ogilvie *et al.*, 1963) and its mathematical analysis was carried out in detail by Patlak *et al.* (1963). It appears that the mechanism of local osmosis proposed by Diamond (1964) to explain the isotonic water transport across epithelial layers may be considered as a limiting case of the two membrane models for $\sigma_2 = 0$; here, too, water is transported by a local osmotic gradient into a restricted space, presumably in the cell membrane, and then expelled by hydrostatic pressure through some broader channels, replacing membrane 2 in Fig. 4.7. The osmolarity of the fluid thus transported corresponds to compartment A.

House (1964) studied the phenomenon of endogenous osmosis across frog skin (usually the term nonosmotic flow is used for this tissue). He succeeded in explaining the nonosmotic flow by a two-membrane model, the outer membrane having a smaller reflection coefficient for sodium than for potassium ions and the situation being reversed on the inner membrane. An important feature of this model is its ability to explain the small inward water movement which persists when no sodium is transported in this direction across the skin.

If mechanisms similar to those described above do not operate directly in cell membranes, equations like eq. (*111*) may be sufficient for the description of water movement and steady-state distributions between the cell and its surroundings. All possible sources of hydrostatic pressure should be considered, which may be of importance not only in plant cells and microorganisms posessing elastic walls but also in animal cells, where a contractile mechanism was suggested by Kleinzeller (1965) as a probable explanation of ouabain-insensitive water and electrolyte extrusion. If, on the other hand, mechanisms similar to the two-membrane system described above are present in cell membranes, this source of possibly uphill (though not active, being not *directly* coupled with the energy-supplying metabolism) transport would have to be also considered.

5. ACTIVE AND COUPLED UPHILL TRANSPORT

5.1. ACTIVE TRANSPORT

5.1.1. Criteria

In chapters 3 and 4, transport mechanisms that permit substances to pass across membranes in order to reach states of chemical or electrochemical equilibrium were treated. In other words, the transfer of substances, whether uncharged or ionic in character, by such systems need not be directly coupled to a source of metabolic energy, all the movement being provided for by the simple physicochemical forces (concentration, electrical potential gradients) operating in the system.

Many cases are known, however, in which molecules or ions move across cell membranes against an obvious concentration or electrochemical potential difference, remaining in solution at both sides of the membrane. It follows from elementary thermodynamics that such a process requires a source of energy other than the random thermal movement.

The above-mentioned uphill transport of molecules and ions gave rise to a number of definitions attempting to distinguish between an active and a passive process. The definitions rest mainly on the involvement or the lack of it of a metabolic, free-energy-producing reaction in the transport. Thus Conway (1954) defined active transport as movement of solutes dependent on the activity or energy change of another system. Rosenberg (1954) conceived active transport as movement of a substance which is influenced by forces other than the chemical or analogous potential gradient of this substance.

It appears that the most satisfactory definition of active transport is due to Kedem (1961) who based it on the formalism of thermodynamics of irreversible processes and who stated that a transport is active if there is a demonstrable interaction between the transmembrane flow of solute and a metabolic reaction, the experimental proof of such a transport consisting either in demonstrating an increase of metabolic activity (oxygen uptake and the like) concomitant with the transport or, better still, an effect of metabolic inhibitors (preferably uncouplers of oxidative phosphorylation) thereon.

We shall now attempt to use Kedem's arguments for defining active transport processes as rigorously as possible. It will be remembered that in the thermodynamics of irreversible processes each flux in a system is expressed as a linear function of all forces operating in the system. Following Kirkwood (1954), Kedem (1961) chose another, but equivalent, set of phenomenological relations in which each force operating in the system is a linear function of all flows in the system. Thus for a system with a membrane in which fluxes of various solutes, Φ_j's ($j = 1, 2, \ldots, n$) in moles crossing a unit area of the membrane per unit time, occur together with a chemical reaction flux Φ_r which may be expressed as the number of moles of oxygen consumed by the unit area of the membrane per unit time, the following phenomenological relations are written:

$$-\Delta \bar{\mu}_i = \sum_{j=0}^{n} R_{ij} \Phi_j + R_{ir} \Phi_r$$

$$-F_r = R_{rr} \Phi_r + \sum_{i=0}^{n} R_{ri} \Phi_i$$

$$(1)$$

Here $\Delta \bar{\mu}_i$'s are the differences of the chemical or electrochemical potentials of the respective solutes across the membrane and F_r is the affinity of the chemical reaction (the change of Gibbs free energy, for example, per mole of oxygen consumed). R's are the generalized resistance coefficients obeying Onsager's relations:

$$R_{ij} = R_{ji}$$

$$R_{ir} = R_{ri}$$

$$(2)$$

Equations (1) are written in an integrated form, the integration involving assumptions analogous to those for the integration of phenomenological relations carried out in section 4.2.

To make clear the physical meaning of the coefficients, Kedem (1961) rewrote eq. (*1*) in the form

$$\Phi_i = -\frac{1}{R_{ii}} \left(\Delta\bar{\mu}_i + \sum_{\substack{j=0 \\ j\neq i}}^{n} R_{ij}\Phi_j + R_{ir}\Phi_r \right)$$

$$\Phi_r = -\frac{1}{R_{rr}} \left(F_r + \sum_{i=0}^{n} R_{ri}\Phi_i \right)$$

(3)

The quantity $1/R_{ii}$ may serve as a certain measure of the permeability of the membrane toward the ith ion. Actually, in the case of a charged solute, it may be seen to be closely related to the "integral conductance" of the membrane discussed in section 4.1. The coefficients R_{ij} express the drag or entrainment effects between two different fluxes. If flux Φ_j produces or enhances flux Φ_i in the same direction, R_{ij} may be seen from eq. (3) to be a negative quantity.

For deriving the criterion of active transport, the coefficient R_{ir} is of greatest importance. If $R_{ir} \neq 0$, the flow of i is coupled to a chemical reaction in the membrane or, in other words, it is actively transported by the membrane. The part of the flux of i given by the term $-(R_{ir}/R_{ii})\,\Phi_r$ is defined as the active transport of i.

Due to coupling with the chemical reaction a net flow of i may take place, even if the chemical or electrochemical gradient of this ion and various drag forces predicted movement in the opposite direction. Some descriptions of active transport used in the past were based on this fact but the new definition is clearly more satisfactory. The movement of solute produced by coupling with a metabolic reaction is then called active transport, whatever the direction of potential or chemical gradients and drag forces in the system may be. Nothing is predicted by the thermodynamics of irreversible processes as to the mechanism of coupling with the chemical reaction (in other words, the mechanism of the pump).

An extension of the above definition of active transport was suggested by Kedem herself when she stated (1961) that: "R_{ij} is different from zero in the most typical case of active transport. With $R_{ij} \neq 0$, i may still be transported actively if $R_{ij} \neq 0$ and $R_{jr} \neq 0$. In this case, the flow of i is not coupled directly to the reaction but it is taken along with a 'carrier' that is itself transported by the reaction." It may be expedient to limit this extension to cases in which the species j, acting as the carrier, is a component of the membrane; otherwise a great number of flows produced due to gradients of actively transported substances would have to be denoted as active (among others, the solute-linked movement of water, discussed in section 4.2, countertransport caused by gradients of actively transported ions or molecules, etc.). Important distinctions would thus be lost from sight. If, on the other hand, the extension is limited to the carriers as components of the membrane, the flux coupled to the flow of the actively transported carrier may appear indistinguishable from a direct coupling with a metabolic reaction.

The various criteria of active transport were expressed by Kedem (1961) in terms of the resistance coefficients as follows.

Let us consider a simple system in which the membrane separates two solutions of an uni-univalent salt. The flow of cations (denoted as 1) and the flow of anions (denoted as 2) are not coupled in the membrane apart from electrostatic coupling accounted for by the use of electrochemical potentials ($R_{12} = R_{21} = 0$). The flow of water (denoted as 0) is not coupled with metabolism ($R_{0r} = 0$).

Equations (3) for this simple system are written in the form

$$\Phi_0 = -\frac{1}{R_{00}} (\Delta\mu_0 + R_{10}\Phi_1 + R_{20}\Phi_2)$$

$$\Phi_1 = -\frac{1}{R_{11}} (\Delta\bar{\mu}_1 + R_{10}\Phi_0 + R_{1r}\Phi_r)$$

$$\Phi_2 = -\frac{1}{R_{22}} (\Delta\bar{\mu}_2 + R_{20}\Phi_0 + R_{2r}\Phi_r) \tag{4}$$

$$\Phi_r = -\frac{1}{R_{rr}} (F_r + R_{1r}\Phi_1 + R_{2r}\Phi_2)$$

Criteria for an active transport of ion 1 or 2 are provided by measurements in which a nonzero value of either R_{1r} or R_{2r} can be established. Of these, the *ability of the membrane to produce a short-circuit current* (defined as a criterion by Ussing and Zerahn (1951) and described in detail in chapter 12) is a valuable criterion for the active nature of a transport of a charged particle and is expressed by Kedem as follows.

The short-circuit current is given by

$$I = F(\Phi_1 - \Phi_2) \tag{5}$$

where F is the Faraday constant. $\Delta\bar{\mu}$'s are zero under the conditions of short-circuit current experiments, so that

$$\Phi_1 = -\frac{R_{1r}}{R_{11}} \Phi_r \qquad \Phi_2 = -\frac{R_{2r}}{R_{22}} \Phi_r \tag{6}$$

and

$$I = -F(R_{1r}/R_{11} - R_{2r}/R_{22}) \Phi_r \tag{7}$$

The criterion fails only if the expression in parentheses on the right-hand side of the equation is zero due to the fact that the cation and the anion are pumped at equal rates.

Another criterion discussed by Kedem is the *influence of solute flow on a chemical reaction*. From the last of the set of eq. (4)

$$\left(\frac{\partial \Phi_r}{\partial \Phi_1}\right)_{\Phi_{21}F_r} = -R_{1r}/R_{rr} \tag{8}$$

$$\left(\frac{\partial \Phi_r}{\partial \Phi_2}\right)_{\Phi_{11}F_r} = -R_{2r}/R_{rr} \tag{9}$$

If, say, the oxygen consumption is found to depend on a flux of an ion (the fluxes of other *actively* transported ions being kept constant) the ion is said to be actively transported.

Finally, let us turn our attention to an important characteristic of the transport of a substance across a membrane which may occasionally serve as a criterion of the active nature of transport. This is the so-called *flux ratio* mentioned already on p. 70.

It may be remembered that in section 4.1 of this book the fluxes of various ions were expressed under the constant-field assumption by eqs. (15), (16), and (17) which were of the form

$$\Phi = P \frac{zFE_m/RT}{e^{-zFE_m/RT} - 1} (c_i - c_o e^{-zFE_m/RT}) \tag{10}$$

The net flux of an ion given by eq. (10) may be considered as a difference of two unidirectional fluxes, the efflux Φ_{ex} and the influx Φ_{in}, where

$$\Phi_{ex} = P \frac{zFE_m/RT}{e^{-zFE_m/RT} - 1} c_i \tag{11}$$

$$\Phi_{in} = P \frac{zFE_m/RT}{e^{-zFE_m/RT} - 1} c_o e^{-zFE_m/RT} \tag{12}$$

The ratio of the two, $f = \Phi_{in}/\Phi_{ex}$ called the flux ratio, is then given by

$$f = \frac{c_o}{c_i} e^{-zFE_m/RT} \tag{13a}$$

Using the definition of the electrochemical potential $\bar{\mu} = \mu_0 + RT \ln c + zF\psi$ and of the membrane potential $E_m = \psi_i - \psi_o$ the flux ratio may be written as

$$RT \ln f = \bar{\mu}_o - \bar{\mu}_i \tag{13b}$$

or

★
$$f = e^{(\bar{\mu}_o - \bar{\mu}_i)/RT} \qquad (13c)$$

This relation, which may be shown to be valid generally for passive and independent movements of ions, was derived by Ussing (1949) and Teorell (1949). For noncharged solutes moving by simple diffusion the flux ratio is of course reduced to the ratio of the concentrations of the solute. Various factors may cause eq. (13) not to be satisfied. One of them is, obviously, the active transport of the solute for which the equation is used, i.e., in the sense of the present definition an interaction with a chemical reaction in the membrane. Another is an interaction with movements of other substances across the membrane, e.g., solvent drag encountered if the flux of water and the flux of the solute proceed by the same pathways. Finally, there is a class of phenomena resulting from the interactions among the movements of the particles of the studied substance themselves. Some of these particles may (or must, if fluxes are measured) be labeled isotopically but this will not affect appreciably the nature of these interactions. It should be borne in mind that interactions of this kind are not artifacts caused by the isotopes and that the same phenomena take place when no recognizable isotopes are present. For this reason, the term "isotope interaction," sometimes used to denote these phenomena, does not appear to be suitable. Various models of such interactions have been described. One of them invokes the presence of a saturable mechanism, transporting the ion without energy expenditure across the membrane. A mechanism of this kind was suggested by Levi and Ussing (1948) and was called exchange diffusion (cf. the discussion of this term on p. 79). When such mechanisms are operating in the membrane, lower ion flux ratios than those predicted by eq. (13) are found. It should be noted that for an uncharged solute the flux ratio under these saturation conditions is equal to 1.

Another example of interaction, this time influencing the flux ratio in the opposite sense (flux ratios higher than predicted are found) is the single-file diffusion or long-pore effect described by Hodgkin and Keynes (1955 b). All these factors may be accounted for in a sufficiently general derivation of the expression for the flux ratio. Before proceeding to this, however, a special case of eq. (13) may be of interest.

If the ion is in flux equilibrium, $f = 1$, so that eq. (13 a) reduces to

$$\frac{c_i}{c_o} = e^{-zFE_m/RT} \qquad (14)$$

which is equivalent to eq. (4) of chapter 4. If a flux equilibrium can be achieved, this equation represents a valuable criterion for active transport, a ratio of concentrations of the permeating ion greater than corresponds to the membrane potential suggesting active transport (or coupling with the flow of another substance). If the ratio of concentrations is smaller than predicted by the membrane potential, mediated diffusion is indicated.

The flux ratio, as already mentioned, can be derived in a more general way when all the possible interactions mentioned above can be taken into account. This is based again on the formalism of the thermodynamics of irreversible processes and the derivation was done first by Hoshiko and Lindley (1964) and, more recently, in a still more general way, by Kedem and Essig (1965).

Kedem and Essig demonstrated that various interactions (apart from the interactions among the movements of the particles of the species tested) are accounted for if the flux ratio is expressed by a formula of the form

★ $$f = e^{(\bar{\mu}_o - \bar{\mu}_i - \sum_{j \neq i} R_{ij}\Phi_j)/RT} \qquad (15)$$

The family of terms $R_{ij}\,\Phi_j$ may include $R_{ir}\,\Phi_r$ known from eq. (3) to express the effect of active transport on the flux of a solute.

Likewise, interactions between particles of the studied species were formally accounted for by

★ $$f = e^{[(R_{ii} - R_{ik})/R_{ii}](\bar{\mu}_o - \bar{\mu}_i - \sum_{j \neq i} R_{ij}\Phi_j)/RT} \qquad (16)$$

Coefficient R_{ii} was already encountered in eq. (3); coefficient R_{ik} is an expression of intraspecies interactions. When there is mutual drag between particles (as in single-file diffusion) R_{ik} is negative; when the particles mutually diminish their transport (e.g., by competition for a carrier) R_{ik} is positive. Thus the first kind of interactions is seen to increase, the second to diminish the flux ratio.

5.1.2. Kinetics

5.1.2.1. Regular Model

In the following we shall present a simple kinetic analysis of an active transport of nonelectrolytes, intentionally omitting uptake mechanisms by intracellular adsorption, firstly because these do not seem to play a major role in general, and secondly because unequal distribution due to intracellular adsorption is kinetically uninteresting, being simply a matter of affinity and capacity of the intracellular binding sites with respect to the solute in question (cf. the sorption theories described in chapter 7).

Let it be stated at the very beginning that any active transport mediated by a membrane component (carrier) involves a different effective affinity of the system at each side of the membrane. Thermodynamically, this change in affinity is the energy-requiring step of the process while kinetically the difference accounts for the unequal distribution of the solute in steady state, there being a greater concentration of solute in equilibrium with a transport system of lower affinity (generally inside the cell) than with a higher-affinity system (generally facing the external medium).

Let us proceed in the analysis from a model implying the conversion of one carrier form C to another with lower affinity for substrate which we shall designate Z. It may be assumed that either the conversion of C to Z intracellularly or that of Z to C extracellularly requires coupling with metabolic energy (shown by heavy arrows in the model), the reversion being spontaneous. Alternatively, both conversions may be metabolically coupled. Nothing is said here about the mechanism of the $C \rightarrow Z$ or $Z \rightarrow C$ conversions. The reactions may involve the carrier as such but also the enzyme, if there is one, catalyzing the binding of substrate to the carrier.

The model to be treated kinetically is as follows:

$$
\begin{array}{ccccc}
\text{I} & & \mathrm{CS_I} \underset{}{\overset{n}{\rightleftharpoons}} \mathrm{CS_{II}} & & \text{II} \\
& & a \updownarrow b \qquad e \updownarrow f & & \\
& & + \ \ \mathrm{C_I} \overset{m}{\rightleftharpoons} \mathrm{C_{II}} \ + & & \\
\mathrm{S_I} & & i \updownarrow j \qquad l \updownarrow k & & \mathrm{S_{II}} \\
& & + \ \ \mathrm{Z_I} \overset{m}{\rightleftharpoons} \mathrm{Z_{II}} \ + & & \\
& & d \updownarrow c \qquad h \updownarrow g & & \\
& & \mathrm{ZS_I} \underset{}{\overset{n}{\rightleftharpoons}} \mathrm{ZS_{II}} & &
\end{array}
$$

The model, although more complete than any of those described in the literature, still contains several simplifications.

(1) The rate constants of the transmembrane movement of both carrier forms, C and Z, and similarly, of CS and ZS, are considered equal but the loaded carrier constants are different from the free carrier ones.

(2) Only the free carrier forms are interconvertible ($C \rightleftharpoons Z$ but not $CS \rightleftharpoons ZS$).

(3) An equilibrium approach to the reaction with substrate is used.

The model is thus analogous to Model II of the mediated diffusion treatment (p. 75) but contains, in addition, relatively slow reactions interconverting C and Z.

Using the steady-state conditions when

$$\frac{d(c_I + cs_I)}{dt} = -mc_I + mc_{II} - ncs_I + ncs_{II} + iz_I = 0 \qquad (17a)$$

$$\frac{d(c_{II} + cs_{II})}{dt} = -mc_{II} + mc_I - ncs_{II} + ncs_I - kc_{II} = 0 \qquad (17b)$$

$$\frac{d(z_I + zs_I)}{dt} = -mz_I + mz_{II} - nzs_I + nzs_{II} - iz_I = 0 \qquad (17c)$$

$$\frac{d(z_{II} + zs_{II})}{dt} = -mz_{II} + mz_I - nzs_{II} + nzs_I + kc_{II} = 0 \qquad (17d)$$

and the relations that

$$K_{CS} = c_I \cdot s_I / cs_I = c_{II} \cdot s_{II} / cs_{II} \qquad (18a)$$

and

$$K_{ZS} = z_I \cdot s_I / zs_I = z_{II} \cdot s_{II} / zs_{II} \qquad (18b)$$

we can express the concentrations of the carrier forms at side I in terms of c_I and those at side II in terms of c_{II}. The carrier conservation equation can be written as

$$c_{tI} + c_{tII} =$$

$$= c_I \left[1 + \frac{s_I}{K_{CS}} + \frac{k}{i} \frac{\left(m + \frac{ns_I}{K_{CS}}\right)}{\left(m + k + \frac{ns_{II}}{K_{CS}}\right)} + \frac{k}{i} \frac{s_I}{K_{ZS}} \frac{\left(m + \frac{ns_I}{K_{CS}}\right)}{\left(m + k + \frac{ns_{II}}{K_{CS}}\right)} \right]$$

$$+ c_{II} \left[1 + \frac{s_{II}}{K_{CS}} + \frac{k}{i} \frac{\left(m + i + \frac{ns_I}{K_{ZS}}\right)}{\left(m + \frac{ns_{II}}{K_{ZS}}\right)} + \frac{k}{i} \frac{s_{II}}{K_{ZS}} \frac{\left(m + i + \frac{ns_I}{K_{ZS}}\right)}{\left(m + \frac{ns_{II}}{K_{ZS}}\right)} \right]$$

$$= c_I(1 + \alpha + \beta + \gamma) + c_{II}(1 + \delta + \varepsilon + \zeta) = c_I A_I + c_{II} A \qquad (19)$$

The carrier flux cycle is defined here as

$$\Phi_C = m(c_I - c_{II}) = m(c_{tI}/A_I - c_{tII}/A_{II}) \qquad (20a)$$

$$\Phi_{CS} = n(cs_I - cs_{II}) = n(c_{tI}/A_I\alpha - c_{tII}/A_{II}\delta) \qquad (20b)$$

$$\Phi_Z = m(z_I - z_{II}) = m(c_{tI}/A_I\beta - c_{tII}/A_{II}\varepsilon) \qquad (20c)$$

$$\Phi_{ZS} = n(zs_I - zs_{II}) = n(c_{tI}/A_I\gamma - c_{tII}/A_{II}\zeta) \qquad (20d)$$

and, since the sum of the fluxes must be equal to zero, $c_{tI} \neq c_{tII}$. Actually, as a simple calculation shows,

$$c_{tI} = 2c_t \frac{A_I X_{II}}{A_{II} X_I + A_I X_{II}} \tag{21a}$$

and

$$c_{tII} = 2c_t \frac{A_{II} X_I}{A_{II} X_I + A_I X_{II}} \tag{21b}$$

where $X_I = m(1 + \beta) + n(\alpha + \gamma)$ and $X_{II} = m(1 + \varepsilon) + n(\delta + \zeta)$. Then

$$\bigstar \quad v_S = \Phi_{CS} + \Phi_{ZS} = \frac{2c_t\, n}{A_{II} X_I + A_I X_{II}} [X_{II}(\alpha + \gamma) - X_I(\delta + \zeta)] \tag{22}$$

The *initial rate*, when $s_{II} = 0$, can be derived in a simple form for two limiting cases:

(1) For very small s_I, when

$$v_{S_0} = 2c_t m \frac{s_I}{\dfrac{2mK_{CS} K_{ZS}(im + ik + km)}{niK_{ZS}(m + k) + kmnK_{CS}} + s_I} \tag{23}$$

The formula is analogous to the simple formula for mediated equilibrating transport but with a complex term corresponding to K_T.

(2) For very large s_I, when

$$v_{S_0} = \frac{2c_t\, nm}{n + K_{ZS}} \frac{s_I}{\dfrac{K_{ZS}\, m}{n + K_{ZS}} + s_I} \tag{24}$$

which is analogous to the simple Model II formula (p. 76).

The *unidirectional flux* of S which is expressed by $(\vec{\Phi}_{CS} + \vec{\Phi}_{ZS})$ is defined by:

$$\vec{\Phi}_S = 2c_t n \frac{(\alpha + \gamma)[m(1 + \varepsilon) + n(\delta + \zeta)]}{A_{II}[m(1 + \beta) + n(\alpha + \gamma)] + A_I[m(1 + \varepsilon) + n(\delta + \zeta)]} \tag{25}$$

and is thus seen to depend on s_{II} in a complicated manner. It may be calculated, however, that the unidirectional flux (say, inward) is increased by preloading of cells with S, this phenomenon being partly analogous to that described for Model II of mediated diffusion where $D'_{CS} > D'_C$ but partly containing a novel element. Even for $D'_{CS} = D'_{ZS} = D'_C = D'_Z$, preloading with substrate will increase the unidirectional flux (or an extra-polated initial rate of uptake) of substrate (*cf.* Silverman and Goresky, 1965).

In the way of example, let $c_t = 1$, $k = 2$, $i = 1$, $K_{CS} = 10^{-4} M$, $K_{ZS} = 10^{-2} M$, $m = 1$, $n = 1$ and $s_I = 10^{-4} M$. Then $\vec{\Phi}_s$ for $s_{II} = 0$ will be equal to 0.375, for $s_{II} = 10$ it will be equal to 0.5.

A quantity which has not appeared in the treatment of mediated diffusion is the *accumulation ratio* of substrate in the uphill-transporting system $(s_{II}/s_I = \omega)$.

It may be seen that in the present model the net movement of substrate ceases when

$$\frac{1 + \varepsilon}{1 + \beta} = \frac{\delta + \zeta}{\alpha + \gamma}$$

At low s_I values

★
$$\omega = \frac{K_{ZS}i(m + k) + K_{CS}km}{K_{CS}k(m + i) + K_{ZS}im} \tag{26}$$

Hence, for low substrate concentrations, the plot of s_{II} against s_I is linear. For relatively small transmembrane rate constants, moreover, eq. (26) becomes

★
$$\omega = K_{ZS}/K_{CS} \tag{27}$$

At high concentrations, on the other hand, $\omega = 1$ (Fig. 5.1).

Similarly, if an inhibitor acts to suppress the $C \rightarrow Z$ reaction so that $k \cong 0$, the steady-state distribution becomes equal to unity as intuitively predictable.

The present model accounts satisfactorily for a number of uphill-transporting systems (amino acids in tumors, sugars in bacteria, etc.). It possesses all the features of a mobile carrier system of mediated diffusion and, in addition, predicts the asymmetry of solute distribution in steady state. The mobility of the carrier is again documented by the occurrence of countertransport and related phenomena but it should be borne in mind that the countertransport minimum observed readily in most mediated diffusion processes will be less pronounced in uphill transport for the following reason. As substrate S is added during steady state of R (this being perhaps the labeled form of S) the total substrate concentration is greatly increased and hence the s_{II}/s_I ratio attained at this increased concentration will be lower than prior to adding S (*cf.* Fig. 5.1). Thus the intracellular radioactivity will drop to a lower level, the decrease being in the same direction as the decrease due to countertransport. Therefore, as is

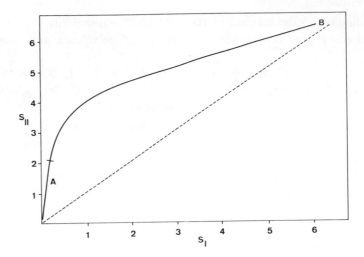

FIG. 5.1. Theoretical distribution of solute across the cell membrane in an uphill transport. In the region marked with **A**, the distribution ratio $\omega = K_{ZS}/K_{CS}$; at **B**, ω tends toward unity.

shown in Fig. 5.2 obtained in a computer using probable values of quantities involved, the countertransport minimum may all but disappear.

A similar consideration (with movements of S and R in the opposite direction) holds for the countertransport according to Miller (1965*b*).

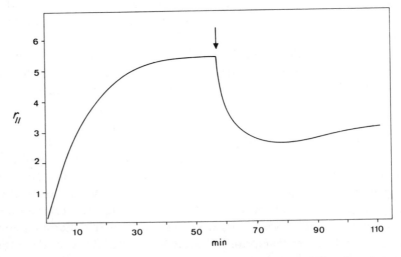

FIG. 5.2. Theoretical prediction of countertransport in an uphill-transporting carrier system (details are found in the text).

5.1.2.2. Self-regulating Model

It was found in several instances that the s_{II} rather than the s_{II}/s_I tends toward a constant at high values of s_I or else is found to increase only very little. This phenomenon can be explained either by limited solubility of the substrate in the intracellular water (but then it should appear in a crystalline form inside the cells) or by a self-regulating mechanism which effectively pumps the solute out of cells at high solute concentrations.

Using the present model, we may envisage a kind of excess-substrate inhibition of the $C \to Z$ conversion resulting in an enhancement of the conversion rate constants j and l and in a suppression of i and k.

Thus the steady-state equations ($17a$–d) will acquire different last terms, as follows:

$$iz_I \quad \text{will be replaced by } iz_I/s_I - js_Ic_I \tag{28a}$$

$$-kc_{II} \qquad \text{by } -kc_{II}/s_{II} + lz_{II}s_{II} \tag{28b}$$

$$-iz_I \qquad \text{by } -iz_I/s_I + js_Ic_I \tag{28c}$$

$$kc_{II} \qquad \text{by } kc_{II}/s_{II} - lz_{II}s_{II} \tag{28d}$$

Addition of eqs. ($28a$) and ($28b$) or of ($28c$) and ($28d$) yields

$$js_Ic_I - iz_I/s_I = lz_{II}s_{II} - kc_{II}/s_{II}$$

which, for very large s (where the effect is of interest), gives

$$c_I/z_{II} = ls_{II}/js_I \tag{29}$$

Digestion of the resulting equations in a manner analogous to that described above, making use of the fact that in this consideration s_I, $s_{II} \gg K_{CS}$, K_{ZS}, results in an expression for the ω ratio equal to

★
$$\frac{nj + ln + ljK_{CS}}{nj + ln + ljK_{ZS}} \tag{30}$$

i.e., a constant. If, moreover, the process rate is limited by a low value of n,

★
$$\omega = K_{CS}/K_{ZS} \tag{31}$$

Hence, in a system of this type (shown in Fig. 5.3), the slope of the s_{II}/s_I curve at low s_I will be K_{ZS}/K_{CS} and at high s_I it will be K_{CS}/K_{ZS}. This is roughly in agreement with experimental relationships obtained, for example, in the monosaccharide transport in some *Candida* species.

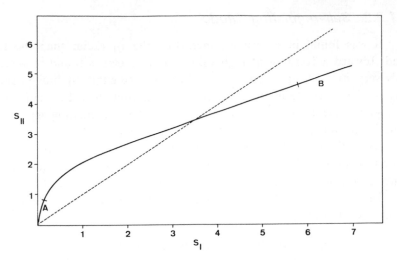

FIG. 5.3. Theoretical distribution of solute in a self-regulating uphill transport. In the region marked with A, $\omega = K_{ZS}/K_{CS}$; in region B, $\omega = K_{CS}/K_{ZS}$.

5.1.2.3. Energetics

While the transport of one mole of substance against a concentration difference of $(s_{II} - s_I)$ is given by the simple formula

$$E = RT \ln s_{II}/s_I \tag{32}$$

the result being in cal/mole, the energy required for maintaining an asymmetric steady state of a solute must include also a time factor and is calculated from the unequal distribution of the various carrier forms C, Z, CS, and ZS. For the sake of simplicity, let us assume that in our model the rates of movement of all the species are equal $(m = n)$ which does not qualitatively affect the results obtained. Then, using D' for the mobility, we have

$$E = D'(c_I - c_{II})RT \ln(c_I/c_{II}) + D'(cs_I - cs_{II})RT \ln(cs_I/cs_{II})$$
$$+ D'(z_I - z_{II})RT \ln z_I/z_{II} + D'(zs_I - zs_{II})RT \ln zs_I/zs_{II} \tag{33}$$

If, under steady-state conditions, $cs_I + zs_I = cs_{II} + zs_{II}$ and $c_I + z_I = c_{II} + z_{II}$, one obtains a simpler form of eq. (33), viz.,

$$E = D'(cs_I + c_I - cs_{II} - c_{II})RT \ln(z_{II}c_I/c_{II}z_I) \tag{34}$$

which becomes meaningful when two limiting cases are considered:

(1) For very low s,

$$\bigstar \qquad E = D' \left(\frac{1}{1 + z_I/c_I} - \frac{1}{1 + z_{II}/c_{II}} \right) RT \ln(z_{II}c_I/z_Ic_{II}) \qquad (35a)$$

(2) For very large s,

$$\bigstar \quad E = D' \left(\frac{K_{ZS}/K_{CS}}{K_{ZS}/K_{CS} + z_I/c_I} - \frac{K_{ZS}/K_{CS}}{K_{ZS}/K_{CS} + z_{II}/c_{II}} \right) RT \ln(z_{II}c_I/z_Ic_{II}) \quad (35b)$$

Since, in a functional uphill-transporting system, $z_I < c_I$ and $z_{II} > c_{II}$ and, also, $K_{ZS} > K_{CS}$, the expression in parentheses of eq. $(35a)$ will be greater than the same in eq. $(35b)$ and, hence, the energy requirement for maintaining a gradient at low concentrations of solute will be greater than at high ones. This semiquantitative prediction actually fits in with observations on, for instance, amino acid transport in ascites tumor cells.

5.1.2.4. Pump-and-Leak Models

The above-treated regular carrier model can be drastically simplified and still, for many purposes, will yield satisfactory results. The models obtained belong to the group of pump-and-leak mechanisms, designed first for the transport of amino acids by ascites tumor cells and employed also in considerations of cation transports in various cells. The simplifications consist mainly in omitting the outward carrier-mediated flux and substituting for it a uni- or ambidirectional diffusion process (leak). The model with the unidirectional leak is inconsistent with what we know about membranes but the one with diffusion proceeding both ways represents a relatively good approximation to what has been found experimentally. Thus,

$$\bigstar \qquad \frac{ds_{II}}{dt} \equiv v_s = V \frac{s_I}{s_I + K_T} + K_D(s_I - s_{II}) \qquad (36)$$

where K_D is a diffusion constant.

Since at high s_I the first term on the right tends toward a constant it may be denoted Y. Integration between 0 and t yields then

$$s_{II}/s_I = (Y/K_D s_I)(1 - e^{-K_Dt}) + (1 - e^{-K_Dt}) \qquad (37)$$

and a plot of $1/s_I$ against s_{II}/s_I is linear with an intercept on the s_{II}/s_I axis equal to $(1 - e^{-K_Dt})$ and on the $1/s_I$ axis equal to $-K_D/Y$. K_D is easily calculated from the ordinate intercept and hence the last term of eq. (36)

is known for any particular s_I and s_{II}. Subtraction of this term leaves us with the expression $Y = Vs_I/(s_I + K_T)$ which, in a reciprocal plot of $1/Y$ against $1/s_I$ will produce the kinetic constants V and K_T.

It will be seen that after a long incubation time when t is very large, the distribution ratio

$$\omega = s_{II}/s_I = [V/K_D(s_I + K_T)] + 1 \tag{38}$$

If $s_I \ll K_T$,

★ $$\omega = (V/K_D K_T) + 1 \tag{39a}$$

If, on the other hand, $s_I \gg K_T$,

★ $$\omega = (V/K_D s_I) + 1 \tag{39b}$$

which, for s_I tending to infinity, tends toward unity, as is observed with many uphill transports.

Similarly, making the simplifying assumption that the inward flux is mediated by two carriers, one of them coupled to a metabolic reaction M so that

$$\Phi_{in} = V_\alpha \frac{s_I}{s_I + K_\alpha} M + V_\beta \frac{s_I}{s_I + K_\beta} \tag{40a}$$

while the outward flux

$$\Phi_{ex} = -V_\beta \frac{s_{II}}{s_{II} + K_\beta} \tag{40b}$$

at steady state

$$V_\beta \left(\frac{s_I}{s_I + K_\beta} - \frac{s_{II}}{s_{II} + K_\beta} \right) = V_\alpha \frac{s_I}{s_I + K_\alpha} M \tag{40c}$$

If one can estimate the unidirectional Φ_{in} (by a tracer technique) we can write

$$\frac{s_{II}}{\Phi_{in}} = K_\beta/V_\beta + s_I/V_\beta \tag{41}$$

from which both K_β and V_β can be estimated.

5.2. COUPLED TRANSPORT

There are many instances among the existing data on biological transport when the uphill transport of one substance, usually an inorganic ion, drives the uphill or even downhill transport of another substance, either in the opposite or in the same direction. The best-known examples of such coupling are that of sodium activation of monosaccharide transport in the intestine and that of amino acid transport in tumor cells.

It should be observed that the "coupled" transport, even if proceeding uphill, is not strictly speaking an active transport (following the definition of Kedem) since it is not directly coupled with a metabolic reaction and since the activator which *is* transported actively can be simply added to one (the driving) side of the membrane and thus the uphill transport of the substance in question pronouncedly altered.

The kinetics of such coupled transport activation are again formally similar to activation of ordinary enzyme reactions and we shall discuss them briefly to assist the reader in orientating himself in the field.

Let us assume, in concord with common practice, that both the substrate S and the activator A are in equilibrium with the carrier C and that, in general, both the CS and the CSA complexes are able to move across the membrane. For the present purpose, let us further examine only the unidirectional flow of S across the membrane so that

$$C + S \underset{k_{-1}}{\overset{k_1}{\rightleftharpoons}} CS \xrightarrow{D'_{CS}} \qquad K_{CS} = c \cdot s/cs \tag{42a}$$

$$C + A \underset{k_{-2}}{\overset{k_2}{\rightleftharpoons}} CA \qquad\qquad K_{CA} = c \cdot a/ca \tag{42b}$$

$$CS + A \underset{k_{-3}}{\overset{k_3}{\rightleftharpoons}} CSA \xrightarrow{D'_{CSA}} \qquad K_{CSA} = cs \cdot a/csa \tag{42c}$$

$$CA + S \underset{k_{-4}}{\overset{k_4}{\rightleftharpoons}} CAS \xrightarrow{D'_{CAS}} \qquad K_{CAS} = ca \cdot s/cas \tag{42d}$$

$$= K_{CSA} \cdot K_{CS}/K_{CA}$$

Let us assume that $D'_{CSA} = D'_{CAS}$.

We do not consider here the possibility that S and A might combine with each other before being bound to the carrier (enzyme). The total flux of S will then be

$$\Phi_S = D'_{CS}\, cs + D'_{CSA}(csa + cas) \tag{43a}$$

or, if the binding of activator is essential for the movement of S,

$$\Phi_S = D'_{CSA}(csa + cas) \tag{43b}$$

The maximum rate for $s = \infty$ and $a = \infty$ is

$$V_a = D'_{CSA}\, c_t \tag{44a}$$

while for $s = \infty$ and $a = 0$ (if activation is not essential)

$$V_s = D'_{CS}\, c_t \tag{44b}$$

The general flux equation for an activated substrate transport is then

★
$$\Phi_{in} = \frac{V_s\, K_{CSA}\, s + V_a\, a\cdot s}{K_{CSA}\, s + a\cdot s + K_{CAS}\, K_{CA} + K_{CAS}\, a} \tag{45a}$$

or, for essential activation,

★
$$\Phi_{in} = V_a \frac{a\cdot s}{s(K_{CSA} + a) + K_{CSA}(K_{CA} + a)} \tag{45b}$$

If the path of formation of the activated complex is compulsory, i.e., either via reactions (42a) and (42c) or via reactions (42b) and (42d), eq. (45b) simplifies further. In the first case $K_{CAS} = 0$ and $K_{CA} = \infty$ so that

$$\Phi_{in} = V_a \frac{a\cdot s}{s(a + K_{CSA}) + K_{CSA}\, K_{CS}} = V_a \frac{s}{s + K_{CSA}(s + K_{CS})/a} \tag{46a}$$

while in the second case $K_{CSA} = 0$ and $K_{CS} = \infty$ so that

$$\Phi_{in} = V_a \frac{a\cdot s}{a(s + K_{CAS}) + K_{CAS}\, K_{CA}} = V_a \frac{s}{s + K_{CAS}(a + K_{CA})/a} \tag{46b}$$

It should be noted that in the compulsory path described by eq. (42a) the activator makes it possible for the carrier–substrate complex to be formed (in the form of CAS) while in that described by eq. (42b) the activator allows the CS complex to migrate across the membrane (as CSA).

According to eq. (42a) the rate of transport will show saturation kinetics with respect to s as well as to a. If $K_{CSA} \gg a$, V will be a linear function of a but the half-saturation constant will be independent of it.

According to eq. (42b) the rate of transport is again saturated at high s but the half-saturation constant is a function of a. In fact, if $K_{CA} \gg a$, the half-saturation constant is inversely proportional to a. On the other hand, V is independent of a.

There is an intrinsic similarity between these two compulsory paths of activation and those of noncompetitive and competitive inhibition, respectively. While in the inhibition formulae, V is divided or K_T multiplied by $(1 + i/K_i)$, in the activation formulae, V is divided and K_T is multiplied by $(1 + K_{CA}/a)$.

In either of the two activated transports described above one can derive the steady-state ratio $\omega = s_{II}/s_I$ by setting $\Phi_{in} = \Phi_{ex}$ (Φ_{ex} being in this treatment, where S and A can enter only by the mechanisms considered, formally identical with Φ_{in}). It follows that in a steady state according to mechanism (42a)

★
$$\omega = a_I/a_{II} \tag{47a}$$

while in that according to mechanism (42b)

★
$$\omega = a_I/a_{II} \frac{K_{CA} + a_{II}}{K_{CA} + a_I} \tag{47b}$$

In the transports considered here, where A may be driven by a specific pump the ω ratio will be different from unity and will change as the amount of activator in the system is changed.

If the presence of the activator does not influence the binding of substrate to the enzyme $K_{CS} = K_{CAS}$ and $K_{CA} = K_{CSA}$ so that

$$\Phi_{in} = V_a \frac{a \cdot s}{(K_{CA} + a)(K_{CAS} + a)} \tag{48}$$

It should be realized that the equilibrium conditions assumed in the above derivations need not obtain, in other words, the D'_{CS} and/or D'_{CSA} may be of the same order of magnitude as the rate constants for reactants C, S, and A. Then the pertinent equations become more complex. Proceeding from the steady-state considerations when $dcs/dt = 0$, $dca/dt = 0$, and $dcsa/dt = 0$, one obtains complex expressions which are not readily analyzed. However, if the activator is essential and the paths leading to the formation of the CSA complex are compulsory, the following expressions can be derived:

$$\Phi_{in} = V_a \frac{k_2 \, a \cdot s}{s(k_2 a + D'_{CSA}) + K_{CAS} k_2 (K_{CA} + a)} \tag{49a}$$

for the path (C + A) + S and

$$\Phi_{in} = V_a \frac{k_1 K_{CS} \, a \cdot s}{k_1 K_{CS} \, s(K_{CSA} + a) + K_{CA}(k_1 K_{CS} K_{CSA} + D'_{CSA} \, a)} \tag{49b}$$

for the path (C + S) + A.

A comparison with the equilibrium relationship for this last case (eq. (46)) shows that while from the equilibrium consideration at very high a the carrier reaction proceeds at its maximum rate regardless of s (if $s \geq c_t$), the steady-state treatment shows this to be true only if D'_{CSA}/k_1

$\ll K_{\text{CS}}K_{\text{CSA}}/a$. The practical use of these equations will be described in chapter 11.

The situation may arise that two activators, A and B, compete for the same binding site of the carrier system or that they are bound to two different sites (this may be true again in cases of sodium and/or potassium activation of various processes or transport). Here an additional set of equations must be employed, namely

$$C + B \underset{k_{-5}}{\overset{k_5}{\rightleftarrows}} CB \qquad\qquad K_{\text{CB}} = c \cdot b / cb \qquad (50a)$$

$$CS + B \underset{k_{-6}}{\overset{k_6}{\rightleftarrows}} CSB \xrightarrow{D'_{\text{CSB}}} K_{\text{CSB}} = cs \cdot b / csb \qquad (50b)$$

$$CB + S \underset{k_{-7}}{\overset{k_7}{\rightleftarrows}} CBS \xrightarrow{D'_{\text{CBS}}} K_{\text{CBS}} = cb \cdot s / cbs \qquad (50c)$$

and again $D'_{\text{CSB}} = D'_{\text{CBS}}$.

Consequently, the maximum rate is

$$V_s = D'_{\text{CS}}\, c_t \text{ for nonessential activation} \qquad (51a)$$

$$V_a = D'_{\text{CSA}}\, c_t \text{ for essential activation by A} \qquad (51b)$$

and

$$V_b = D'_{\text{CSB}}\, c_t \text{ for essential activation by B} \qquad (51c)$$

The flux is then expressed by

$$\Phi_{in} = \frac{s(V_a\, a/K_{\text{CSA}} + V_s + V_b\, b/K_{\text{CSB}})}{s(1 + a/K_{\text{CSA}} + b/K_{\text{CSB}}) + K_{\text{CS}}(1 + a/K_{\text{CA}} + b/K_{\text{CB}})} \qquad (52)$$

Various limiting cases can be readily derived from this equation, such as $V_s = 0$ (essential activation), $V_s = V_a = V_b$ (nonessential activation with identical maximum velocities), types where substrate can react with either CA or with CB only or where the modifiers have no effect on the enzyme-substrate affinity, etc.

Another special case arises if the sum of concentrations of the activators is kept constant (which may be of importance in studies with animal cells or tissues where the osmolarity must not vary). For a detailed treatment of this situation, consult Semenza (1967).

6. KINETICS OF TRACER EXCHANGE

6.1. INTRODUCTION

It was shown several times before (p. 67, p. 145) that the kinetics of transmembrane movement are greatly simplified if the process studied is first-order, i.e., if the rate of the movement of the substance studied is directly proportional to its concentration. It was also shown that, in addition to conditions of very low saturation of the transport systems when the movement proceeds as an approximation to a diffusion process, first-order kinetics can be attained by the use of tracers. The use of tracers provides an additional bonus in that it makes it possible to follow unidirectional fluxes of a substance under conditions when the substance moves across a membrane in both directions.

A few words concerning the justification of measuring unidirectional fluxes with tracers are in order here to prevent possible misunderstanding. Suggestions that such measurements are incorrect, even when using kinetically indistinguishable isotopes, were made in the literature and the situation may appear somewhat confusing. Thus it was stated (Nims, 1962) that "... the flow of the tracer bears little relation to the flow of the substance concerned" in a paper which has drawn attention to important mutual interactions between the movements of particles, the transport of which is studied. (Such interactions are discussed in chapter 5 of the present book.) Other authors (Curran *et al.*, 1967 *b*) state that the previous author "... suggested that the general approach of estimating unidirectional fluxes with tracers is incorrect" and on the basis of both a theoretical analysis and experimental data they concluded that "... flow interactions do not affect the interpretation of tracer measurements very markedly in simple systems at relatively low solute concentrations, so that suggestions that the tracer technique is invalid in general because of interactions is unwarranted." Nims' mathematical reasoning (1962), however, appears to prove that there is, in fact, no simple relation between the tracer flux and the *net* flux of the substance concerned, whether there are interactions or not, and thus, even if some measurements of unidirectional fluxes are critically mentioned by the author, they have little bearing on the relation between the flux of a kinetically indistinguishable tracer and the *unidirectional* flux. The view held in this book is that the relation between a tracer flux from one compartment to another, and the unidirectional flux between the two compartments, is simple and always the same: the tracer flux divided by the specific activity in the first compartment is equal to the unidirectional flux. The unidirectional flux, a concept useful in theoretical analysis (see, for instance, the discussion of the flux ratio in section 3.2 and chapter 5) may be defined as the number of moles of the substance which, *being originally present in the first compartment*, penetrate per unit time across a unit area of the barrier into the other compartment. The above calculation is then seen to involve the assumption that if some particles which we are able to recognize (e.g., as they happen to disintegrate during the time of our radioactivity measurement) penetrate from the first compartment into the second, the kinetically indistinguishable particles in the first compartment which we cannot recognize (since they will disintegrate later or not at all, belonging to a nonradioactive isotope) will do the same in proportion to their relative participation in the first compartment (apart from statistical fluctuation) whatever the forces and interactions encountered en route may be.

The principal aim of this section is to discuss the simple mathematical principles underlying the interpretation of tracer experiments and the determination of the magnitudes of the fluxes (or related parameters like rate constants) involved. At first, however, some of the concepts important in these calculations will be defined and some of the general ideas of the field of the kinetics of tracer exchange will be discussed.

To represent satisfactorily the behavior of unlabeled substances, the labeled ones must resemble them as much as possible. The method of choice is labeling with a radioactive isotope or, where no such isotope is available, with a heavy stable isotope of the element studied ^2H, ^{15}N, ^{18}O). Certain errors can arise due to the use of tracers since the diffusion and reaction velocities generally decrease as the weight of the labeling isotope increases. However, as the difference in velocities depends mostly on the percent difference in molecular weights of the substance studied, the effect will be appreciable only with very simple compounds of the lightest elements (such as ordinary, heavy, and tritiated water).

Two conditions must be fulfilled to achieve reliability of tracer studies: (i) The analytical concentration of the tracer must be negligibly low as compared with the concentration of the nonlabeled substance so that addition of the tracer to one of the compartments under consideration brings about no change in total concentration. (ii) The substance studied must be labeled uniformly, i.e., the molecules containing recognizable isotopes must be well mixed with the unlabeled substance, the behavior of which they are to represent.

By a "physiological compartment" (Solomon, 1960 a) we shall understand a spatial part of a biological system showing a certain homogeneity with respect to the substance considered; there are no important barriers inside such compartment such as would influence the distribution of the substance, and its concentration is therefore the same throughout the compartment. The total amount of substance leaving the compartment per unit time may be called the total outflow of the substance; the amount entering it during the same time the total inflow. When the area of the surface surrounding the compartment is known, the flows per unit area of the membrane (the fluxes) may be calculated and the properties of different membranes, irrespective of their dimensions, may be compared.

The experimental conditions for which the mathematical description of the temporal changes of the tracer content is simplest are actually the conditions which are most often encountered in biological systems and which have a special significance for them. First, it is the steady state of the compartment considered, i.e., the state when the total content of the substance (irrespective of its labeling) in the compartment and also the volume of the compartment are constant. The amount of the substance in a compartment does not change with time whenever its inflow equals its outflow, this condition being satisfied with a good approximation rather frequently. The so-called net flows of substances connected with the growth of living cells or with the gradual decay of surviving tissues are often too small to invalidate seriously the assumption of the equality of the inflow with the outflow. As stressed by Robertson (1957), this equality is the only condition of a steady state rather than the temporal constancy of the individual flows. This temporal constancy is, however, the second condition required by the kinetic equations to become as simple as they are in the present treatment. Also this second condition is often satisfied in biological systems; it is evident that permanent equality of inflow and outflow is attained most simply when neither of them changes with time. Under conditions of steady state and when the individual flows are constant the observable behavior of the *labeled substance* is independent of the mechanism by which the studied substance permeates through the membrane and may be described by simple linear differential equations, analogous to those by which first-order chemical reactions, the simplest electrical circuits, radioactive decays, or the first phases of organic growth are described. By way of illustrating this statement, let us return to the treatment of carrier-mediated flows on p. 63. There, using the generally employed symbolism, it was shown that the rate of change of intracellular concentration of R (a tracer) in the presence of a constant amount of S is described by $dr_{II}/dt = K(r_I - r_{II})$ which is a first-order rate equation formally identical with eq. (9) for the simple case of an equilibrating transport (no uphill movement) where $c_i = c_o$. This can be easily understood: In a steady state the concentration of the labeled substance in a compartment changes in direct proportionality to the flow of the labeled substance across the boundaries of the compartment. If the total flow of the substance, labeled as well as unlabeled, is always the same, the flow of the tracer depends only on the relative abundance of the labeled substance in this flow, i.e., on the specific activity in the compartment from which this flow originates. For this reason, in a steady state and when the flows are constant, the differences in the specific activities act as driving forces of the observable changes in the concentrations of labeled substance. As soon as an isotopic equilibrium is achieved in a system in steady state (i.e., the specific activity is the same throughout the system) changes in concentrations of labeled substances cease to be measurable.

The following symbols will be used in the calculations:

c_i^*,   concentration of labeled substance in compartment i; in counts·cm^{-3} × min^{-1};

c_i,   chemical (i.e., total) concentration of the studied substance in compartment i; in mmol·liter^{-1}; c_i^*/c_i is then the specific activity in compartment i;

Φ_{io},   flux of the substance studied from compartment i into compartment o; in mmol·cm^{-2}·sec^{-1};

V_i,   volume of compartment i; $V_i c_i^*$ is then the total amount of the labeled substance in compartment i;

A_i,   area of the membrane across which the exchange with compartment i takes place; $\Phi_{io} A_i$ is then the total flow of the studied substance from compartment i into compartment o.

6.2. COMPARTMENT IN A STEADY STATE, COMMU-
NICATING WITH ONE OTHER COMPARTMENT

As a fundamental case for which a differential equation can be written
—a case which will be later solved for various special conditions—we may
consider the system of two compartments in steady state communicating
only with each other across a membrane of area A by time-independent
fluxes $\Phi_{io} = \Phi_{oi} = \Phi$ (Fig. 6.1).

For the change of concentration of a labeled substance in compartment i the following equation may be written:

★
$$\frac{dc_i{}^*}{dt} = \frac{\Phi_{oi}}{V_i} A \frac{c_o{}^*}{c_o} - \frac{\Phi_{io}}{V_i} A \frac{c_i{}^*}{c_i} \tag{1}$$

The rate of change of concentration of the labeled substance is directly
proportional to the total flows of the labeled substance (i.e., to the fluxes
of the studied substance, multiplied by the specific activity in the compartments from which the fluxes originate, and multiplied by the area of the
membrane) and indirectly proportional to the volume of the compartment.

An analogous equation applies to compartment o:

$$\frac{dc_o{}^*}{dt} = \frac{\Phi_{io}}{V_o} A \frac{c_i{}^*}{c_i} - \frac{\Phi_{oi}}{V_o} A \frac{c_o{}^*}{c_o} \tag{2}$$

On comparing eqs. (1) and (2) it may be seen that a third equation must
be valid:

$$V_i \frac{dc_i{}^*}{dt} = - V_o \frac{dc_o{}^*}{dt} \tag{3}$$

Since the compartments communicate only with each other, an increase
in the amount of labeled substance in one compartment equals its decrease
in the other compartment.

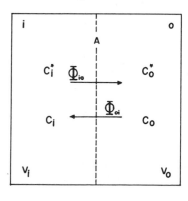

FIG. 6.1. System of two compartments.
Explanation in the text.

6.2.1. Compartment and a Reservoir

For further simplification of equations such as (*1*), the concept of a reservoir may be introduced. A reservoir is understood as a physiological or experimental compartment in which not only the chemical (i.e., total) concentration of the studied substance but also the concentration of the labeled substance remains constant. In proctice, such a situation is most easily achieved when the reservoir is much larger than the other compartments and the amount of the tracer exchanged may be therefore neglected with respect to the total content of the tracer in the reservoir, or when the medium in the reservoir is continuously replaced with a solution containing the original concentration of the labeled substance.

1. Labeled Substance in the Compartment. A special case of the reservoir defined above is a reservoir in which the concentration of the labeled substance equals zero. A compartment with a known initial concentration of the labelled substance $c_{it=0}^*$ will be considered, communicating only with a reservoir where the specific activity of the substance studied equals zero (Fig. 6.2). Experimental conditions corresponding to this assumption are obtained rather simply in the so-called washing-out experiments (see p. 208).

The concentration of the labeled substance in the medium c_o being practically zero, eq. (*1*) will take the form

$$\frac{dc_i^*}{dt} = - \frac{\Phi_{io}}{V_i} A \frac{c_i^*}{c_i} \qquad (4)$$

Under the conditions specified above (steady state, time-invariant fluxes) four of the five magnitudes on the right-hand side may be included in a

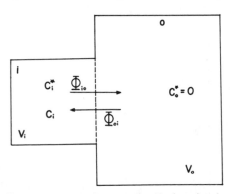

Fig. 6.2. Compartment and reservoir. Explanation in the text.

single constant k, called the rate constant

$$k = \frac{\Phi_{io}}{V_i} A \frac{1}{c_i}$$

and

$$\frac{dc_i{}^*}{dt} = - kc_i{}^* \tag{5}$$

Perhaps it may be useful to stress here again that the rate constant k as determined in this analysis does not tell us anything about the mechanism of the movement across the membrane barrier. The magnitude and character of flux Φ which is included in the constant actually depends greatly on the complexity of the mechanism involved. Thus, even in simple carrier transport, it would include the number of carrier molecules participating in the process, their affinity for substrate, and their mobility in the membrane. To obtain information on these parameters, different approaches to the problem must be chosen (cf. p. 73).

The very simple differential equation (5) can be solved by separating the variables and by integrating between zero time, when the concentration of the labeled substance inside the compartment is $c_{it=0}^*$ and time t, when the concentration is $c_i{}^*$:

$$\int_{c_{it=0}^*}^{c_i{}^*} \frac{dc_i{}^*}{c_i{}^*} = - k \int_0^t dt \tag{6}$$

Solving, we get

$$\ln \frac{c_i{}^*}{c_{it=0}^*} = - kt \tag{7a}$$

Equation (7a) may be also written in the exponential form:

★ $$c_i{}^* = c_{it=0}^* e^{-kt} \tag{7b}$$

where e is the basis of natural logarithms ($e = 2.71828...$). The concentration of the labeled substance inside the compartment thus decreases exponentially (Fig. 6.3); equation (7b) can be graphically represented as shown in Fig. 6.4.

For practical purposes, however, it is much more convenient to plot the values of $c_i{}^*/c_{it=0}^*$ against t of eq. (7a) semilogarithmically, since then the relationship is a straight line (Fig. 6.5). The slope of this straight line (i.e., the change of the variable plotted on the vertical axis per unit time) corresponds to the rate constant k. It can be read easily from the graph, and if the total amount of the studied substance in the compartment $V_i c_i$ is

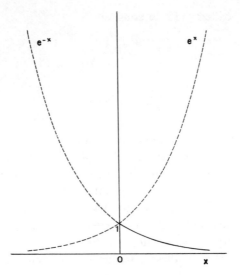

FIG. 6.3. The exponential function.

known the rate constant can be multiplied by it and the total unidirectional flow $\Phi_{io}A_i = \Phi_{oi}A_i$ calculated. If the area of the membrane across which the exchange takes place (A_i) is known, the fluxes per unit area may be determined.

Alternatively, an exponential process may be characterized by its half-time $t_{0.5}$, i.e., by the time during which the concentration of the labeled substance inside a compartment (more generally, the concentration of the molecules of the studied substance present there at any specified time)

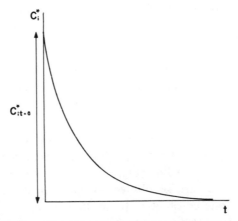

FIG. 6.4. Exponential decrease of activity. Explanation in the text.

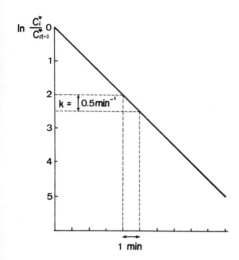

FIG. 6.5. Semilogarithmic plot of the exponential time course. Explanation in the text.

decreases to one-half. From eq. (*7a*)

$$\ln \frac{1}{2} = -\ln 2 = -kt_{0.5}$$

i.e.,

★
$$t_{0.5} = \frac{\ln 2}{k} = \frac{0.693}{k} \tag{8}$$

It is not always convenient to work with natural logarithms and the use of a "semilogarithmic paper" may not be precise enough. It is then possible to plot the variable concentrations of the labeled substances in the usual Briggsian logarithms. Since $\ln x = 2.303 \log x$, the rate constant k may be found if the slope read from such a graph is multiplied by the modulus 2.303.

2. Labeled Substance in the Reservoir.

This simple case is analogous to the previous one but the positions of the labeled and the unlabeled substance are reversed. The reservoir contains the labeled substance in the time-invariant concentration $c_o{}^*$ and the compartment communicating with it contains at the beginning of the experiment no labeled substance. Equation (*1*) where $c_o{}^*$ is a constant and $\Phi_{io} = \Phi_{oi} = \Phi$ is relevant:

$$\frac{dc_i{}^*}{dt} = \frac{\Phi}{V_i} A_i \frac{c_o{}^*}{c_o} - \frac{\Phi}{V_i} A_i \frac{c_i{}^*}{c_i}$$

Using again the notation

$$\frac{\Phi}{V_i} A_i \frac{1}{c_i} = k$$

the equation will take the form

$$\frac{dc_i^*}{dt} = k\left(\frac{c_i}{c_o} c_o^* - c_i^*\right) \tag{9}$$

Equation (9) may again be easily solved by separating the variables and integrating within appropriate limits

$$\int_{\frac{c_i}{c_o} c_o^*}^{\frac{c_i}{c_o} c_o^* - c_i^*} \frac{dc_i^*}{\frac{c_i}{c_o} c_o^* - c_i^*} = k \int_0^t dt$$

i.e.,

$$- \ln \frac{\frac{c_i}{c_o} c_o^* - c_i^*}{\frac{c_i}{c_o} c_o^*} = kt \tag{10a}$$

which may be written

★ $$\frac{c_i^*}{c_i} = \frac{c_o^*}{c_o} (1 - e^{-kt}) \tag{10b}$$

From the graphical form of the function $(1 - e^{-kx})$ (Fig. 6.6) it follows that the specific activity in the compartment approaches asymptotically the specific activity in the reservoir (Fig. 6.7).

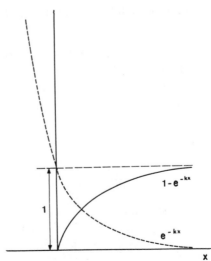

FIG. 6.6. Plot of $(1 - e^{-kx})$.

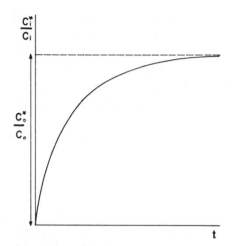

FIG. 6.7. Exponential build-up of specific activity. Explanation in the text.

6.2.2. Closed System of Two Compartments

Let us now discuss the case when neither of the two compartments can be considered as a reservoir with a time-invariant concentration of the labeled substance; the concentration of tracer will change appreciably in both. At zero time the labeled substance is present in compartment o at concentration $c_{ot=0}^*$. An equation may be introduced expressing the fact that no tracer is lost from the closed system; the labeled substance present at zero time in compartment o may be found during the experiment in one or the other compartment:

$$c_i^* V_i + c_o^* V_o = c_{ot=0}^* V_o \qquad (11)$$

Combining (11) with (1) when, moreover, $\Phi_{io} = \Phi_{oi} = \Phi$, a differential equation describing our system is obtained:

$$\frac{dc_i^*}{dt} = \frac{\Phi}{V_i} A_i \left(\frac{c_{ot=0}^*}{c_o} - \frac{V_i c_i + V_o c_o}{V_o c_i c_o} c_i^* \right) \qquad (12)$$

After separating the variables the equation may be integrated and the result expressed in a logarithmic (13a) or exponential (13b) form

$$\frac{V_o c_i c_o}{V_i c_i + V_o c_o} \ln \left(1 - \frac{V_i c_i + V_o c_o}{V_o c_i} \frac{c_i^*}{c_{ot=0}^*} \right) = - \frac{\Phi}{V_i} A_i t \qquad (13a)$$

$$\frac{c_i^*}{c_i} = \frac{c_{ot=0} V_o}{V_i c_i + V_o c_o} \left(1 - e^{- \frac{(V_i c_i + V_o c_o) \Phi}{V_o c_i c_o V_i} A_i t} \right) \qquad (13b)$$

It may be seen that the system approaches exponentially the state where the specific activity in compartment i is the same as in the other compartment and equals the total amount of the labeled substance in the system divided by the total amount of the studied substance, labeled as well as unlabeled, in both compartments. The rate constant, i.e., the coefficient in the exponent of the exponential function, is already a rather complicated expression, the value of which depends on the amounts of the studied substance in each compartment. The calculation of flux Φ under these conditions is more difficult; it is therefore advisable, wherever possible, to use experimental conditions where one of the compartments behaves as a reservoir.

6.3. COMPARTMENT IN A STEADY STATE, COMMU-NICATING WITH TWO OTHER COMPARTMENTS

In the above calculations a special case of the steady state was considered: the inflow of a substance from a reservoir or from another compartment was exactly balanced by its outflow into the same reservoir or compartment. In the following, we shall analyze compartments in which the inflow of the studied substance from one reservoir (e.g., from one with a constant concentration of the labeled substance) will be exactly balanced by the outflow into another reservoir (e.g., into one with a negligibly low concentration of the labeled substance).

6.3.1. Compartment with Irreversible Outflow

An especially simple situation arises when the outflow of the studied substance from the compartment is irreversible; in this case it need not even flow into a reservoir with a negligibly low concentration of the labeled substance, since neither the labeled nor the unlabeled substance returns to the studied compartment (see Fig. 6.8). The situation is described by

$$\frac{dc_i^*}{dt} = \frac{\Phi_{1i}}{V} A_1 \frac{c_1^*}{c_1} - \left(\frac{\Phi_{i1}}{V} A_1 + \frac{\Phi_{i2}}{V} A_2 \right) \frac{c_i^*}{c_i} \tag{14}$$

If it is assumed that the system is in a steady state and the fluxes are independent of time, we can write

$$\frac{dc_i^*}{dt} = k_1 c_1^* - k_2 c_i^* \tag{15}$$

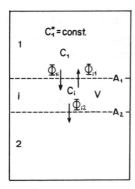

FIG. 6.8. Compartment with irreversible outflow. Explanation in the text. (According to Janáček, 1967.)

The first term on the right-hand side is a constant, the second term contains a constant coefficient. After separating the variables one can integrate

$$\int_{k_1 c_1^*}^{k_1 c_1^* - k_2 c_i^*} \frac{dc_i^*}{k_1 c_1^* - k_2 c_i^*} = \int_0^t dt \qquad (16a)$$

$$-\frac{1}{k_2} \ln \frac{k_1 c_1^* - k_2 c_i^*}{k_1 c_1^*} = t \qquad (16b)$$

$$1 - \frac{k_2}{k_1} \frac{c_i^*}{c_1^*} = e^{-k_2 t} \qquad (16c)$$

$$c_i^* = c_1^* \frac{k_1}{k_2} (1 - e^{-k_2 t}) \qquad (16d)$$

Because of the prevailing steady state the total inflow of the substance must be equal to its total outflow:

$$\Phi_{1i} A_1 - \Phi_{i1} A_1 = \Phi_{i2} A_2 \qquad (17)$$

which is equivalent to

$$\frac{k_1}{k_2} = \frac{c_i}{c_1} \qquad (18)$$

and, therefore, eq. (16d) may be written as

★ $$\frac{c_i^*}{c_i} = \frac{c_1^*}{c_1} (1 - e^{-k_2 t}) \qquad (19)$$

The unlabeled substance disappears exponentially from the compartment with irreversible outflow and the specific activity there approaches the specific activity in the reservoir from which the substance arrives. As an example of such a situation we might cite a compartment into which a substrate arrives to be metabolized and in this way leaves the compartment irreversibly (*cf.* p. 244 where the steady-state intracellular level of a metabolic substrate is discussed).

6.3.2. Compartment with Reversible Outflow

Another important case of a compartment communicating with two other compartments is the compartment with reversible outflow of the substance studied. An epithelial layer separating two media is likely to behave in this way, e.g., the intestinal wall dividing an experimental perfusion solution from blood plasma or the isolated frog skin placed between two aerated solutions. In contrast with the previous case, the specific activity in such a compartment never reaches the specific activity of the reservoir from which the substance arrives.

For the sake of simplicity let us again consider the case where the studied substance comes from a reservoir with a constant concentration of the labeled substance and leaves for a reservoir with a negligibly low concentration of the labeled substance (Fig. 6.9).

As in the preceding case, no labeled substance returns from the second reservoir (flux Φ_{2i} is not zero, but the specific activity in reservoir 2 is practically nil) and hence the same differential equation (*14*) is applicable, giving the same solution in terms of rate constants (eq. (*16d*)) but the steady-state condition will be different:

$$(\Phi_{1i} - \Phi_{i1})A_1 = (\Phi_{i2} - \Phi_{2i})A_2 \qquad (20)$$

The ratio of the rate constants is thus not simply equal to the ratio of the concentrations of the studied substance in the examined compartment and in the first reservoir, but the specific activity in the compartment can be

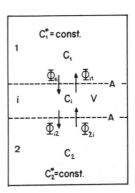

FIG. 6.9. Compartment with reversible outflow. Explanation in the text. (According to Janáček, 1967).

expressed in terms of fluxes, areas, volumes, and concentrations as follows:

$$\bigstar \qquad \frac{c_i^*}{c_i} = \frac{\Phi_{1i}A_1}{\Phi_{i1}A_1 + \Phi_{i2}A_2}\frac{.c_1^*}{c_1}(1 - e^{-k_2 t}) \qquad (21)$$

The specific activity in our compartment communicating with two reservoirs thus approaches exponentially a value which is less than the specific activity. in the reservoir; which fraction of this activity it will reach is determined by the ratio of the individual flows. This characteristic of a compartment with reversible outflow of the studied substance communicating with two reservoirs was discussed in detail by Morel (1959) in connection with sodium transport across frog skin. The flux Φ_{i2} from the cells to the unlabeled medium corresponds in the case of frog skin to sodium transported actively by the sodium pump. The value of this flux cannot be calculated by dividing the amount of the labeled sodium appearing in the medium at the inner surface of the skin per unit time by the specific activity of the medium at the outer surface. It would be necessary to divide it by the specific activity inside the cells and this activity is always lower, the labeled sodium there being continually diluted by the unlabeled one entering from the medium bathing the inner surface.

Let us consider at this stage the characteristics which all the systems discussed so far had in common. The concentration of the labeled substance was followed always in one compartment, communicating either with one other compartment or with one or two reservoirs. Equations describing the concentration changes of the labeled substance were always of the form

$$c^* = c_{t=0}^* e^{-kt} \qquad \text{or} \qquad c^* = c_{t=\infty}^*(1 - e^{-kt})$$

Each equation contained only one time-dependent exponential term, the concentration constants $c_{t=0}^*$, $c_{t=\infty}^*$, and the rate constants k, the meaning of which was defined by the theoretical analysis of the system. In a logarithmic form, the equations were equations of a straight line

$$\ln \frac{c^*}{c_{t=0}^*} = -kt \qquad \text{or} \qquad \ln c^* = (\ln c_{t=0}^*) - kt$$

and

$$\ln \left(1 - \frac{c^*}{c_{t=\infty}^*}\right) = -kt$$

Thus in the first case one can plot either the logarithm of the ratio of the concentration of the labeled substance to its concentration at zero time against time or else the logarithm of the concentration itself so that the original concentration may be found as the ordinate at zero time (the intercept on the vertical axis). In the second case the difference between one and the ratio of the concentration of the labeled substance at time t to its final concentration is plotted against time. Straight lines are thus obtained and the rate constants may be calculated from their slopes.

In most cases the situation is not so simple; the changes of concentrations in the compartments studied suggest an exponential character, but after a semilogarithmic plotting curvilinear dependences rather than straight lines are found. It may be shown, however, that in many cases these curves may be represented as the logarithm of a sum of a small number of exponential terms. This situation is suspected to occur if in the semilogarithmic plot both a curved and a linear part are seen; for sufficiently long times the curve turns into a straight line.

6.4. TWO COMPARTMENTS AND A RESERVOIR

Let us examine now why in some simple systems the measured concentration of a labeled substance changes as a sum of several exponential terms. It is shown on p. 241 how one can resolve the experimental curve by graphical analysis and express it by an equation containing several exponential terms. The meaning of this equation will be made clear by theoretical calculations.

6.4.1. Two Compartments in Parallel

In the first and simplest example discussed above the labeled substance flowed from a single physiological compartment into a reservoir where the concentration of the labeled substance was negligibly low; the whole process was described by a single exponential term. Now let us assume that not only one, but two such compartments with different properties are placed into such a reservoir without labeled substance; neither the amounts of the studied substance in the compartments nor their rate constants are identical. Such situation may arise if, for instance, two kinds of cells are dispersed in a medium and a labeled substance flows from them into a reservoir without returning, the reservoir being either very large or the medium in it often replaced. The concentration of the labeled substance in the first compartment, i.e., in the first kind of the cells, decreases in agreement with eq. (22)

$$\frac{c_1{}^*}{c_{1t=0}^*} = e^{-k_1} \tag{22}$$

while in the other type of cells it decreases according to the analogous equation (23)

$$\frac{c_2{}^*}{c_{2t=0}^*} = e^{-k_2 t} \tag{23}$$

If it were experimentally feasible to measure the activity in the single compartments one would simply follow the concentration of the labeled substance in each of them and the evaluation of the experimental data would not require any new principles. If, on the other hand, there is no possibility of separating the two kinds of cells, the individual concentrations of the labeled substances cannot be directly estimated. It may be, however, convenient to measure the ratio of the total amount of the labeled substance in both the compartments at time t to the total amount present at time zero. If the volumes of the two compartments are V_1 and V_2 this ratio will be equal to

$$\frac{c^*}{c_{t=0}^*} = \frac{c_1{}^* V_1 + c_2{}^* V_2}{c_{1t=0}^* V_1 + c_{2t=0}^* V_2} = \frac{c_{1t=0}^* V_1}{c_{1t=0}^* V_1 + c_{2t=0}^* V_2} e^{-k_1 t} + \frac{c_{2t=0}^* V_2}{c_{1t=0}^* V_1 + c_{2t=0}^* V_2} e^{-k_2 t} \tag{24}$$

The right-hand side of the equation contains two exponential terms. The coefficients of these terms are the fractions of the content of the studied substance present in each of the compartments. The coefficients of the exponents are rate constants, related in the way already discussed to the fluxes of the studied substance across the membranes limiting the compartments.

6.4.2. Two Compartments in Series

It was shown how the parallel arrangement of two compartments is characterized by an experimental curve, which can be analysed graphically (see p. 241) as a sum of two exponential terms. Now it will be seen that a quite different arrangement of two compartments, one in series, is also described by a sum of two exponential terms. The interpretation of the coefficients in their exponents will be different as is shown by a theoretical consideration. It may be of importance to realize that if an experimental curve obtained in such a system with reservoir can be broken down to two exponential parts the presence of two compartments is suggested but nothing is learned about their mutual arrangement.

Two steady-state compartments in series will be now considered. The specific activity will be the same in both at the beginning of the experiment.

At time zero they will be transferred to a reservoir where the concentration of the labeled substance is negligibly low. It is often assumed that a piece of nerve or muscle tissue in steady state and isotopic equilibrium with a radioactive medium which is then transferred to a nonradioactive medium of the same composition behaves in this way. The two compartments in series in this case are the cells and the extracellular space of the tissue; the exchange of substances between the extracellular space and the external medium is a diffusion rather than a permeation process, but still it may be approximately described by a single exponential. The system of two compartments in series is shown in Fig. 6.10.

The rate of change of the label concentrations in the two compartments is described by the following differential equations:

$$\frac{dc_1^*}{dt} = \frac{\Phi_{21}A_{12}}{V_1}\frac{c_2^*}{c_2} - \frac{\Phi_{12}A_{12}}{V_1}\frac{c_1^*}{c_1} \tag{25}$$

$$\frac{dc_2^*}{dt} = \frac{\Phi_{12}A_{12}}{V_2}\frac{c_1^*}{c_1} - \left(\frac{\Phi_{21}A_{12}}{V_2} + \frac{\Phi_{20}A_{20}}{V_2}\right)\frac{c_2^*}{c_2} \tag{26}$$

For the sake of simplicity the constant coefficients may be considered as rate constants:

$$\frac{dc_1^*}{dt} = k_1 c_2^* - k_2 c_1^* \tag{27}$$

$$\frac{dc_2^*}{dt} = k_3 c_1^* - k_4 c_2^* \tag{28}$$

These are two first-order equations, each containing two unknowns. From each of them one unknown may be expressed in terms of the other, its derivative calculated and these values inserted into the other equation.

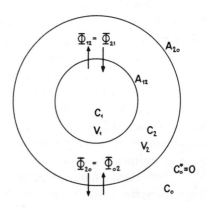

FIG. 6.10. Two compartments in series. Explanation in the text.

Two second-order equations with constant coefficients, without a constant term and each containing only one unknown are thus obtained; moreover, they contain the same coefficients:

$$\frac{d^2c_1^*}{dt^2} + (k_2 + k_4)\frac{dc_1^*}{dt} + (k_2k_4 - k_1k_3)c_1^* = 0 \qquad (29)$$

$$\frac{d^2c_2^*}{dt^2} + (k_2 + k_4)\frac{dc_2^*}{dt} + (k_2k_4 - k_1k_3)c_2^* = 0 \qquad (30)$$

Such differential equations have solutions of the form

$$c_1^* = Ae^{r_1t} + Be^{r_2t} \qquad (31)$$

$$c_2^* = Ce^{r_1t} + De^{r_2t} \qquad (32)$$

where r_1 and r_2 are the roots of the auxiliary equation—a quadratic equation with the coefficients of the differential equations:

$$r^2 + (k_2 + k_4)r + k_2k_4 - k_1k_3 = 0$$

$$r_{1,2} = -\frac{k_2 + k_4}{2} \pm \sqrt{\frac{(k_2 + k_4)^2}{4} - k_2k_4 + k_1k_3}$$

The constants A, B, C, and D can be calculated from the initial conditions

$$t = 0 \qquad c_1^* = c_{1t=0}^* \qquad c_2^* = c_{2t=0}^*$$

$$\left(\frac{dc_1^*}{dt}\right)_{t=0} = k_1c_{2t=0}^* - k_2c_{1t=0}^* \qquad \left(\frac{dc_2^*}{dt}\right)_{t=0} = k_3c_{1t=0}^* - k_4c_{2t=0}^*$$

Derivatives of the solutions (31) and (32) are

$$\frac{dc_1^*}{dt} = r_1Ae^{r_1t} + r_2Be^{r_2t}$$

$$\frac{dc_2^*}{dt} = r_1Ce^{r_1t} + r_2De^{r_2t}$$

At time zero (since $e^0 = 1$):

$$c_{1t=0}^* = A + B$$

$$c_{2t=0}^* = C + D$$

$$r_1A + r_2B = k_1c_{2t=0}^* - k_2c_{1t=0}^*$$

$$r_1C + r_2D = k_3c_{1t=0}^* - k_4c_{2t=0}^*$$

In this way four equations for four unknowns are available and the values of the constants can be calculated:

$$A = \frac{k_1 c_{2t=0}^* - (k_2 + r_2)c_{1t=0}^*}{r_1 - r_2} \qquad C = \frac{k_3 c_{1t=0}^* - (k_4 + r_2)c_{2t=0}^*}{r_1 - r_2}$$

$$B = \frac{k_1 c_{2t=0}^* - (k_2 + r_1)c_{1t=0}^*}{r_2 - r_1} \qquad D = \frac{k_3 c_{1t=0}^* - (k_4 + r_1)c_{2t=0}^*}{r_2 - r_1}$$

The ratio of the total amount of the labeled substance in both compartments to its total amount at time zero (this ratio is experimentally measurable) is equal to

$$\frac{c^*}{c_{t=0}^*} = \frac{c_1^* V_1 + c_2^* V_2}{c_{1t=0}^* V_1 + c_{2t=0}^* V_2} = \frac{V_1 A + V_2 C}{c_{1t=0}^* V_1 + c_{2t=0}^* V_2} e^{r_1 t} + \frac{V_1 B + V_2 D}{c_{1t=0}^* V_1 + c_{2t=0}^* V_2} e^{r_2 t} \tag{33}$$

where the constants A, B, C, D, r_1, r_2, k_1, and k_2 have the meaning specified above.

It may be seen that the solution of the series model is also composed of two exponential terms. In contrast with the parallel arrangement, however, the coefficients of these terms and of their exponents, which may be determined by graphical analysis, are not characteristic for the individual compartments. How the amount of the labeled substance present in the inner compartment ($c_{1t=0}V_1$ in our nomenclature) can be calculated if the movement in time of the labeled substance from the system to a reservoir is described as a sum of two exponential terms, was shown in a most elegant derivation by Huxley (1960).

6.5. NONSTEADY-STATE COMPARTMENTS

Studying systems in steady state simplifies greatly the kinetics of the exchange of labeled substances. In some cases, however, no steady state can be achieved, such as with an isolated tissue that does not survive so as to preserve its steady state. Keynes and Lewis (1951 a) studied the exchange of potassium in surviving crab nerve and their equations were published by Solomon (1960 a). At the beginning of the experiment the surviving tissue without labeled potassium was transferred to a large reservoir with a constant concentration of labeled as well as of unlabeled potassium. The total potassium content of the nerve (irrespective of its labeling) decreases during the experiment, this drop being approximately exponential. In our nomenclature we can write

$$c_i = c_{it=0}e^{-\beta t} \tag{34}$$

The rate of change of the total concentration of potassium is therefore described by

$$\frac{dc_i}{dt} = \frac{\Phi_{oi}A}{V_i} - \frac{\Phi_{io}A}{V_i} = -c_{it=0}\beta e^{-\beta t} \tag{35}$$

Setting

$$\Phi_{io} = k_{io}\frac{V_i}{A}c_i$$

we can write

$$-c_{it=0}\beta e^{-\beta t} = \frac{\Phi_{oi}A}{V_i} - k_{io}c_{it=0}e^{-\beta t}$$

$$\frac{\Phi_{oi}A}{V_i} = c_{it=0}(k_{io} - \beta)e^{-\beta t}$$

The rate of change of the concentration of the labeled substance is described by eq. (1):

$$\frac{dc_i^*}{dt} = \frac{\Phi_{oi}A}{V_i}\frac{c_o^*}{c_o} - \frac{\Phi_{io}A}{V_i}\frac{c_i^*}{c_i}$$

Combining the above equations and rearranging the terms we have

$$\frac{dc_i^*}{dt} + k_{io}c_i^* = c_{it=0}\frac{c_o^*}{c_o}(k_{io} - \beta)e^{-\beta t}$$

which is an equation of the form $y'(x) + ky(x) = g(x)$ with the solution

$$y = e^{-kx}\left[\int g(x)e^{kx}\,dx + C\right]$$

The value of the integration constant C is defined by the initial condition

$$t = 0 \qquad c_i^* = 0$$

and the time course of the concentration of the labeled potassium in the nerve is described by the equation

$$c_i^* = c_{it=0}\frac{c_o^*}{c_o}(e^{-\beta t} - e^{-k_{io}t})$$

i.e., by a difference of two exponential terms.

This example shows some of the principles useful in considering the kinetics of the tracer exchange in nonsteady state systems. Difficulties in these theoretical calculations arise from the fact that values like c_i, V_i, and Φ are not time-independent constants. The differential equations containing them can be solved if their time dependence is expressed by empirical functions.

It is obvious that theoretical calculations like those above may be carried out also for a greater number of compartments in various arrangements; some of these derivations may be found in comprehensive articles by Robertson (1957) and Solomon (1960 a). Understanding of the simple basic cases described above should be of assistance when more complicated cases are considered. Solutions of equations describing such systems contain a greater number of exponential terms and experimental curves obtained when studying such systems can be analyzed graphically, as explained on p. 241. The accuracy of the data obtainable with biological material restricts the validity of this procedure—otherwise theoretically unlimited—to experimental curves which can be analyzed into two or at most three exponential terms. If a greater number of exponential terms (each of them containing two adjustable constants!) is used it is often possible to represent satisfactorily experimental curves which are intrinsically not composed from exponentials and where one would then fruitlessly look for the corresponding number of compartments in the experimental object. It may be seen that the study of tracer exchange and the related compartmental analysis have their limitations and that they are ideally applicable only to very simple systems.

7. SPECIAL TYPES OF TRANSPORT

7.1. SORPTION THEORIES

It was stated in the introduction to the section on mediated diffusion that the role of plasma membranes in the transport of substances into and out of cells is incontestable and that intracellular adsorption cannot explain a number of phenomena of kinetic nature. It definitely has no justification in the case of transport of organic substances by single cells, since these cells (like erythrocytes and some bacteria) can be completely "emptied" and closed again and the remaining plasma membranes still effect a practically unimpaired transport into the cell.

However, the fact cannot be overlooked that a certain fraction of cell ions can be bound intracellularly, being osmotically inactive and practically unexchangeable, the binding being rather specific. This has been suggested for a part of K^+ of kidney cortex cells (Kleinzeller, 1961) and for a fraction of Na^+ in frog skin (Rotunno *et al.*, 1967) and in muscle (Lev, 1964 *a*; Cope, 1965). Thus, although intracellular sorption does not belong among membrane-mediated processes such as comprise the major part of this book, a few words should be said about the sorption theory.

Theories of ion sorption by cell constituents date back to Nasonov and Aleksandrov who, in 1943 and 1944, suggested that a major portion of electrolytes in the protoplasm are bound and do not contribute to bioelectric potentials. It is only upon injury that some of the ions are mobilized and create a potential difference between cell and medium (injury potential). The distribution of ions is then governed by the adsorption properties of the protoplasm, the specificity of the sorption being provided by metabolic energy. Although the Soviet transport school represented by Troshin (e.g.,

173

1958) has adhered to the sorption theory for years, its strongest advocate has been G. N. Ling (e.g., 1962; Ling and Ochsenfeld, 1966) who drew much on the physicochemical treatment of equilibrium ionic specificity by Eisenman (e.g., 1961).

The sorption theory in its most up-to-date form, bearing the name of association–induction hypothesis, bases most of its evidence on studies of muscle ion content and transport and, indeed, is capable of accounting for some of the phenomena observed there better than a simple membrane theory could. The association–induction hypothesis can be characterized as follows:

(1) The living cell is actually a protein fixed-charge system, the proteins forming a three-dimensional network to which ions (and perhaps other molecules) are adsorbed (Fig. 7.1).

(2) The specificity of adsorption is determined by the arrangement of the fixed charges in the protein network, the arrangement being only semi-permanent, allowing for changes of specificity due to disturbance by external factors (binding of Na^+ instead of K^+).

(3) An important parameter of the binding interaction is the quantity denoted as c, expressed arbitrarily in Å, which measures the magnitude of interaction between the fixed anion and the external countercation. It is defined as

$$c = r_f - \frac{1}{\eta/D_f r_f + \sum\limits_{i=1}^{n} Z_i/D_i r_i}$$

where Z_i is the valence of the fixed anion, D_i the effective dielectric constant for the interaction, r_i the shortest distance between the fixed anion and the center of the cation, D_f is the effective dielectric constant within a distance r_f from the center of the cation, and η is a positive number.

The constant c is positive if the excess electrons of a fixed charge (due to mesomeric or resonance effects) are displaced toward the cation, and negative if they are displaced away from the cation. The value of c becomes more negative as the binding affinity increases (it increases with the disso-ciation energy of the ion–counterion pair). An analogous c' value has been derived for fixed cationic sites.

(4) The cell potential is due to the unequal distribution of ions in the interstitial spaces of the protein network (not the fixed charges!) and in the free external solution. The equation describing this potential is analogous to the Nernst equation eq. (4.4) but it refers to an interface separating two different phases rather than to two similar dilute ionic solutions.

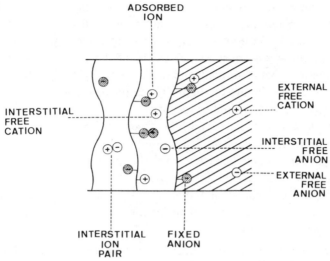

FIG. 7.1. Diagram of a fixed-charge system in contact with a free solution phase. (According to Ling and Ochsenfeld, 1966.)

(5) The competitive character of uptake of similar ions, like K^+ and Rb^+, is explained by competition for the fixed adsorption sites and can be derived from a Langmuir isotherm.

Some of the arguments supporting the sorption theory are strong, e.g., the lower requirement of energy for maintenance of unequal distribution of ions than in the case of a membrane-located pump, or the practical non-exchangeability of at least a part of intracellular ions. Other arguments are weaker, e.g., the fact that Donnan equilibria do not adequately explain the unequal distribution of some ions between the medium and the cell.

Without attempting to pronounce a final judgment we may conclude by saying that intracellular adsorption may play a greater or lesser role in the uptake and interactions of cations in some types of cells, particularly those with an ordered structure of their proteins, such as muscle cells. However, no modern theory can ignore the rather direct evidence for the function of membrane systems in the transport of substances into cells, both nonionic (sugar permeases) and ionic (Na,K-ATPases) in character.

7.2. PINOCYTOSIS AND PHAGOCYTOSIS

Unlike the tenets of the sorption theory where the function of the plasma membrane is disregarded altogether as far as transport of substances is concerned, the concepts of pinocytosis and phagocytosis invest the plasma membrane with an additional capacity, that of physical movement of whole sections of the membrane. Many time-lapse cinematographic observations of cells, particularly those of protozoans and of some higher animal cells, indicate a more or less constant movement of the membranes, infoldings and invaginations being formed and destroyed, especially when food or substrate are placed in the medium. These processes are associated with what we call phagocytosis and pinocytosis.

Phagocytosis is a process by which specialized cells, the so-called scavenger cells of higher organisms (e.g., granular leukocytes) and some unicellular organisms engulf particles from the environment. These particles may be bacteria (Fig. 7.2), bits of organic debris, or artificially administered objects like particles of India ink. One is compelled to analyze the process by morphological rather than kinetic means, three types of phagocytosis having been distinguished.

(1) Formation of pseudopodia which trap the object, followed by fusion of their peripheral portions and separation of the engulfed object surrounded by the pseudopodial membrane from the cell membrane.

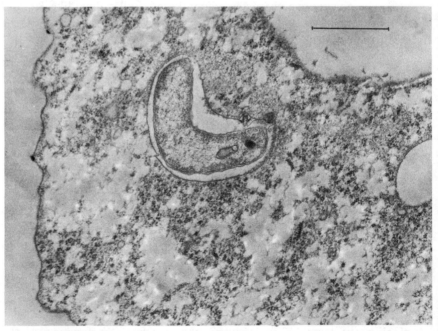

FIG. 7.2. Ultrathin section of *Entamoeba histolytica* showing a food vacuole containing a phagocytized bacterium. Scale line 1 μ. (Courtesy of Dr. J. Ludvík, Institute of Microbiology, ČSAV, Prague.)

(2) Invagination of cell membrane, following the formation of an indentation upon contact with the object of interest which becomes eventually fused at the cell periphery.

(3) Opening of the cell-surrounding membrane connected with endoplasmic reticulum, permitting entry of particles into the cell through canaliculi.

The phagocytotic vacuole surrounding the foreign body after its trapping by the cell either survives the trapped particle which is eventually digested or it gradually merges with other vacuoles and subcellular components.

The term *ultraphagocytosis* has been applied to the process whereby particles of submicroscopic dimensions are ingested, including colloids, vital dyes, etc.

Pinocytosis is a process through which the cell surface engulfs small droplets of the outside medium containing substances that are of use to the cell and, inversely, through which droplets of secretory materials are liberated from the cell. Sometimes it is difficult to draw a boundary line between pinocytosis and phagocytosis but, generally, the latter term is reserved for the engulfment of solid particles with special emphasis on the sanitary functions of leukocytes. Pinocytosis is known to occur in amoebae, blood capillary walls, macrophages, leukocytes, intestinal and kidney tubule cells, and possibly even yeast. Most animal cells, however, have not been found to display it. The role of pino- and phagocytosis may not be without significance in the development of immunological reactions.

The process of pinocytosis is initiated by contact of the cell surface receptors with a stimulant (Holter, 1964). The selectivity of the stimulus is not universally defined; on the contrary, it may be very broad, ranging from molecules of thorium dioxide to proteins (*Amoeba proteus*), or rather narrow, discriminating between thrombin and prothrombin (liver Kupffer cells; Gans *et al.*, 1968). At any rate, the mechanism of pinocytosis represents a vehicle for the uptake of large molecules which otherwise could not penetrate the cell. The stimulation of the receptors appears to be invariably associated with adsorption on the outer face of the plasma membrane as is handsomely shown by cells possessing filamentous projections on this membrane. Only then, by a discrete process (it has a different temperature dependence from that of the adsorption), will the plasma membrane begin to invaginate, form canaliculi or be pinched off in globular vesicles (Fig. 7.3) and, with its content of the outer medium, penetrate deeper into the cell.

The extent to which pinocytosis proceeds is evidently limited by the

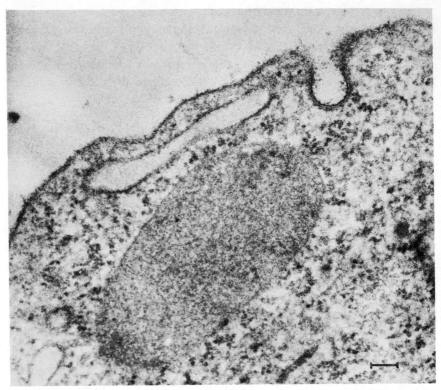

FIG. 7.3. Ultrathin section of *Trichomonas vaginalis* showing a peripheral part of the cell with a pinocytotic vesicle. Scale line 0.1 μ. (Courtesy of Dr. J. Ludvík, Institute of Microbiology, ČSAV, Prague.)

amount of plasma membrane available. It should be recognized that during the invagination of the plasma membrane its surface is actually enlarged, this being probably made possible by a conversion of the globular or granular conformation of the membrane to the bilayer lipid leaflet which exhibits greater area per amount of material. Thus, an amoeba placed in a medium with nutrients that it absorbs by pinocytosis completes its pinocytotic cycle within some 30 min and then it has to "rest" for about two or three hours before new material and energy are made available for a new cycle to set in.

In animal cells, however, the plasma membrane is apparently continuously synthesized during the process, as suggested by an increased turnover of phosphatides in a cell engaged in pinocytosis.

The further fate of pinocytotic vesicles seems to be connected with the appearance of *lysosomes*, a name given to vesicular cell inclusions or granules containing a number of hydrolytic enzymes which digest the material presented to them probably by fusion with pinocytotic or phagocytotic vesicles. The lysosomes, then called *phagosomes* or digestive vacuoles, may then either contain indigestible remnants of the engulfed particle for a very long time or else release it actively from the cell (in amoebae). The lysosomes and their modifications are probably functionally associated with the Golgi apparatus where the process of pinocytosis may actually reach its turning point after which secretion sets in. The reverse pinocytosis is perhaps even more important for the smooth function of a cell specializing in the production of hormones and other secretions.

It has been suggested by some authors that a kind of micropinocytosis may occur by a continuous synthesis of plasma membrane at the sites where an invagination projects into the cell, accompanied by continuous breakdown at the site where the invagination ends with a tip. If the membrane binds small molecules, they are actually transported inward without the necessity of formation of pinocytotic vesicles.

Kinetically, the transport by pinocytosis does not fit any of the models described in this book, as its time uptake curve would be characterized by the following features: A rapid rise due to the binding to the plasma membrane surface (perhaps a few seconds), a plateau of some 10 min or more before pinocytotic vesicles begin to migrate inward, a slow rise as more of the transported solute is adsorbed on the surface and transferred inward. Upon transfer to a solute-free medium there would be again either an instantaneous drop as the solute would be desorbed (unless bound irreversibly), followed by a slow decline (again unless the substance is retained intracellularly and reverse pinocytosis cannot occur). However, wherever transport curves were studied in detail, they were either smooth and could be fitted to an integrated equation of facilitated diffusion or active transport or else showed irregularities of a quite different type (oscillations, initial peaks, and the like).

It should be also observed that pinocytosis cannot explain any type of uphill transport unless other concentrating systems (such as a water-extrusion pump) are operative.

Pinocytosis and phagocytosis represent only special types of movement and morphological alterations of the plasma membrane whereby substances can be transported in and out of the cell. It should be realized that various plasma membranes can be considerably specialized to engage in cell-to-cell or surface-to-organelle transport, the processes being sometimes analogous to pinocytosis as described above. For a very informative treatise on this subject, see Fawcett (1962).

Molecular Aspects

8. MOLECULAR BASIS OF TRANSPORT

8.1. NATURE OF TRANSPORTING MOLECULES

In the preceding chapters the assumption was implicit that the membranes of very diverse (and perhaps all) cells contain binding sites for various solutes and that it is by virtue of the movement of these binding sites from one side of the membrane to the other that most substances of biochemical interest are transported into cells. What do we actually know about the nature of these binding sites and, if they are mobile, about their movement?

One of the more obvious characteristics of both facilitated diffusion and active transport is their substrate specificity; in fact, it may be, although mostly is not, as narrow as to accept only one substrate or a few related analogues (e.g., some of the amino-acid-transporting systems of microorganisms). Hence it may be deduced that the binding site is of protein character as no other type of substance is known to possess this type of selective behavior, this being best exemplified by the distinction between D- and L-configurations of transported substrates.

Evidence for the protein character of the binding molecules is supported by the effect on transport of some inhibitors which are known to react either with SH-groups (e.g., N-ethylmaleimide) or with the N-terminus of peptide chains (e.g., dinitrofluorobenzene).

However, although protein is very probably involved in transmembrane transport, the specificity pattern of a given transport may be determined not by the mobile membrane component itself but by an enzyme catalyzing the binding of substrate to the carrier and hence containing binding loci for both. Kinetic evidence is available for the existence of such enzymes in particular from the inhibition of sugar transport in erythrocytes by phloretin and its analogues (*cf.* Rosenberg and Wilbrandt, 1962). If this is actually the transport mechanism for a given substrate there is no evidence as to the nature of the carrier itself, if such an entity exists separately from the binding enzyme. The carrier could then easily be a small molecule, perhaps a peptide or even a lipid which would move back and forth in the membrane.

All carrier-mediated transports display a saturation effect at high concentrations of substrate, but some, in addition, show unusual features at different substrate concentrations. One of these is a higher order of reaction toward substrate, indicating two or more binding sites of the same or similar type existing on one molecule of carrier, which then show cooperative effects (e.g., amino acid transport in the intestine, possibly sugar transport in human erythrocytes). The cooperation may be of the "allosteric" type where binding of a substrate molecule to the carrier system increases the affinity or perhaps probability due to steric reasons that another substrate molecule will be bound; or it may consist in a positive effect of additional molecules bound to the carrier on the mobility of the carrier (these two possibilities may be difficult to distinguish experimentally).

While a homologous cooperation of substrate molecules in enzyme reactions is a widely observed phenomenon (*cf.* Monod *et al.*, 1965) some transports show even a heterologous cooperation which may—but need not—turn into inhibition at other concentrations of the two types of substances. This may be the case with sugars and amino acids in baker's yeast (Poncová and Kotyk, 1967) and probably with sugars, amino acids, and ions in the intestine (Alvarado, 1966a). After all, the stimulation or even essential activation of transport of sugars and amino acids by the presence of alkaline cations and anions, observed in a variety of objects ranging from bacteria to tumor cells, is indicative of two or more heterologous binding sites on the carrier system. Without endeavoring to generalize the findings, it may be stated at the present state of knowledge that many carrier systems appear to be heteropolyvalent and considerations of the carrier size and mobility in such cases need not take into account small molecules that could shuttle back and forth across the membrane.

8.2. MOVEMENT OF THE CARRIER

How then is the movement, which still is the essential element of carrier function, achieved? If we rule out translational diffusion as highly improbable both for steric and for structural reasons, we are left with oscillation, rotation, change of conformation, and perhaps several other, more subtle, varieties of these types of movement.

The *oscillating* protein molecule can perform its function by being localized in a hydrophilic "pore" or by taking the place of lipid in one of the globules postulated by the modern membrane structure theories (chapter 1) and by exposing the binding site for substrate either to the one or to the other face of the membrane. This oscillatory movement does not require any energy beyond that for thermal movement and may thus easily operate in facilitated diffusion. In a metabolically coupled transport, one would have to postulate the presence of a modifier or energy transducer which would change the carrier affinity at one of the membrane faces. On the whole, the diameter of the globules which such a carrier-plus-modifier system might occupy is easily 50–100 Å which is in agreement with the generally observed plasma membrane thickness.

We can envisage the *rotation* of the carrier in two ways. One follows the suggestion of Danielli's (1954) where the membrane carrier turns about a pivot and performs the same function as the oscillating carrier, exposing the substrate binding site alternately to the one or to the other membrane face. To the present author, this type of movement does not appear to be supported by any analogous evidence. It would require the existence of a peculiar mechanism which would cyclically break the bonds of the rotating protein with its surroundings and re-establish them with rather unexplainable regularity.

Another variety of the rotating protein is the *rolling carpet* hypothesis propounded by Janáček, where the membrane proteins are assumed to move in resemblance to a conveyor belt about the micelles in the plasma membrane. This hypothesis, although assuming again the feasibility of an ordered continuous movement on the molecular level, has the merits of postulating large (and potentially polyvalent) carrier molecules which can carry an array of enzymes and, moreover, can account for the turnover of the carrier proteins as they temporarily appear intracellularly, without affecting much the function of the relatively large rolling system.

The theory of *invaginating membrane* (Crane, 1966 b) has some features in common with the rolling carpet in that parts of the membrane surface receptors with substrate slide inward through pores and thus transport their substrate molecules into the cell. They may, however, return the same way instead of completing the movement about a central pivot.

Some resemblance to this model is displayed by the *sliding membrane* of Booij (1962) according to which the membrane is in a continuous movement along the surface of cytoplasmic blocks bounded by invaginations.

Thus, in one of the invaginations it is sliding outward, carrying small ligands like Na⁺, in the next invagination it is moving inward, carrying perhaps K⁺ and the like. This hypothesis is already much like the micropinocytosis described on p. 179.

The possible role of conformational change in the protein involved in transmembrane movement of substrate (Vidaver, 1966) is actually identical with the oscillating carrier, the only difference being that here only a part of the protein molecule moves, exposing its binding site to the left or to the right.

It should be pointed out at this stage that what has been termed the essential feature of carrier transport, the carrier mobility, is not actually the proper criterion by which the abundant carrier systems are characterized. It is rather the requirement that the binding site of the carrier be exposed alternately to the one and to the other face of the membrane but never to both at the same time, since then the substrate on both sides would equilibrate and phenomena such as countertransport could not be observed.

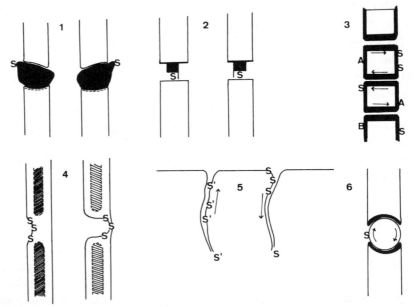

FIG. 8.1. Schematic representation of various carrier models showing the movement of substrate S from left to right. 1 Oscillatory carrier filling a globular vesicle in the membrane; 2 gate-type carrier, identical with the conformation-change mechanism; 3 "rolling carpet" carrier with sites for binding substrate and different modifiers; 4 invaginating membrane, with a lipoid core, the fabric of which may change during transport; 5 sliding membrane transporting S′ outward and S inward; 6 rotating carrier.

(The same situation would arise if the binding of substrate to the carrier were very slow relative to carrier movement; *cf.* p. 81.) Hence the carrier need not move as such but a mechanism must be present for exposing its binding site alternately to the two membrane sides. Such a mechanism underlies the hypothesis advanced by Patlak (1957) where a gate-like mechanism achieves the task. Thus not the carrier but another membrane component is mobile.

Recently, at least one sound model has been proposed that does away with a mobile carrier altogether and still retains the typical characteristics of carrier transport, such as countertransport, competitive acceleration, etc. (Heckmann, 1965). It is based on assuming a S_N2 (substitution-bimolecular-nucleophilic) reaction taking place between the binding site (with substrate attached) and another substrate molecule at the protein–lipid interface in the membrane. Parallel to this reaction, there may proceed a more or less free diffusion that can bring the substance across the membrane without the necessity of undergoing the S_N2 reaction. In this way, the model also accounts for flux acceleration by preloading (see p. 76).

Figure 8.1 shows the models of facilitated diffusion as discussed above.

8.3. UPHILL TRANSPORT SYSTEMS

The coupling of the carrier with energy-yielding reactions on the molecular level presents even greater difficulties and is perhaps even less understood than the operation of the carrier itself.

The *permease* concept due to the Institut Pasteur group (Cohen and Rickenberg, 1956; Rickenberg *et al.*, 1956) has received wide support from genetic studies where it was shown that a certain genetically controlled protein is required for the operation of an active transport of various solutes in microorganisms. Without this protein, called the permease, the cell can at best use a system of facilitated diffusion (mostly rather inefficient in the cells studied) to transport the substance inward. The term permease has been criticized both justly and unjustly, the main objection being that the ending *-ase* implies an enzymic transformation of substrate and this, generally speaking, is not the case with permeases. Likewise, it has been argued that there is no clear definition of what a permease is and that it has never been isolated. Substitute terms have been suggested, like *transportase* (Sols and de la Fuente, 1961) or *transfor* (Pardee, 1968). The fact remains, however, that the permease concept has contributed substantially to further investigation of transport phenomena and that, at the present stage of research, the terminological question has acquired a completely new dimension once some of the transport proteins have been actually isolated. The role of the classical permease (like that for β-galactosides in *Escherichia coli*) may consist of a change in the carrier configuration leading to a decreased affinity at the intracellular side of the membrane rather than in acting as the carrier itself. Alternatively, the permease might make the movement of the carrier possible by fitting into a proper vacancy in the transport complex. This might be the case with sugars in yeasts where, prior to induced formation of a membrane transport protein, no transport of a given solute is possible at all.

The term permease in the broad sense may embrace all the proteins which were recently found to operate in the transport of sugars in a number of microorganisms (*E. coli, Staphylococcus aureus, Aerobacter aerogenes, Salmonella typhimurium, Pseudomonas aeruginosa*), the whole permease concept (at least in the case of sugars) thus acquiring a novel and unifying meaning.

The significant discovery in this context is that of Kundig and co-workers (1964) who isolated a bacterial phosphotransferase system operating as shown in the accompanying scheme:

$$
\begin{array}{ccc}
\text{Phospho-}enol\text{-pyruvate} & \text{HPr} & \text{Sugar phosphate} \\
\text{Enzyme I, Mg}^{2+} & & \text{Enzyme II, Mg}^{2+} \\
\text{Pyruvate} & \text{P-HPr} & \text{Sugar}
\end{array}
$$

While Enzyme I is probably a rather nonspecific protein catalyzing the phosphorylation of HPr, a heat-stable protein, the other components of the system display remarkable specificities. Two different heat-stable proteins have been isolated from *E. coli* (Anderson *et al.*, 1968), both having a molecular weight of 10,900 daltons, both accepting one phosphate group linked to N_1 of a histidine imidazole residue. The two HPr's differ in their amide content and enzyme activity. Moreover, they are different from the HPr found in *S. aureus*.

Even more specific is probably the enzyme II which was found to consist (at least in *A. aerogenes*; Hanson and Anderson, 1968) of two components: a large protein "apoenzyme" and an inducible specifier protein whose function it is to increase the affinity of enzyme II for a given sugar. The whole system is localized in the membrane, apparently in an orientated manner, and achieves an effective translocation of a number of mono- and disaccharides as well as some polyols from the outside medium into the cell, albeit in the form of phosphorylated derivatives.

The position of the phosphate group on the sugar is not universally established. In *A. aerogenes*, fructose-1-phosphate is formed (Hanson and Anderson, 1968) and so is lactose-1-phosphate in *Staphylococcus aureus* (Hengstenberg *et al.*, 1968). However, 2-deoxy-D-glucose-6-phosphate is found during transport in *E. coli* (Ghosh and Ghosh, 1968) and other sugars also appear to cross the membrane as 6-phosphates.

The question might arise as to which of the components of the system may be identified with the classical permease but the argument is largely academic. If genetic control is taken as the principal characteristic of permeases, either HPr or enzyme II might be a candidate. If the interaction with the transported substrate is considered as significant, only enzyme II is acceptable since HPr has no affinity for the transported substrate. What comes closest to the classical permease protein is the specifier of enzyme II which is under genetic control, is inducible, and comes into direct contact with the transported substrate.

The HPr system promises to be fairly widespread among bacterial species and it might be shown to be involved in the transport of practically all sugars. The fact that many sugars (the best-known example is lactose in *E. coli*) occur free rather than phosphorylated in the cell may be explained by the action of sugar phosphatases at the inner side of the plasma membrane (Lee *et al.*, 1967).

However, there remain uphill transports of compounds where the HPr system does not operate, such as in the uptake of amino acids in a variety of cells and even of sugars in yeast and fungal species. Moreover, even in cells where the HPr system is operative, there are at least two objections against its uniqueness and indispensability. One is of a kinetic nature and has to do with the fact that if all the sugar were transported in the phosphorylated form one could not observe phenomena such as countertransport or competitive acceleration, found commonly in the cells in question. This might indicate that there is a parallel route for the free sugars across the plasma membrane.

The other objection is somewhat utilitarian in nature. It is considered definitely wasteful for the cell to transport a substrate by means of phosphorylating it (at the expense of metabolic energy), then to dephosphorylate it inside the cell and immediately rephosphorylate it (this again at the expense of metabolic energy) to break it down metabolically. Be it as it may, the existence of the highly complicated bacterial system involved in the transport of sugars emphasizes the need of caution when trying to fit the various transport observations to simple mobile carrier models. In a system such as the HPr, any of the reactions (plus the physical movement of the enzyme–sugar complex) may be rate-limiting. The persistence in

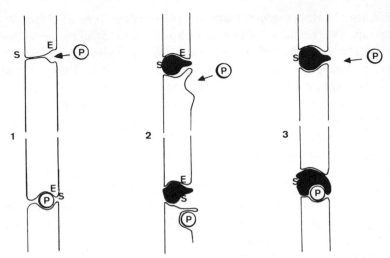

Fig. 8.2. Different possibilities of permease **P** function in transport of substrate **S** from left to right. **1** Permease acting as carrier itself (enzyme **E** catalyzes the change of affinity of permease for substrate); **2** permease increasing the mobility of the carrier (black), the affinity change being brought about by enzyme **E**; **3** permease altering the affinity of the carrier for its substrate (possibly allotopically), the effect being apparent at one membrane face only.

bacteria of facilitated diffusion when the HPr system is not operating and the occurrence of countertransport and acceleration by trans concentration (*cf.* Horecker *et al.*, 1961) indicate the existence of an integrated system of transport independent of the superimposed active component.

Figure 8.2 shows some of the feasible models of metabolically coupled transport.

8.4. TRANSLOCATION BY METABOLIC ENZYMES

The direct involvement of metabolic enzymes in the transport of substances has been a matter of controversy for years. Apart from the Kundig–Roseman system discussed above there are suggestions for a direct role of a metabolic enzyme in the translocation of the transported substrate. The concept may be due to Mitchell (1957) and has found support of several authors who conjecture that, for example, glucose is transported in liver cells by glucose-6-phosphatase (Siekevitz, 1961) or 2-deoxy-D-glucose in baker's yeast by a phosphokinase (van Steveninck, 1968). Other suggestions of this type are based on the fact that uranyl ions which are clearly active only on the yeast membrane surface inhibit differently the anaerobic and the oxidative breakdown of glucose in yeast as if the corresponding enzymes were accessible from the outside (Rothstein, 1954). Although the existence of such a dual function of metabolic enzymes has not been completely rejected there is no straightforward evidence for it.

8.5. NUMBER OF CARRIER SITES PER CELL

Before attempts were made to isolate the transport proteins from cell membranes several assessments of the amount of various carriers in the membrane were published.

When studying the inhibition of the glucose transporting system in human erythrocytes by dinitrofluorobenzene, Stein (1964) reached a value of 1.3×10^6 sites per cell. Recalling now the globular micelle model of Lucy (p. 21) one can estimate the number of such globules to be of the order of 10^7 per cell so that every tenth globule or so would have to fulfil the function of a glucose carrier. This is a rather high estimate which may be due to dinitrofluorobenzene being bound in a similar manner to proteins other than the glucose carrier.

Other estimates of the amount of glucose binding sites in erythrocytes are due to van Steveninck and associates (1965) who reached a value of 7×10^5 sites per cell and to LeFevre (1961 b) who obtained 2×10^5 sites per cell.

The rather high figure of 10^7 uranyl binding sites per yeast cell was obtained by Rothstein et al. (1948) but here it should be realized that more than one uranyl ion can be bound to each protein molecule and that the involvement of all these in glucose transport is not altogether obvious.

A much lower estimate, in this case for the leucine-transporting protein from E. coli, was obtained by Penrose and co-workers (1968) who postulated a turnover of the protein equal to 200/min and reached a value of 10^4 molecules per cell. However, the dimensions of E. coli being less than either of yeast cells or of erythrocytes, there may be only $5-10 \times 10^5$ of the Lucy model globules so that perhaps every 50–100th globule would be involved in leucine transport, a fairly high figure again.

8.6. ISOLATION OF CARRIER MOLECULES

It is only quite recently that attempts were made, some of them already crowned with success, to isolate the protein molecules involved in the transmembrane movement of substances. Some of the methods were highly sophisticated, others were simple fractionations, the latter having been generally more successful.

At least three approaches are possible and have been used by one author or another: (1) Double labeling of the transport protein in case it is inducible. (2) Binding of inhibitors or substrate analogoues to the transport protein. (3) Search for binding activity among membrane subfractions.

8.6.1. Double Labeling of Inducible Transport Proteins

The credit for pioneering in this field goes to Stein and his co-workers (Kolber and Stein, 1966, 1967) who cultivated *E. coli K12* in the absence of inducer with ^{3}H-labelled L-phenylalanine and another batch of the same bacterium in the presence of the gratuitous inducer of the *lac* operon enzymes, β-thiomethylgalactoside, with ^{14}C-labeled L-phenylalanine. After growth, they washed and mixed the cultures, prepared spheroplasts, burst them with dilute Tris buffer, and applied the soluble extract to a DEAE–cellulose column. Radioactivity due to the two isotopes in the emerging fractions was estimated and regions of enrichment with ^{14}C were analyzed (Fig. 8.3). Three such regions were found, corresponding to thiogalactoside transacetylase, β-galactosidase and, probably, β-galactoside permease. The protein responsible for the permease enrichment was subsequently found to be of low molecular weight and probably to constitute only a component of the permease system. However, an enrichment probably corresponding to the galactoside permease was later found among proteins solubilized by 1 M NaI from isolated plasma membranes.

The same procedure was used by Haškovec and Kotyk (1969) to isolate the inducible galactose carrier (or permease) from baker's yeast membranes and a peak of unequal labeling presumed to correspond to the carrier was found in DEAE–cellulose elution diagrams of solubilized membrane proteins both from the wild strain and from a galactokinase-less mutant of *Saccharomyces cerevisiae*. The membrane proteins isolated by Stein's and by the Prague group have not yet been purified as homogeneous entities.

On the other hand, an ingenious approach applied to the same problem by Fox and Kennedy and co-workers (Fox and Kennedy, 1965; Fox *et al.*, 1966) resulted in a molecular isolation of at least one protein component of the transport system. Proceeding from the observation that β-galactoside transport in *E. coli* is blocked by N-ethylmaleimide but that the binding is prevented by protecting the site with β-galactoside, they treated their bacteria (both induced and uninduced) with unlabeled N-ethylmaleimide in the presence of thiodigalactoside and exposed the induced cells to ^{14}C-N-ethyl-

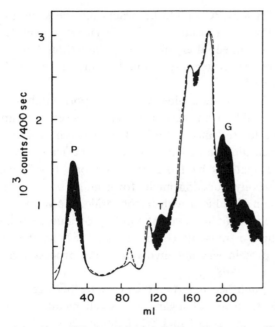

FIG. 8.3. Radioactivity due to ^{3}H-phenylalanine (broken line) and to ^{14}C-phenylalanine (full line) incorporated into *Escherichia coli* proteins during growth without and with inducer of the *lac* operon enzymes, respectively. The black areas correspond to β-galactosidase (**G**), thiogalactoside transacetylase (**T**), and, possibly, a component of the β-galactoside permease (**P**). (According to Kolber and Stein, 1966.)

maleimide and the uninduced cells to ^{3}H-N-ethylmaleimide, thus increasing the specificity of labeling of the transport proteins. Solubilization of the membranes with Triton X-100 led to the separation of a ^{14}C-enriched protein which was termed the *M*-protein and which showed binding affinity for β-galactosides. The protein has since been isolated and its molecular weight shown to be 31,000 daltons (Jones and Kennedy, 1968).

8.6.2. Binding of Inhibitors to the Transport Protein

Here again the work of Stein is of importance. He made use of the fact that dinitrofluorobenzene (DNFB) blocks glucose transport in human erythrocytes in proportion to the square of its concentration. By treating one part of erythrocytes with 1 mM ^{3}H-DNFB for 10 min and another part with 10 mM ^{14}C-DNFB for 1 min, mixing the suspension, preparing ghosts, and analyzing a papain digest as to unequal labeling of proteins he could localize the carrier since, for linearly reacting proteins, the amount of label with ^3H and ^{14}C should be practically the same (a first-order reaction with a low rate constant) (Stein, 1964). Small peptides with unequal labeling were obtained after suitable fractionation but so far no exact identification of the protein responsible for glucose binding by erythrocytes has been reported.

The use of substrate analogues or transport inhibitors which might be bound irreversibly to the carrier protein deserves more attention than it has received so far (*cf.* Baker, 1967). The question of a chemical reaction (if any) between glucose and the carrier during transport in erythrocytes provoked investigations by Langdon and Sloan (1967) who found that the glucose carbonyl function might form imine bonds with lysine of the transport protein if reduced with borohydride. However, reduction of the cells and the concomitant covalent binding of glucose resulted in no depression of glucose transport (Evans *et al.*, 1967) so that it is likely that the transport protein was not involved in the process (see also LeFevre, 1967; Rose *et al.*, 1968).

Recent unpublished reports indicate that other agents might be useful in blocking the transport of sugars across cell membranes (succinic anhydride and glucose transport in Ehrlich tumor cells).

8.6.3. Search for Binding Affinity

The least complicated and the most straightforward of the approaches used for isolating membrane transport proteins has been the examination of binding properties of various membrane protein fractions toward the substrate in question. The first complete isolation of a transport protein has been achieved in this manner and is due to Pardee's team (Pardee and Prestidge, 1966; Pardee *et al.*, 1966; Pardee, 1967). They proceeded from osmotically lyzed or sonicated *S. typhimurium* which accumulates sulfate (and thiosulfate) against very high concentration differences. Applying their soluble fractions to a sulfate-binding ion-exchange resin they were able to separate one which liberated sulfate from the resin, demonstrating its high affinity for the anion. Eventually, the protein responsible for this activity was crystallized and thoroughly analyzed (Fig. 8.4). There appear to be 10^4 molecules of the protein per bacterium. Its properties are summarized in Tables 8.1 and 8.2. The protein undoubtedly represents a part of the sulfate-transporting system but it is present even in transport-negative mutants and is exposed on the cell membrane surface (Pardee and Watanabe, 1968), suggesting that it requires the presence of another (more or less intracellular) component for operation.

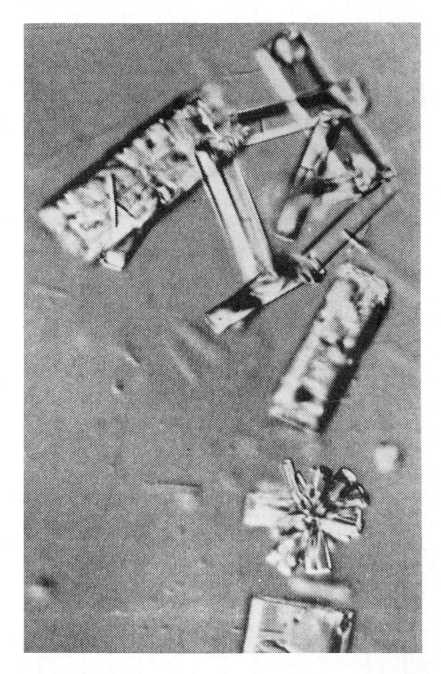

FIG. 8.4. Crystals of the isolated sulfate-transporting protein from *Salmonella typhimurium*. (Taken with kind permission from Pardee, 1967.)

Table 8.1. Properties of Membrane Transport Proteins

Source	Transport substrate	Molecular weight	Sites per molecule	Crystals obtained	K_{diss} (mM)	K_T in vivo (mM)	Reference
Salmonella typhimurium	Sulfate	32,000	1	+	0.03	0.004	Pardee (1967); Pardee et al. (1966)
Escherichia coli	β-Galactosides	31,000	1	–	?	0.6	Winkler and Wilson (1966); Jones and Kennedy (1968)
Escherichia coli	L-Leucine	36,000	1	+	0.0003	0.001	Piperno and Oxender (1966); Penrose et al. (1968)
Escherichia coli	L-Leucine	36,000	1	+	0.002	0.001	Anraku (1968a,b,c)
Escherichia coli	D-Galactose	35,000	1	–	0.001	0.004 (0.0005)	Anraku (1968a,b,c); Rotman and Radojkovic (1964)
Escherichia coli	Phospho-enol-pyruvate	10,900	1	–	Covalent	?	Anderson et al. (1968)
Escherichia coli	Glucose-6-phosphate	?	?	–	?	?	Dietz et al. (1968)
Escherichia coli	L-Arabinose	35,000	?	–	0.002 (0.006)	0.05	Hogg and Englesberg (1969); Schleif (1969)
Pseudomonas natriegens	L-Arabinose	?	?	–	?	?	Rhodes and Payne (1968)
Streptococcus faecalis	Na+, K+	300,000	?	(+)	–	–	Schnebli and Abrams (1969)
Saccharomyces cerevisiae	D-Glucose	?	1	–	1.1	6	Azam and Kotyk (1969)
Saccharomyces cerevisiae	D-Galactose	?	1	–	0.6	3	Haškovec and Kotyk (1969); Cirillo (1967)
Chick duodenum	Ca^{2+}	28,000	1	–	0.004	?	Wasserman et al. (1968)
Guinea-pig brain	Na+, K+	670,000*	?	–	–	–	Medzihradsky et al. (1967)
Human erythrocyte	D-Glucose	?	1	–	5–20	12	Bobinski and Stein (1966)

Table 8.2. Amino Acid Composition (in Residues per Molecule) of Some Isolated Transport Proteins *

Amino acid	Sulfate (S. typhimurium)	L-Leucine (E. coli)	D-Galactose (E. coli)	Ca²⁺ (chick)
Alanine	23	42	42	15
Aspartic acid	30	38	49	28
Arginine	8	7	6	5
Cysteine	0	1	—	3
Glutamic acid	21	36	28	32
Glycine	17	33	22	15
Histidine	4	4	3	3
Isoleucine	12	18	15	9
Leucine	16	22	24	24
Lysine	19	28	30	20
Methionine	1	4	6	4
Phenylalanine	9	10	7	11
Proline	9	14	9	3
Serine	13	11	13	9
Threonine	11	16	12	8
Tryptophan	5	3	4	—
Tyrosine	9	12	6	8
Valine	17	23	29	6

* After Pardee (1968).

Two laboratories vied for the priority in isolating transport proteins for galactose and for leucine from *E. coli*: that of Oxender and that of Anraku. Both proteins were isolated in the pure state, the one for leucine existing in a crystalline form (Penrose *et al.*, 1968), their properties being again described in Tables 8.1 and 8.2. Anraku, in a brilliant series of papers, described the release by an osmotic shock of proteins binding L-leucine and D-galactose which were purified by precipitation with protamine sulfate, ammonium sulfate and chromatography on DEAE–cellulose, on hydroxyl-apatite, and on DEAE-Sephadex. The proteins alone do not restore the transport capacity to shocked cells, requiring the presence of another protein factor of the shock fluid which shows no affinity for either leucine or galactose, reminiscent of the HPr factor. The binding of galactose to the transport protein was unaffected by a number of ions and inhibitors but was suppresed by 3 mM urea and 10% sucrose while the binding of leucine to its transport protein was slightly (20–30%) depressed by Ca^{2+}, Zn^{2+}, Mn^{2+}, Cu^{2+}, EDTA, N-ethylmaleimide, sodium azide, and urea. It is noteworthy that 1 mM NaCl, 10% sucrose, and 30% ethanol strikingly increased the binding of leucine (Anraku, 1967, 1968a,b,c).

The L-arabinose-binding protein has been highly purified from *E. coli* by two teams (Hogg and Englesberg, 1969; Schleif, 1969) and its molecular properties have been defined.

In all these isolations use was made of the so-called shock fluid of *E. coli*, obtained (as described by Heppel, 1967) by treating young cells with 0.5 M sucrose containing dilute Tris buffer and 10^{-4} M EDTA. The mixture is centrifuged and the pellet is rapidly dispersed in cold 5×10^{-4} M MgCl$_2$. After another centrifugation, the supernatant solution is found to contain 3.5% of cell proteins including a number of enzymes (ribonuclease, 5′-nucleotidase, and many phosphatases) and transport factors.

In addition to these purifications from microorganisms much work has been done on the isolation of the transport protein for Ca^{2+} from chick duodenum (Wasserman *et al.*, 1968) (Tables 8.1 and 8.2).

The common feature of all the transport proteins isolated so far is their ability to recognize the transported substrate and hence they are apparently an important component of the respective transport systems. However, in no single case can it be stated with certainty what precise function the protein has in transport and it has not been possible so far to cause the isolated proteins to change transport properties (or rather the impermeability) of artificial lipid membranes. The isolated proteins may even be associated with the chemotactic behavior of the cells. It is only too obvious that much will have to be done before at least one transport system is isolated in its entirety.

Rather promising in this respect is the work done on the isolation of transport ATPase. In Post's laboratory a great deal has been done to elucidate the molecular mechanism of ATPase function and it is due to their efforts that we now recognize the existence of two phosphorylated intermediates (acyl phosphates) in the reaction (Bader *et al.*, 1966; Sen and Post, 1966). Unfortunately, the solubilization of the whole ATPase complex proved extremely difficult as activity was lost upon disintegration of the plasma membrane. However, using stabilization with ATP and Na$^+$ and K$^+$, Hokin's group (Medzihradsky *et al.*, 1967) succeeded in isolating the membrane ATPase from guinea pig brain microsomes and in determining its molecular weight. The isolated lipoprotein shows an affinity for ATP, Na$^+$, and K$^+$ as predicted by its physiological function.

An ATPase was isolated from the membranes of *Streptococcus faecalis* (Schnebli and Abrams, 1969), the molecular weight being less than that of the microsomal ATPase (300,000 daltons). Under the electron microscope, the ATPase shows a hexagonal arrangement of six globules of 40 Å in diameter.

A special mention should be made of the work of Semenza's group on the isolation of the monosaccharide transport system from the intestine (e.g., Semenza, 1969b). The observation that the glucose moiety of sucrose is absorbed by the mucosa more readily than free glucose led to the assumption of a close spatial relationship between intestinal sucrase and the monosaccharide carrier. Treatment of brush-border membranes of the rabbit intestine with papain solubilized a complex which was subsequently resolved into subunits, one of which shows sucrase activity but does not bind glucose, the other binds glucose but shows no hydrolytic activity toward sucrose. This glucose fixation is sodium-dependent like the transport of glucose *in vivo* and it is not affected by Tris (in contrast with sucrase which is blocked by Tris buffer). The protein complex which includes the binding sites for sucrose, glucose, and Na^+ has been named *hyphezyme* by Semenza to indicate its fabric or mesh structure.

8.7. FINAL COMMENT

When reviewing the available data on the transport of various substances one cannot escape the unpleasant feeling that there are actually a great number of mechanisms involved in one way or another in membrane transport. The variety is appalling in contrast with such universally functioning biochemical systems as the genetic code and protein synthesis, photosynthesis, and fundamental metabolic reactions. Considering the fact that of the million plus animal species and the 300,000 or so plant and bacterial species only about a hundred have been studied in some detail as to their membrane transport properties, one might expect an even greater assortment of transport mechanisms as research extends to other representatives of the biosphere. Although some mechanisms are known which have a broader validity extending over the highest taxonomic categories, such as the transport ATPase, it appears at present that we are simply ignoring some fundamental component of the transport process which would unify the present conjectures and hypotheses and which would permit us to reduce the differences which appear to be qualitative in nature to a mere variability of quantitative characteristics.

Methodological Aspects

9. INCUBATION AND SEPARATION TECHNIQUES

9.1. INCUBATION

9.1.1. Types of Incubators and Flasks

The great majority of transport studies *in vitro* are performed by incubating the cells in suspension or the pieces of tissue in a saline solution at the appropriate temperature in the presence of the solute, the transport of which is investigated.

Several varieties of constant-temperature incubators are available on the market, there being three principal types as are described below.

The *first* of these is a simple water bath, represented by a glass jar or a metal box with adjustable clamps or racks for tubes or flasks, provided with a contact thermometer, a heater, and a stirrer. The heating and stirring functions are usually accomplished by a compact unit, easily transferable, which can be attached to a chemical stand and used for practically any vessel that happens to be available (Fig. 9.1). The problem of stirring the suspension in the flask is usually overcome by placing into it the tip of a pipet through which gas under pressure is gently introduced. The passage of gas fulfils a dual function, that of agitating the flask contents and that of providing the appropriate gaseous phase for the incubation.

The *second* type of incubator is the one introduced by Dubnoff and is based on the principle of a reciprocal shaker which is placed in a constant-temperature bath. The tray with clamps for holding flasks (usually the Erlenmeyer type) is supplied for several flask sizes and is easily exchangeable (Fig. 9.2). Stirring of the flask contents is assured by the movement of the tray (a range of speeds is usually provided). If a special gaseous phase is

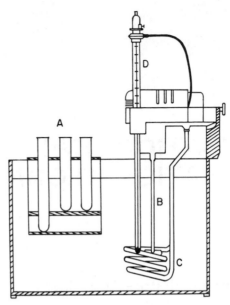

Fig. 9.1. Self-contained unit for incubation of biological samples in tubes (**A**), containing a stirrer (**B**), a heater (**C**), and a constant-temperature thermometer (**D**).

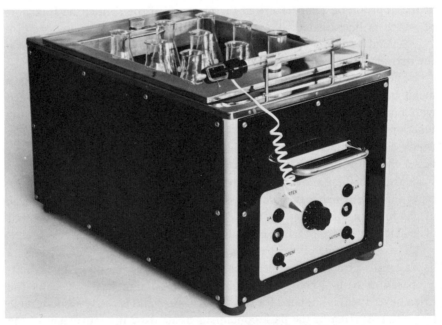

Fig. 9.2. A commercial Dubnoff-type constant-temperature incubator. (Courtesy of the Development Workshop, Institute of Microbiology, ČSAV, Prague.)

required it can be supplied in two ways: (i) The flasks are filled before incubation with the required gas and stoppered (the filling operation must be repeated after each sample withdrawal). (ii) The incubator is provided with a lid covering the tray with flasks which remain open, the whole space under the lid being flushed with the appropriate gas (this arrangement practically precludes rapid sample withdrawal). Hence, for transport studies requiring rapid sampling, particularly in an atmosphere different from air, the first type of incubator is to be preferred.

The *third* general type of incubator is one based on forced circulation of the constant-temperature liquid around the experimental object (Heppler or Wobser type). This is a versatile device applicable even to the most demanding operations. A pump forces either water or cooled saline (for low-temperature experiments) through a system of tubing or into a reservoir which serves as a jacket for a set of tubes or flasks or for a chamber in which the experimental object (e.g., frog skin) is fixed, or even for an optical cuvette if a transport process is followed directly in a (spectro)photometer (see p. 230). The circulation system can either function as a closed unit, the heating (cooling) liquid remaining out of contact with the atmosphere, or else the thermostatic jacket is provided with an overflow through which the liquid returns to the pump.

Stirring of the experimental suspension or medium must be provided either by bubbling a gas through it (as in the first type of incubator) or with the aid of a magnetic or other stirrer.

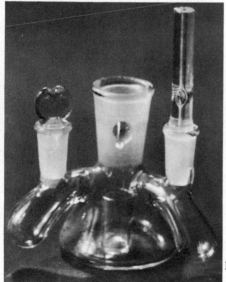

FIG. 9.3. A Warburg-type flask with two side-arms and a central well.

The tubes or flasks used for incubation in any of the arrangements are either glass or plastic of different types, there being no general preference for any material or design. (Beware, however, of using polymethacrylates, such as Perspex, plexiglas, or lucite, for studies where absorption of oxygen might be detrimental, and ordinary glass if extremely low concentrations of alkali metal ions are of importance.) For some experiments, it is convenient to use flasks with one or more side arms (of the Warburg manometer type) where the investigated solutes and/or inhibitors are placed, to be tipped into the main compartment at the appropriate time. If, in addition, the flask has a central well compartment (Fig. 9.3) it can be used for special reagents like oxygen absorbers.

It might be added here that for a variety of experiments with objects which are not particularly sensitive to temperature changes (poikilotherm tissues) one may do completely without a constant-temperature bath, the work being carried out at the ambient temperature of the room.

9.1.2. Gaseous Phases Used for Incubation

The gases used for membrane transport studies are of two general types, those for aerobic conditions and those for anaerobic experiments. Aerobic incubations are most frequently carried out with air as the gaseous phase but sometimes, if the oxygen requirement is high or if the thickness of the experimental object is too great as to impair gas diffusion to the deeper lying cell layers, pure oxygen or oxygen with a certain amount of carbon dioxide (0.5–5%) are advisable. These gases are obtained in steel cylinders already mixed to order.

Anaerobic incubation requires the removal of oxygen from the experimental system (both the gaseous and the liquid phase). To this end, air is displaced from the flask or tube with another gas, different types having been used by different investigators: hydrogen, helium, nitrogen, argon, and carbon dioxide. The use of some of these is distinctly discouraged: hydrogen, because of its inflammability and lightness; helium, because of its lightness and relative costliness; carbon dioxide, because of its potential metabolic effects and the usual high content of impurities. Nitrogen has been used most often but argon is the gas of choice for anaerobic experiments for several reasons: (i) it can be supplied at a high degree of purity, (ii) it is not prohibitively expensive, (iii) it appears to have no metabolic effects (unlike nitrogen which, even in microorganisms not involved in nitrogen fixation, may affect metabolic processes), (iv) it is heavier than air and hence, when using it for flushing open flasks, it does not escape from the liquid surface as quickly as nitrogen does.

If it is essential for the success of the experiment to remove even the least traces of oxygen one is compelled to use one of a number of oxygen-absorbing agents.

More information on the subject may be found in Kleinzeller (1965).

9.2. SEPARATION OF CELLS AND TISSUES

For an analysis of transport processes it is usually necessary (for exceptions see the methods on p. 229) to be able to manipulate cells and tissues separately from the medium surrounding them during incubation.

9.2.1. Handling of Macroscopic Objects

If pieces of tissue are investigated the problem of separation is rather simple. The tissue can be removed from the incubation medium with forceps or "fished out" with a small plastic or stainless-steel net, blotted gently with a lint-free filter paper, and analyzed. For experiments where repeated transfers of tissue at short time intervals are necessary, it is useful to incubate the tissue in a modified tube provided with a nylon mesh bottom

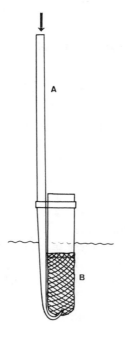

FIG. 9.4. Tube with nylon-mesh bottom (**B**) for incubating tissue slices or chunks, permitting flushing with the appropriate gas through a rigid tube (**A**) serving also for holding the device in position.

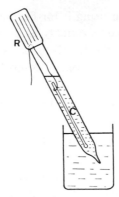

Fig. 9.5. Tube for large algal cells. The cell (**C**) is held in position by a cotton thread fixed at the rubber squeezer (**R**). (According to Dainty and Hope, 1959*b*.)

(Fig. 9.4) where the tissue is simultaneously flushed with the appropriate gas (Kleinzeller *et al.*, 1962). This type of tube proved invaluable in tracer-exchange studies.

For handling large single cells (e.g., the algae *Hydrodictyon, Chara, Nitella, Rhodymenia*), a pipet according to Dainty and Hope (1959*b*) was found to be useful (Fig. 9.5). A constant-temperature device for handling algal cells was described by McRobbie and Dainty (1958 *b*).

9.2.2. Handling of Microscopic Objects

For separating small single cells from the incubation medium several techniques of three main types, centrifugation, filtration, and chemical termination, are available.

9.2.2.1. Centrifugation Techniques

The procedures to be described here are based on centrifuging a suspension sample after removing it from the incubation flask, either directly or across a liquid layer. Direct centrifugation is simpler but requires subsequent washing of the sediment to remove that part of the original incubation medium which is trapped in the sediment, and recentrifugation. In this way, the procedure becomes laborious, time-consuming and, what is most important, causes losses of intracellular solutes due to the washing. In some cases, however (e.g., where the uptake of a solute is irreversible or where the extracellular medium is to be analyzed) the method is applicable and, therefore, it might be useful to consider some practical aspects of the technique.

First of these is the performance of the centrifuge expressed in terms of g, the acceleration due to gravity. The formula to be used here has been derived from elementary mechanics.

$$G = 1.12 \times 10^{-5} \, (rpm)^2 r, \tag{1}$$

where G is the multiple of the value of acceleration due to gravity, rpm is the number of revolutions of the rotor per minute, r is the radius of centrifugation in cm.

For most transport studies with single cells, accelerations where $G < 5000 \, g$ will be sufficient but, to increase the rate of sedimentation, higher multiples of g may be used.

The second useful formula connected with centrifugation is the one expressing the time required for a cell to reach the bottom of the centrifuge tube. Using an integrated form of the Stokes Law for the sedimentation of particles we obtain after suitable substitutions the following expression:

$$t = 843 \eta K^{-1} \, (rpm)^{-2} D^{-2} (d_p - d_l)^{-1} \log(r_{max}/r_{min}) \tag{2}$$

where η is the viscosity of the liquid in poises ($g \cdot cm^{-1} \cdot sec^{-1}$), r_{max} and r_{min} are the radii (in cm) of rotation of the bottom and of the meniscus of the liquid, respectively, K is the particle shape factor (0.222 for a sphere, 0.19 for a disc, 0.06 for a cylinder), D is the particle diameter in cm, d_p the particle and d_l the liquid density in $g \cdot cm^{-3}$. The resulting time is obtained in sec.

The formula can be used for assessing the time required for a given centrifugation run but, because of many interfering factors which are poorly defined, it is advisable at least to double the calculated time to achieve satisfactory results.

To speed up the centrifugation procedure and to circumvent the necessity of washing the sedimented pellet, Werkheiser and Bartley (1957) suggested the use of a three-layer liquid system in the centrifuge tube (Fig. 9.6). A suspension sample is pipetted on top of a silicone oil layer which, in its turn, rests on a layer of a heavier liquid (e.g., 10% trichloroacetic acid). Spinning in a centrifuge will "pull" the cells from the top layer through the oil to the bottom. During passage through the oil layer, the cells are separated from the aqueous medium and can be analyzed in the bottom layer without any "intercellular" liquid. The technique can be simplified by omitting the bottom layer so that the cell pellet after centrifugation is surrounded by the oil. One must then carefully remove the top aqueous layer, wash the tube walls with water, remove the silicone layer, and analyze the cells.

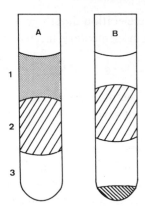

FIG. 9.6. Filtration of cell suspension (1) through a layer of silicone oil (2) into a solution of deproteinizing agent (3). **A** before, **B** after filtration.

For the technique to be useful one must employ a silicone oil of a density intermediate between that of the incubation medium and that of the cells. Most frequently used is the H-oil with $d = 1.05$ which is a polymer containing dimethylsiloxy and methyl-H-siloxy groups in a ratio of 3:1 with a molecular weight of 5–15,000 daltons. If the necessity arises to reduce the density of the oil layer, the H-oil can be mixed with a lighter silicone; for increasing the density, chloroform may be added.

The technique has several drawbacks which apparently prevent it from being used more extensively. First, the tubes must be filled carefully to eliminate formation of globules of water in oil. Second, a certain amount of medium may be carried down with the cells when they migrate in the form of droplets along the tube walls. This "leak" can be greatly reduced if the tube walls are coated with silicone prior to the experiment. Third, the silicone oil is resistant to general glass-washing procedures. It is best to remove it either in a hot concentrated detergent (not all commercial products will do) or in an approximately 10% solution of sodium hydroxide in ethanol.

The drawbacks of the silicone method are overcome in the technique introduced by Ju Lin and Geyer (1963). Its principle is the same but, instead of the silicone oil, a relatively long column of an aqueous solution is used (Fig. 9.7). However, this technique is again not an ideal one: The device needed for the separation of cells is somewhat cumbersome and difficult to fix in the centrifuge head; moreover, the relatively long passage of cells through the aqueous medium may deprive them of some of the substance we are investigating.

It is a common feature of all centrifugation techniques that they require a relatively long time for each sample to be separated. A highly skilled worker, using the silicone technique (which is the fastest of all) and a high-speed miniature centrifuge (e.g., Danish Ole Dich or West German Ecco-

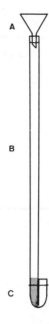

FIG. 9.7. Filtration of cell suspension introduced in funnel **A**, through a column of solution **B** to form a sediment at the bent end of the polyethylene tube **C**.

Quick) can take serial samples of, say, erythrocytes or yeast cells, in about 50-sec intervals while separation of bacteria will require as much as 3 min.

9.2.2.2. Filtration Techniques

The most frequently employed procedure whereby cells are separated from the incubation medium is filtration, aided by water- or oil-pump vacuum, of suspension samples through suitable filters which may be followed by washing the pellet on the filter with water or buffer. The technique was first developed for filter paper discs but now cellulose nitrate filters are used practically exclusively. They are made in several countries, the best known brand being probably the Millipore filter. According to the size of cells or particles to be separated, one may choose from a range of pore sizes. The accompanying table (Table 9.1) shows the filtration times of water for the various filter classes.

Naturally, a cell suspension will filter more slowly than water, while application of more powerful suction (or pressure from the top side) will accelerate the flow.

It should be noted that there are substantial differences between the rates of filtration of cells even of roughly the same size, depending on the surface characteristics and rigidity. To achieve satisfactory filtration rates for cells like erythrocytes it is helpful to use a paper filter and cover it with

Table 9.1. Filtration Rate of Water across Standard Membrane Filters *

Pore diameter, cm^{-4}	Rate of flow, ml/min
8	2120
3	1000
1	500
0.5	160
0.2	50
0.1	5
0.01	0.5

* Filter area 2.5 cm², water-pump vacuum corresponding to a pressure difference of about 740 Torr.

a layer (about 1 mm thick) of Hyflo Super Cel or a similar material, introduced as a suspension in buffer before applying vacuum. Alternatively, a glass prefilter (like the Millipore AP-20) may be used on top of the filter to prevent clogging of the membrane filter. The recently offered polycarbonate filters (such as Nuclepore) appear to have faster filtration rates and are less easily clogged by protein material.

One of the filter supports in use is shown in Fig. 9.8. On this type of device serial samples of suspensions can be withdrawn at much shorter intervals than is possible in any of the centrifugation techniques (Table 9.2). A fast-flowing pipet must be used and its contents emptied quickly so that the cells are distributed evenly over the filter area.

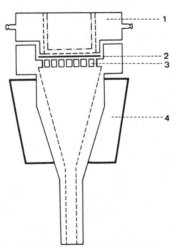

Fig. 9.8. Attachment for membrane filter separation of cells from medium. Heavy cover (**1**) holds down the plastic membrane filter (**2**) onto a perforated or porous disc (**3**), the whole device being fixed in a rubber stopper (**4**).

Table 9.2. Rate of Filtration of Various Cell Suspensions through Membrane Filters *

Cell type	Dry weight of cells in a 0.7 ml sample, mg	Pore diameter of filter, cm^{-4}	Time, sec
Saccharomyces cerevisiae	6	0.4	14
Saccharomyces cerevisiae	6	0.8	13
Saccharomyces cerevisiae	6	1.5	13
Saccharomyces cerevisiae	15	0.4	21
Rhodotorula gracilis	6	0.4	22
Rhodotorula gracilis	15	0.4	35
Escherichia coli	2	0.4	45
Ehrlich ascites cells	2	0.4	60
Rhesus kidney cells	2	0.4	60
Human erythrocytes	3	0.4	75
Human erythrocytes	10	0.4	200 (haemolysis)
Human erythrocytes	10	0.4 (+ prefilter)	35

* The times shown are the shortest reasonably attainable and include the filtration, washing of the pellet with 1 ml buffer, removal of the filter with pellet to a tube with a deproteinizing agent, and placing of a fresh filter on the support. The difference in pressure was about 750 Torr, filtration area 2.5 cm².

Several more complicated filtration devices have been designed (e.g., Lassen, 1964) but it is doubtful that they can speed up the procedure significantly.

The washing of cells on the filter is the only somewhat objectionable part of the procedure as it might be argued that some of the intracellular solute is released from the cells by it. However, it seems that in most cases the very brief application of ice-cold washing liquid has no appreciable effect on intracellular solutes. Nevertheless, all measurements of trans-membrane movements should be carried out by several different techniques and, wherever possible, intracellular as well as extracellular samples should be analyzed at least as a control.

9.2.2.3. Chemical Termination

For the sake of completeness, a method of stopping the membrane transport in a cell should be mentioned which does not consist primarily of separating the cells from the medium. It is based on the finding that the transport of some substances is blocked by certain inhibitors, e.g., transport of some monosaccharides in baker's yeast by uranyl ions (Rothstein, 1954) or the same in human erythrocytes by mercuric ions (LeFevre, 1948). To interrupt the transport process the suspension sample is transferred to a solution of the appropriate inhibitor, the procedure allowing withdrawal of serial samples at 3–4 sec intervals. The collected samples are then centrifuged and washed with the inhibitor solution at the end of incubation.

Unfortunately, the method is limited to a few substances and even with these the inhibition of transport may not be complete, so that errors of at least the magnitude incurred in the other techniques can be expected.

10. ESTIMATION OF SOLUTE PENETRATION

10.1. ANALYSIS OF THE INCUBATION MEDIUM

While in the preceding chapter we discussed the techniques employed for incubating cells and tissues in a suitable medium and for separating them from this medium at appropriate time intervals, the present chapter will concern the various methods applicable to the estimation of intracellular solutes, without attempting to come close to being an exhaustive survey of such methods as they constitute the subject of a number of monographs.

It should be made clear at the very beginning that the simplest and most reliable approach is the determination of the solute concentration in the medium. It can be obtained free of cells or tissue by simple operations and analyzed without interference on the part of other cell components. This method is the obvious one for studying outflow of substances from cells and can be used even for studying uptake if the process is an uphill one so that measurable differences in concentration of the medium are likely to occur.

10.2. WATER OF CELLS AND TISSUES

The intracellular concentration of penetrating solute is usually expressed in reference to the water volume of cells or, when this is not known, to the dry weight or to the total cell weight. With single cells this is quite straightforward, the water volume of cells being measurable with the aid of a haematocrit and a nonpenetrating substance like inulin or ovalbumin. The volume of the cell column in the calibrated heamtocrit tube is known (V_s), the dry weight of the cells is determined by drying at 90–110°C and dividing by 1.15–1.25 which appears to be the average density of cell solids, to obtain the volume of the dry solids (V_d), and the intercellular volume (V_i) is determined by estimating the amount of medium of known concentration of the nonpenetrating substance trapped in the column. The cell water volume (V_w) is then given by the simple formula

$$V_w = V_s - V_d - V_i \qquad (1)$$

It may vary between 10 and 90% of the total cell volume for different cells but it is usually about 60–70% for microorganisms and 70–90% for tissue cells.

It should be noted that expressing the intracellular concentrations per total cell water is rather a matter of convention since the value, at its best, represents an average across the whole cell water, the actual local concentrations being apparently rather varied.

10.3. INTERCELLULAR SPACE

If animal or plant tissue slices or blocks are studied and if the intracellular concentration of the solute in question is to be referred to intracellular water, the fact must be taken into account that a part of the tissue volume represents an extracellular space which is believed to be in free communication with the external medium. This extracellular space, or rather the amount of solute present in it, must be subtracted from the total amount of solute present in the tissue. To perform this adequately, the size of the space must be estimated for each individual type of experiment, this being generally done by the use of inulin (preferably labelled with ^{131}I) which is known not to permeate through cell membranes ("inulin space").

The tissue is incubated with inulin for at least 30 min, blotted with filter paper, weighed, and boiled for 3 min in 6N NaOH. The radioactivity of the solution (made up to a mark) is then counted, together with a reference sample prepared from the incubation medium. The inulin space (in $g_{H_2O}/g_{tissue\ weight}$) is obtained as the ratio E of counts/min in 1 g tissue to counts/min in 1 ml medium. The fraction of extracellular tissue water is then equal to $E \times$ tissue wet weight/tissue water, the tissue water being defined as the difference between wet and dry weight.

10.4. ESTIMATION OF INTRACELLULAR COMPOSITION

10.4.1. Preparation of Cells and Tissues for Analysis

Practically all types of chemical analysis require the previous destruction of the cells or tissues examined so that the substance analyzed is present in solution. Depending on the intracellular nature of the substance we may choose from two types of techniques.

(1) If the substance is in solution intracellularly (like chloride, sulfate, nonmetabolized organic compounds) the cells are destroyed either by boiling in water (this is not recommended for inorganic phosphate determination as some high-energy phosphates may be split) or by extraction with a deproteinizing agent. The following are in use: (i) 5–10% trichloroacetic acid, the mixture being centrifuged after some 30 min and the acid removed if necessary by extraction with ether or by neutralization. (ii) 8% perchloric acid, the mixture being centrifuged after 30–60 min and the supernatant neutralized with a mixture of 1 M triethanolamine, 10 M KOH, and 1 M acetic acid (3 : 0.2 : 0.2) to precipitate insoluble potassium perchlorate. (iii) 5% zinc sulfate, followed by the same amount of 0.3 N $Ba(OH)_2$ (titrated previously against the zinc sulfate solution to slightly pink phenolphthalein). The suspension is then centrifuged at a low speed. This technique is particularly suitable for estimations where a salt-free extract is desirable since all ions added to the sample are precipitated as $BaSO_4$ and $Zn(OH)_2$. On the other hand, it is not suited to estimations of, say, phosphate since this anion is mostly carried down with the precipitate. (iv) Mixture of chloroform and methanol which is suitable particularly for estimation of lipids and fatty acids.

(2) If the chemical species we are interested in is not necessarily in the soluble form or if the task is to determine its total amount (this applies particularly to inorganic ions), it is best to ash the cells or tissue, either by actual dry incineration in an electric oven or by wet-ashing in various acid mixtures.

(a) The dry procedure has the advantage of doing away with any extraneous chemicals but it is lengthy, requires a good stock of platinum crucibles or silica vials, and care must be taken that the temperature of the oven does not exceed even locally a certain limit. The optimum temperature for inorganic ions appears to be 400–600°C applied for 6–10 h.

(b) The wet procedure is definitely the recommended one. Here the sample is heated with a mineralization (ashing) mixture until complete decolorization and clearing is achieved. One of the following ashing agents can be used (depending on the element examined): (i) Concentrated nitric acid. (ii) Concentrated sulfuric acid and 60% perchloric acid (3 : 2). (iii) Concentrated sulfuric acid followed by a drop of 30% hydrogen peroxide.

10.4.2. Chemical Analysis

When a proper cell extract or mineralized solution is obtained it can be analyzed by any of a number of analytical methods described in the literature. It would be far beyond the scope of this book to treat in detail the various methods of analysis available. Let us restrict ourselves only to a review of the principles employed.

10.4.2.1. Analysis of Elementary Composition

There appear to be three principal methods widely applicable to the analysis of the elementary composition of the cell interior: flame photometry, atomic absorption spectroscopy, and spectrography. Table 10.1 shows the range of applicability of the three techniques.

Flame photometers, atomic absorption apparatus, and spectrographs are now available commercially but improvements on these have been described by various laboratories, making it possible to estimate at least some elements in samples as small as 10^{-9} liter. While the three methods cover practically the whole periodic system of elements, they are of no use in estimating halogens, inert gases, oxygen, nitrogen, or sulfur. With the exception of the inert gases (which do not come into play in transport studies) there are specific methods for the estimation of all the other elements. Chloride can be estimated potentiometrically, bromide and iodide colorimetrically or polarographically, fluoride by titration with thorium nitrate, etc. Oxygen (if it is at all determined in transport studies) can be estimated by several principally colorimetric methods or as heavy ^{18}O, sulfur can be determined colorimetrically or by polarography, and nitrogen after kjeldahlization to ammonia can be determined by, for instance, Conway's microdiffusion technique.

The appropriate techniques can be found in any of a number of monographs on biochemical analysis (e.g., Oster and Pollister; Glick; Colowick and Kaplan; Florkin and Stotz; as listed in the bibliography).

A method that can be used for detecting minute amounts of elements is the *neutron activation analysis*. The sample (on, say, a' chromatographic paper) is exposed to a flux of slow neutrons in an atomic reactor, the neutrons being trapped by various elements at different degrees of sensitivity. The element atomic mass is raised by the trapping and generally a radio-

Table 10.1. Range of Application of Various Methods of Elementary Analysis
(S = Spectrography, F = Flame Photometry, A = Atomic Absorption)

Element	S	F	A	Element	S	F	A	Element	S	F	A
H	−	−	−	Zn	+	−	+	Nd	+	+	−
He	−	−	−	Ga	+	+	−	Pm	+	+	−
Li	+	+	+	Ge	+	−	−	Sm	+	+	−
Be	+	+	−	As	+	−	−	Eu	+	+	−
B	+	+	−	Se	+	−	−	Gd	+	+	−
C	+	−	−	Br	−	−	−	Tm	+	+	−
N	−	−	−	Kr	−	−	−	Yb	+	+	−
O	−	−	−	Rb	+	+	+	Lu	+	+	−
F	−	−	−	Sr	+	+	+	Hf	+	−	−
Ne	−	−	−	Y	+	−	−	Ta	+	+	−
Na	+	+	+	Zr	+	+	−	W	+	+	−
Mg	+	+	+	Nb	+	−	−	Re	+	−	−
Al	+	−	−	Mo	+	+	+	Os	+	+	−
Si	+	−	−	Tc	+	−	−	Ir	+	−	+
P	+	−	−	Ru	+	−	−	Pt	+	−	+
S	−	−	−	Rh	+	−	+	Au	+	−	+
Cl	−	−	−	Pd	+	+	+	Hg	+	+	+
A	−	−	−	Ag	+	+	+	Tl	+	+	+
K	+	+	+	Cd	+	+	+	Pb	+	+	+
Ca	+	+	+	In	+	+	−	Bi	+	−	−
Sc	+	−	−	Sn	+	+	−	Po	+	−	−
Ti	+	+	−	Sb	+	−	−	At	−	−	−
V	+	+	−	Te	+	−	−	Rn	−	−	−
Cr	+	+	+	I	−	−	−	Fr	+	−	−
Mn	+	+	+	Xe	−	−	−	Ra	+	−	−
Fe	+	+	+	Cs	+	+	+	Ac	+	−	−
Co	+	+	+	Ba	+	+	+	Th	+	−	−
Ni	+	+	+	La	+	+	−	Pa	+	−	−
Cu	+	+	+	Ce	+	+	−	U	+	−	−
				Pr	+	+	−				

active isotope is produced which can then be detected by one of the techniques described below. The sensitivity of neutron activation analysis with respect to various elements of biochemical interest decreases as follows: Mn $(10^{-7}$ g) > Na, Cu, As, Br, I $(10^{-6}$ g) > P, Cl, K, Co, Ni, Se, Rb $(10^{-5}$ g) (Haissinsky and Adloff, 1965).

It will be seen that samples exposed on pure cellulose chromatographic paper will not be contaminated by unwanted isotopes since H, C, and O of the paper cannot be neutron-activated.

10.4.2.2. Analysis of Organic Compounds

Organic compounds are usually estimated by suitable colorimetric or spectrophotometric methods, of which hundreds have been described and for the performance of which some excellent apparatus is on the market (Beckman, Hilger, Eppendorf, Perkin–Elmer, Aminco–Bowman, Unicam, Zeiss, Optica, and others).

10.4.3. Radioactive Isotopes

A technique which has been gaining in importance ever since the work of Schönheimer (1946) is the use of radioactive isotopes of various elements and of compounds labeled with such elements. The merits of this technique are many: (i) Absolute specificity without any effect of contamination by related elements or compounds. (ii) Versatility, the market offering now a whole range of isotopes that can be used in transport studies (Table 10.2). The list of isotopes should be supplemented with a list of several hundred compounds, most of them available commercially, labeled with isotopes such as ^{3}H, ^{14}C, ^{32}P, ^{35}S, ^{131}I, and others. Moreover, some useful methods are available which permit the labeling of many organic compounds with ^{3}H (e.g., Wilzbach, 1956). (iii) Simplicity of assay. It is frequently not necessary to prepare cell or tissue extracts for estimation of radioactivity, in particular if the more penetrating γ emitters are used, such as ^{24}Na, ^{42}K, ^{59}Fe, ^{131}I, ^{203}Hg. (iv) Applicability to steady-state studies (p. 152).

The use of radioisotopes holds a certain danger only if working with greater amounts of penetrating radiation when proper precautions must be taken (lead shields for body protection, storage of isotopes at some distance from the working bench, etc.). Obviously, care must always be taken that no radioactive material is ingested.

Apparatus for counting radioactive samples is available in many varieties (Packard, Tricarb, Nuclear Chicago, Frieseke–Hoepfner). Of these, two are of special importance for transport studies, namely the gas-flow counters with a 2π geometry and the crystal or liquid scintillation detectors where an almost 4π geometry can be attained.

10.4.3.1. Gas-Flow Counters

In the gas-flow counter, samples are placed on planchets made of aluminum, brass, stainless steel, or plastic and moved into the measuring position in a chamber continuously flushed with a gas (e.g., methane) ensuring low quenching of ionizing radiation. In this chamber a tungsten

Table 10.2. Radioactive Isotopes Applicable to Transport Studies

Element	Atomic weight of isotope	Half-life	Mode of decay
H	3 (tritium T)	12.3 years	β^-
C	14	5570 years	β^-
F	18	1.9 hours	β^+
Na	22	2.6 years	β^+, γ
Na	24	15.0 hours	β^-, γ
P	32	14.3 days	β^-
S	35	86.7 days	β^-
Cl	36	300,000 years	β^-
K	42	12.4 hours	β^-, γ
Ca	45	165 days	β^-
Cr	51	27.8 days	γ
Mn	54	291 days	β^-, γ
Fe	59	45 days	β^-, γ
Co	60	(after isomeric transition) 5.3 years	β^-, γ
Ni	63	91.6 years	β^-
Cu	64	12.8 hours	β^-, β^+, γ
Zn	65	245 days	β^+, γ
As	74	18 days	β^-, β^+, γ
Se	75	120 days	γ
Br	82	35.7 hours	β^-, γ
Rb	84	(after internal electron conversion) 33 days	β^-, β^+
Rb	86	18.8 days	β^-
Sr	85	(after isomeric transition) 65 days	γ
Sr	89	50.4 days	β^-
Cd	115	43 days	β^-, γ
Te	132	78 hours	β^-, γ
I	125	60 days	γ
I	131	8.1 days	β^-, γ
Cs	134	(after isomeric transition) 2.2 years	β^-, γ
Ba	133	(after isomeric transition) 7.5 years	γ
Hg	203	47 days	β^-, γ

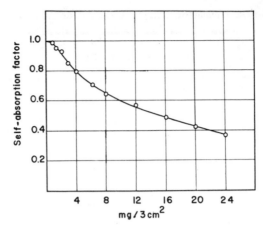

FIG. 10.1. Relative efficiency of counting the radioactivity of a microbial pellet on a membrane filter, showing the amount of self-absorption involved.

wire carrying a high electric potential is suspended. As the gas is momentarily ionized by the emitted radiation, the medium becomes conductive and a "pulse" is registered.

In this arrangement the phenomenon of self-absorption is of importance and, particularly with the soft emitters (^{3}H, ^{14}C, ^{35}S), must be taken into account when comparing samples of unequal weight or distribution on the planchet. Figure 10.1 shows the effect of self-absorption when counting the radioactivity of ^{14}C contained in single cells on a membrane filter in the gas-flow apparatus. Generally, however, proper corrections must be determined empirically (Schlegel and Lafferty, 1961).

10.4.3.2. Crystal Scintillators

The effect of self-absorption is negligible when working with powerful emitters where the method of choice is crystal scintillation. There a crystal of sodium iodide activated with thallium and shaped into a well serves as the receptacle for the sample in a glass or plastic tube. The radiation produced by the sample elicits scintillation of the crystal material (either visible or near ultraviolet), the individual light flashes being properly multiplied to be registered by the counter. However, most of the powerful emitters are rather short-lived and, therefore, if the measurement of radioactivity extends over a considerable period (perhaps several hours) a correction must be taken for the decay of the isotope, correction curves being readily plotted for any given isotope, using a first-order kinetic equation such as on p. 159 and the appropriate half-time of decay.

10.4.3.3. Liquid Scintillators

If very soft and nonpenetrating samples (but not necessarily only those) are to be counted, the modification to be used is the liquid scintillation technique. The sample is placed (either in a minute amount of solution or evaporated) into a tube made of special low-potassium glass or of polyethylene and dissolved in or covered with any of several available organic mixtures (toluene, dioxane, methanol, ethylene glycol) differing among themselves in three points: miscibility with water (higher methanol or glycol contents), freezing point (toluene or methanol depress it), efficiency of counting (toluene or dioxane are best). The mixtures contain substances which, in analogy to the NaI crystal, emit visible or UV light upon interacting with a β particle. There are usually two such compounds included, one as the primary scintillator (most commonly 2,5-diphenyloxazole), the other as the secondary scintillator (usually 1,4-di-2-(5-phenyloxazolyl)benzene). The flashes of light emitted are then again amplified and counted.

While self-absorption should not play a great role in liquid scintillation measurements (unless a heavier residue is counted but that should be only a very exceptional case), the radiation may be quenched by a number of compounds that are present in the sample. A correction for the quenching (mostly built into the modern counters) must always be observed.

10.4.3.4. Doubly Labeled Samples

Because in the scintillation technique the output pulse is roughly proportional to the energy absorbed in the detector (crystal or liquid scintillator) it is possible by suitable screening to count two or more radionuclides in the same sample. This is of special importance for studies where compounds labeled with different isotopes are used, as is becoming rather common in transport studies. There are cases where such a double labeling is indispensable if useful results are to be obtained (e.g., double labeling of transport proteins).

The two or three different radionuclides must be so chosen that their individual pulse height distributions do not overlap very much. This can be achieved, for example, with ^{3}H, ^{14}C, and ^{32}P (Fig. 10.2) by counting in three channels registering three different ranges of energy, e.g., ^{3}H at about 0.002 MeV, ^{14}C at about 0.05 MeV, and ^{32}P at about 1 MeV. However, better results are usually obtained with γ-emitters such as ^{22}Na and ^{42}K in a mixture, since their pulse height ranges are easier to differentiate. The channel width must be somewhat broader so that the steep sides of the peaks do not coincide with the channel limits, since small fluctuations in

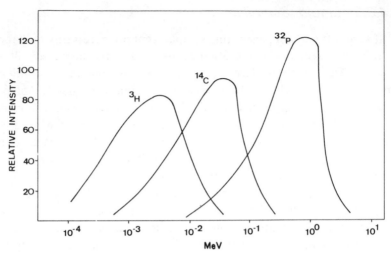

F$_{\text{IG}}$. 10.2. Relative intensity of nuclear disintegrations of three different nuclides over a range of particle energies (in MeV).

amplification could then cause great changes in counting efficiency. The liquid samples are usually refrigerated so as to minimize the thermal emission of the phototube used to register the flashes. To reduce this background further there are two photomultipliers working in coincidence which register only flashes reaching both of them simultaneously so that the random thermal emission is generally not registered.

In evaluating the results we must take into account the radioactivity counted in a given channel caused by the undesirable nuclide (^{14}C in the ^{3}H channel, for instance). Thus, to use an example from an experiment using compounds labeled with ^{3}H and ^{14}C, the actual radioactivity of the individual nuclides is computed as follows:

$$D_C \equiv \text{amount of } ^{14}\text{C} = N_2/C_2 \tag{2a}$$

$$D_H \equiv \text{amount of } ^3\text{H} = (N_1 - C_1 D_C)/H_1 \tag{2b}$$

where N_1 are the counts/min in channel 1 where ^{3}H is preferentially counted, N_2 the counts/min in channel 2 where practically only ^{14}C is counted, C_1 and C_2 are the respective efficiencies of counting ^{14}C in channels 1 and 2 and H_1 is the efficiency of counting ^{3}H in channel 1.

Quenching presents a particular problem in the doubly or triply labeled runs as it is different for each nuclide and due corrections must be made for it. This is usually achieved by counting after each sample a very powerful standard (e.g., ^{137}Cs or ^{133}Ba) radiating through the sample, the relative decrease in count rate compared with standard without sample indicating the amount of quenching, provided the ratio of quenching of the external standard to that of the nuclide tested is known.

Whenever doubt persists as to the quantitative evaluation of doubly labeled samples, one should—if possible—reverse the labeling of the sample and compare the results obtained.

It is perhaps not superfluous to add that no special corrections and advanced equipment are needed if two isotopes are used in a transport study that can be differentiated by their half-time of decay (e.g., ^{22}Na and ^{42}K) when the samples are counted once immediately after preparation and then again after one of the isotopes has practically completely disintegrated (after some 8–10 half-lives).

10.5. CYTOLOGICAL METHODS

For many purposes, a direct localization of the substance in question in the cell or tissue is essential. A number of identification methods are available both for inorganic and organic compounds, based on specific reactions with either a stain or a complexing or precipitating agent. For details on these, the reader is again referred to special monographs as are listed in the bibliography. Mention shall be made here of some special useful methods.

10.5.1. Autoradiography

In autoradiography, one uses radioactive tracers not to determine kinetic rates but the distribution of a substance in a cell or tissue. The amounts of isotopes required are generally greater than those used in kinetic studies. Cells or tissues, after incubation with the labeled compound, fixation (e.g., isopentane cooled to $-180°C$), dehydration (by freeze-drying or chemically), embedding (e.g., paraffin or Vestopal), and sectioning, are placed in contact with an autoradiographic film. One of several techniques can be used, most of the materials being available in convenient form from firms such as Kodak, Ilford, and Gevaert.

(a) The tissue can be placed directly on the emulsion surface on a glass or cellulose slide. Paraffin sections are floated on water and the slide is dipped into the water, or else, if contact with water is to be avoided, mercury is used, and the slide is pressed against the mercury surface.

(b) The object is mounted directly on the microscope slide and coated with melted emulsion in the dark.

(c) Thin sensitive emulsion attached to gelatin is stripped from a glass or celluloid support and floated on the surface of water. When the section has straightened out and swelled, it is picked up on a slide.

The mount with emulsion is then exposed for a period of weeks or even months (but not more than 2 months is recommended because of fading of films and high background due to cosmic radiation). The slides are then developed in a special photographic developer and processed according to instructions for each individual type of film. The objects may then be stained by various microscopic techniques and observed under the optical, phase-contrast, or electron microscope (Fig. 10.3).

For optical observation of stained specimens a dark-ground illumination of the object is of great advantage as the exposed silver grains in the emulsion appear as shining specks. A suitably balanced combination of transmitted and incident lighting will produce satisfactory superimposed pictures of the object with proper location of the radiation sources.

For electron microscopy, mounting the ultramicrotome sections on a slide with a thin membrane of formvar (polyvinyl formaldehyde) or polyvinyl chloride where they can be suitably stained (with uranyl, lead, or so) and coated with a 50–60 Å layer of carbon is recommended. The slide is then dipped in a dilute emulsion of silver halide crystals to form a coat about 1500 Å thick (determined by interference colors). Exposure should take place in the absence of oxygen.

The resolution of an autoradiograph developed under optimal conditions (thin preparation, soft emitter of radioactivity, thin fine-grained emulsion, electron microscope) may be as high as 5×10^{-6} cm.

10.5.1.1. Autoradiography of Soluble Compounds

In transport studies it is of paramount importance to be able to estimate the localization in the cell of water-soluble compounds, such as inorganic ions or small organic molecules which are not incorporated into cell components. As practically all of the generally used techniques include steps in which the mount comes into contact with an aqueous solution special techniques must be used. However, it is still necessary to have the concentration of the isotope inside cells appreciably higher than in the outside medium in order to obtain clear differentiation.

Sections can be prepared either from a tissue or from a frozen suspen-

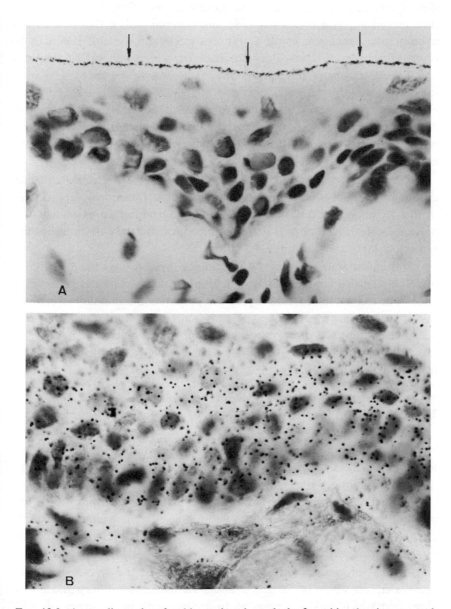

FIG. 10.3. Autoradiography of a thin section through the frog skin, showing accumulation of silver grains formed by disintegration of $^{203}HgCl_2$ at the outer boundary of the epidermis (**A**) and a diffuse distribution of the same isotope inside the epithelial layer (**B**). (Taken with permission from Lodin *et al.*, 1963.)

sion of cells in a cryostat at a temperature well below 0°C. The sections are
then picked up on a preloaded slide and covered with a thin dry film of
photographic emulsion, using a stainless steel loop, or else they can be
placed directly on a layer of dry emulsion on a glass support (if stripping
film is used, its gelatin side should be in contact with the slide). The auto-
radiograms must not thaw at any step of the procedure, unless they have
been freeze-dried instead of liquid-frozen, including the exposure period.
After exposure, the sections are let to thaw, quickly dried in a current
of air and fixed (e.g., with 4% formaldehyde in a pH 7.4 phosphate
buffer).

The danger of artifacts caused by reactions between the chemical
groups of the unfixed cell and the photographic emulsion is appreciable
in this technique; it can be suppressed by interposing an extremely thin
polyvinyl chloride membrane between the object and the emulsion, but this
is a rather laborious technique.

It is unfortunate that as yet no technique for observing the distribution
of soluble labeled compounds by the electron microscope has been developed. The technique of freeze-drying the preparation and fixing in osmium
vapor and of directly applying the emulsion to the section in a gel form
does not ensure good preservation of the biological objects.

More details on the techniques have been given by Rogers (1967),
from which source much of the above information was taken.

10.5.2. Other Techniques

X-Ray absorption spectra show marked discontinuities at certain wave-
lengths, these being characteristic for every element. If a collimated beam
of X rays is allowed to pass through a biological specimen one can record
the attenuation of the X-ray beam at the top and at the bottom wavelength
of the discontinuity, and the amount of the element examined may be
calculated. The object may be scanned by the X-ray beam and the density
of the element actually mapped.

A great promise is held by the application of the so-called electron
probe analyzer which so far has found wide use in metallurgy and miner-
alogy and which makes it possible to analyze with an accuracy of $\pm 1\%$
the elementary composition of a cross section through an experimental
object. It is likely that, in particular for the study of bulk distribution of
elements (with the exception of the very lightest ones), it could be employed
after suitable fixation of the object.

10.6. ESTIMATION OF SOLUTES IN LIVING CELLS

10.6.1. Microelectrodes

Several ions can be estimated inside cells without destroying the cells. This can be carried out by inserting the tip of a cation-sensitive glass micro-electrode into the cell, fixed mechanically in an aqueous medium containing the other (reference) electrode, the measured potential having to be corrected for the membrane potential determined with a calomel electrode. The technique is described in detail elsewhere (p. 257) together with some of its theory. It has the advantage over static postmortem techniques, in that a continuous record of the intracellular concentration changes can be obtained but, unfortunately, it is limited to the estimation of H^+, Na^+, and K^+.

10.6.2. Estimation of Intracellular pH by Acid Dye Distribution

Another technique which allows a determination of intracellular pH to be done is the use of a weak acid with a pK at about 4. Waddell and Butler (1959) used 5,5-dimethyl-2,4-oxazolidinedione but it was later found to be actively transported into cells. In our hands (Kotyk, 1963), superior results were obtained with the acidobasic indicator bromophenol blue.

The principle underlying the method is the fact that ions pass across the cell membrane only with difficulty or are transported by a specific system while the undissociated molecule reaches a practically equilibrium distribution at the two sides of the membrane. As it is used in a minute concentration (about 10^{-6} M) it does not affect the pH of the solution but dissociates according to the pH prevailing in the compartment measured. Proceeding from these premises (making the approximation that the acid has the same pK inside and outside the cells) one can derive the following expression:

$$pH_i = pH_o + \log\left[\frac{c_i}{c_o}(1 + 10^{pK-pH_o}) - 10^{pK-pH_o}\right] \qquad (3)$$

The pH_o in the medium is measured with a pH-meter, the intracellular (c_i) and extracellular (c_o) concentrations of the dye are measured colorimetrically (after raising the pH to convert the dye to its dissociated blue-purple form) and the pK is known to be equal to 4.0. The only drawback of the technique is that in some cells (particularly animal tissues) some of the dye may be adsorbed by cell constituents and this must be subtracted from the measured value to obtain correct c_i.

10.7. ESTIMATION OF MEMBRANE TRANSPORT BY NONSPECIFIC METHODS

All the techniques described in this section are applicable only to cells which respond to a change of osmotic pressure of the medium by a change in their volume in a regular manner, i.e., only single cells, either of animal origin (erythrocytes, tissue culture cells, tumor cells) or microorganisms without their protective walls.

10.7.1. Densitometry

Ørskov (1935) observed that the amount of polychromatic light transmitted through a suspension of erythrocytes decreases as the cells shrink. This seemingly paradoxical observation was confirmed for other cells and has apparently to do with the reflectivity of the cell surface, the amount of scattered light increasing with decreasing cell size. Be it as it may, as a concentrated solution of a substance is added to the cell suspension, transmission increases due to dilution of the suspension (an instantaneous process), the cells begin to lose water to adapt their internal osmotic pres-

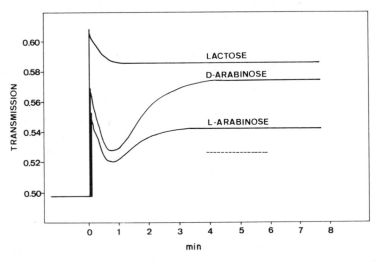

FIG. 10.4. A continuous transmission record of a yeast protoplast suspension at 546 nm. The initial rise of transmission is due to dilution of cell suspension (theoretically to the level of the dashed line) plus an "overshoot" apparently associated with changes of cell surface reflectivity. The subsequent decrease is caused by shrinking of cells due to external hypertonicity which may be permanent if the solute is impermeant (lactose), fully reversible if the solute enters into the cell water freely (D-arabinose) or partly reversible if the solute enters into only a part of the cell water (L-arabinose).

sure, this resulting in decreased transmission (this takes about 30–50 sec both with erythrocytes and yeast protoplasts) and then, if and when the added solute penetrates into cells, their volume increases so that light transmission rises (Fig. 10.4). The whole process is easily recorded on paper when using a sensitive photometer and the uptake curves of many solutes can be thus used for direct evaluation.

However, the method is not suited for fast-penetrating solutes since the beginning of the uptake curve is then distorted by the extrusion of water and, moreover, displays an as yet unexplained anomaly, the initial overshoot in transmission which is related to the solute concentration added. LeFevre and LeFevre (1952) derived an empirical formula for monosaccharide uptake by human erythrocytes, relating the measured transmission L to the transmission of cell suspension without sugar L_o, the extracellular sugar concentration c_o (mol/liter), and the intracellular sugar concentration c_i (mol/liter) in case the cells had been preincubated with sugar:

$$L = L_o(1 + 0.051c_o - 0.024c_i) \tag{4}$$

10.7.2. Indirect Cytolytic Method

This method was originally developed for the uptake of sugars by human erythrocytes (Wilbrandt, 1948) but it can be applied to any osmotically labile cells and to the transport of any solute. It can be characterized as follows.

Osmotically labile cells burst in hypotonic media but display a certain degree of osmotic resistance which is reflected in the fact that the "hypotonicity" to which cells are exposed must attain a certain value before they burst *en masse*. The concentration of the outer medium at which 50% cells undergo lysis (this can be readily estimated photometrically) is greater for higher intracellular contents concentration. As a solute is taken up by the cells the total intracellular concentration rises and from the shift of the concentration of a buffer solution at which the cells burst by 50%, as compared with a control without any permeant, one can calculate the uptake of the solute.

The procedure used in the indirect cytolytic method consists in removing samples (usually four at a time) of cell suspension before and after adding the transported substance (at various time intervals) and mixing them with graded concentrations of ice-cold buffer, e.g., 0.4–0.5–0.6–0.7 isotonic. The buffer concentrations should be so chosen as to give 50% cytolysis in about the middle of the series (Fig. 10.5). The intracellular

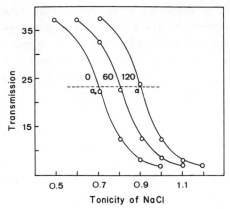

FIG. 10.5. Cytolytic resistance curves of a bovine erythrocyte suspension during penetration of glycerol at 2°C, at 0, 60, and 120 min. For details see the text.

concentration c_i of the penetrating solute at time t is then computed from the following formula:

$$c_i = \frac{n \cdot s + m \cdot a}{n \cdot s_o + m \cdot a_o} \tag{5}$$

Since the experiment is carried out in an isotonic medium (say, 0.154 M NaCl with animal cells) $s_o = 1$; $s = 1 +$ tonicity of the solute (in isotones), a is value of the intersection a in Fig. 10.5 read on the abscissa, a_o is the same without solute (as curve 1 of the figure), n is the volume of sample (generally 0.2 ml), m is the volume of hypotonic buffer into which the sample is pipetted (usually 5 ml).

In addition to their simplicity and rather wide applicability, the "osmotic methods" provide information on whether a tested substance is present in the cells in an osmotically active state or whether it is adsorbed or otherwise immobilized only on the surface.

11. INTERPRETATION OF TRANSPORT DATA

The kinetic principles underlying the various types of membrane transport were discussed in chapters 3–6, together with suggestions as to experimental procedures for arriving at the various transport parameters. In this chapter, some practical information will be given to aid in the interpretation of the various uptake and exit curves.

11.1. INITIAL RATES

The merits of using tracers in such studies were emphasized before but cases may arise when the true initial rate of uptake must be measured. The estimation of the initial rate of a process may be complicated for two principal reasons: (i) The rate of uptake (or exit) is very rapid so that no reliable values can be obtained in the region that is of importance. (ii) The start of the uptake curve can be irregular for various reasons (inadequate mixing, a shock reaction of the cells, dead time of monitoring equipment). The plotting of a tangent to the uptake curve at zero time (as suggested in older books on enzymology) is usually impracticable but an approximative method can be used at least with processes which are not excessively rapid, in other words, where the first experimental point lies well below the third of the final equilibrium (or steady-state) value. The technique is based on the first derivative of Newton's formula for equispaced decreasing differences (presented here in a modified form):

$$v_o = \Delta^1 - \Delta^2/2 + \Delta^3/3 - \Delta^4/4 \qquad (1)$$

233

Its application is apparent from the following sample calculation:

Experimental values spaced at equal time intervals	Δ^1	Δ^2	Δ^3	Δ^4
0				
	16			
16		−4		
	12		1	
28		−3		0
	9		1	
37		−2		
	7			
44				

$$v_0 = 16 + 2 + 0.3 + 0 = 18.3$$

It is not uncommon (particularly with uphill transports) that a good part of the uptake curve is practically linear so that there the estimation of the initial rate presents no problems.

Once the initial rates are known it is usual to plot the reciprocals of the values obtained against the reciprocals of the substrate concentration used as shown in Fig. 11.1. The half-saturation constant and the maximum

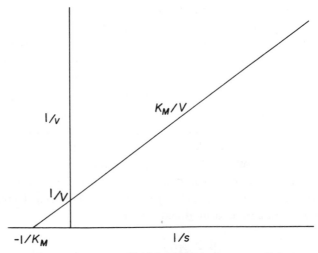

FIG. 11.1. A Lineweaver–Burk plot of initial rate dependence on substrate concentration.

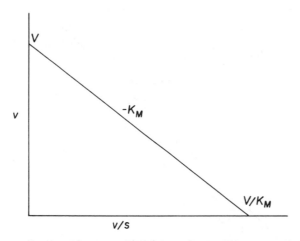

FIG. 11.2. A Woolf–Hofstee plot of v against v/s yielding the maximum rate V and the half-saturation constant K_M.

rate of uptake or exit follow from the plot (method of Lineweaver and Burk, 1934). However, a detailed analysis of the various possible reciprocal methods carried out by Dowd and Riggs (1965) showed that the relative weight of the individual experimental points is best distributed in the plot first described by Woolf (1932) but generally known under the name of Hofstee (1959). The transport parameters are read as shown in Fig. 11.2.

Obviously, as follows from the treatment on p. 65, if the transport is one of simple diffusion both the apparent maximum rate and the half-saturation constants will tend to infinity. Practically, however, infinite values are never obtained due to interactions, even in simple diffusion, of the solute molecules with the barrier or pores through which the process takes place.

11.2. INHIBITION OF TRANSPORT

When using either competing transport substrates or inhibitors of different nature, the above reciprocal plots can be used for determining the constants of the inhibitor as shown in Figs. 11.3, 11.4, 11.5, and 11.6.

The partially competitive or partially noncompetitive inhibitions are more difficult to analyze and require an extrapolation of $V(1 + i/K_i)$ and $K_{CS}(1 + i/K_i)$ to infinite inhibitor concentration. However, to obtain at least qualitative information on the "partial" character of inhibition one can plot $1/v$ against i. If the inhibition is either partially competitive or

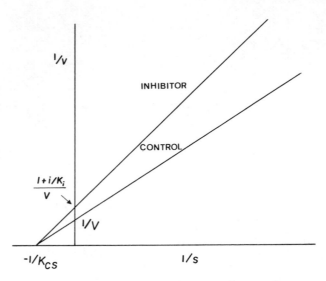

FIG. 11.3. Analysis of noncompetitive inhibition according to Lineweaver and Burk.

partially noncompetitive the plots will not be linear but rather convex upward. Similarly, as used by Thorn (1953), a plot of V/v against i will not yield a straight line if the inhibition is of "partial" nature.

The two types of plot suggested above (as well as a number of other modifications) require a series of measurements with either the substrate or the inhibitor concentration varying, the other remaining constant. For

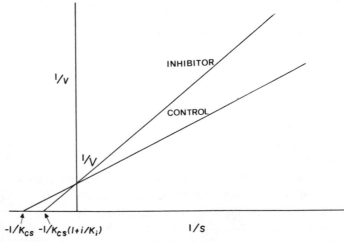

Fig. 11.4. Analysis of competitive inhibition according to Lineweaver and Burk.

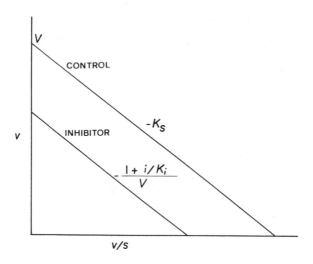

FIG. 11.5. Analysis of noncompetitive inhibition according to Woolf and Hofstee.

cases where a number of experimental data exist such that neither of the concentrations was kept constant it is useful to plot the results according to the method of Hunter and Downs (1945) as shown in Fig. 11.7. In the figure, α stands for the ratio of uptake rates in the presence (v_i) and in the absence (v) of the inhibitor. Instead of $\alpha/(1 - \alpha)$ one can use the expression $v_i/(v - v_i)$.

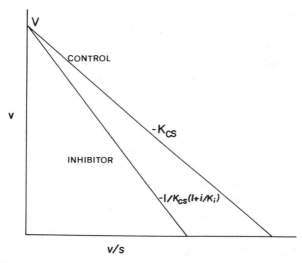

FIG. 11.6. Analysis of competitive inhibition according to Woolf and Hofstee.

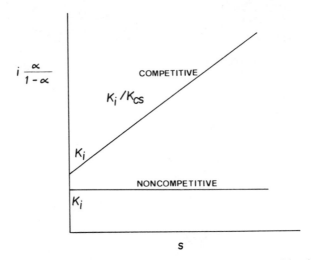

FIG. 11.7. A Hunter–Downs plot of competitive and noncompetitive inhibition.

11.3. ACTIVATION OF TRANSPORT

An analysis of transport activation may prove to be more difficult than that of inhibition. Essential activation, i.e., one where no transport occurs without activator, is more easily analyzed than nonessential, i.e., one where some transport proceeds even without activator. The following rules may be of use in this context.

If curves of $1/v$ against $1/s$ (at different concentrations of the activator) have a common intercept equilibrium conditions of essential activation exist and a compulsory path $(C + A) + S$ is indicated. The intercept on the $1/v$ axis is equal to $1/V_a$, that on the $1/s$ axis to $-a/K_{CAS}(a + K_{CA})$.

If curves of $1/v$ against $1/a$ at different concentrations of substrate have a common intercept, equilibrium conditions of essential activation exist and a compulsory path $(C + S) + A$ is indicated. The intercept on the $1/v$ axis is equal to $1/V_a$, that on the $1/a$ axis to $-s/K_{CSA}(s + K_{CS})$.

If linear plots are obtained for $1/v$ vs. $1/s$ and for $1/v$ vs. $1/a$ but no common intercept is obtained, either the equilibrium defined by eq. (45a) of chapter 5 obtains or else the steady-state treatment defined by eq. (49a) or (49b) of chapter 5 describes the situation.

If, however, $1/v$ vs. $1/s$ is linear but $1/v$ vs. $1/a$ is not, the probable explanation lies in a nonessential activation in equilibrium for any of the compulsory or noncompulsory paths or else in a steady state for the compulsory paths.

If neither the plot of $1/v$ vs. $1/s$ nor that of $1/v$ vs. $1/a$ is linear, the mechanism is one of steady-state activation by noncompulsory pathway, for either essential or nonessential activation (*cf.* Segal *et al.*, 1952).

It is useful in some cases to plot $1/(v - v_-)$ where v_- is the rate without activator present against $1/a$ since these plots (for equilibrium conditions) are always linear. The intercept with the $1/a$ axis is given by $-(1 + K_{CAS}/s)/(1 + K_{CS}/s)K_{CSA}$ which, for the case that $K_{CAS} = K_{CS}$, becomes $-1/K_{CSA}$. The linearity of these plots is preserved even if another modifier B is present.

If two modifiers A and B are present at a constant summary concentration, the plot of $1/v$ vs. $1/a$ will not generally be linear but can become so for essential activation if $D'_{CSB}(a + b) \ll K_{CSB}$. Then the intercept on the $1/v$ axis can be either positive or negative. Again, a plot of $1/(v - v_-)$ vs. $1/a$ is linear for all the cases considered (*cf.* Semenza, 1967).

11.4. SOME LESS COMMON CASES

If a given solute is transported by two types of transport or by two different carriers of widely different affinities and/or maximum velocities of transport the reciprocal plots presented above may show two more or less linear parts which can be extrapolated to intersect with the coordinate axes and yield the appropriate constants (Fig. 11.8).

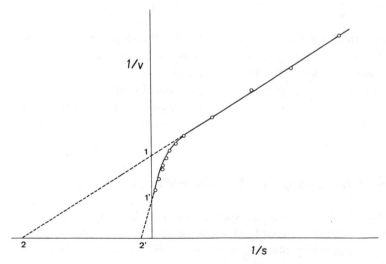

FIG. 11.8. A Lineweaver–Burk plot of data obtained in a transport system containing two carriers with widely differing K_M's and V's. (From Gits and Grenson, 1967.)

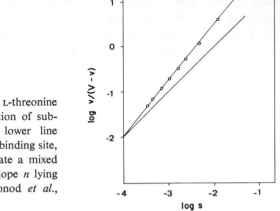

FIG. 11.9. A Hill-type plot of L-threonine deaminase activity as a function of sub-strate concentration s. The lower line would correspond to a single binding site, the experimental points indicate a mixed cooperative interaction, the slope n lying between 1 and 2. (From Monod *et al.*, 1963.)

A useful plot providing an insight into the molecular order of a transport process is based on Hill's empirical relation (1910) for the binding of oxygen to haemoglobin. As is shown in Fig. 11.9, one uses an expression containing the maximum rate of transport and the actual rate of transport at the given concentration of substrate $[\log v/(V - v)]$ and plots it against the logarithm of substrate concentration. The expression is readily related to the transformation of the Michaelis–Menten equation according to Eadie (1942),

$$v = V - K_M v/s \qquad (2)$$

If the system were strictly monomolecular the slope of the line should be equal to unity. If it is greater, more than one binding sites mutually cooperating are indicated (Monod *et al.*, 1963).

An alternative plot is the one developed by Scatchard where $\log v/(V - v)s$ is plotted against v. Curves of negative slope indicate negative interactions; curves with positive slope are indicative of positive cooperative effects.

11.5. ANALYSIS OF EXPONENTIAL CURVES

In a first-order process, such as simple diffusion, carrier transport at very low concentrations of substrate, or movement of tracers, the rate can also be obtained by plotting on a logarithmic scale the difference between the final value of s_{II} and the value at time t against time on a linear scale (Fig. 11.10). A straight line should be obtained, the slope of which, taken

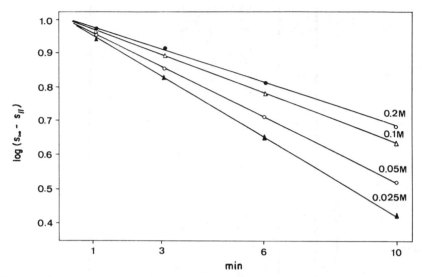

FIG. 11.10. A plot demonstrating the first-order nature of a transport process (uptake of labeled D-ribose in equilibrium of nonlabeled D-ribose). The half-times of uptake range here from 4.8 min at the lowest to 9.3 min at the highest concentration. (From Wilbrandt and Kotyk, 1964.)

with a minus sign, indicates the rate constant of the process. For modifications of this approach, see p. 74.

If the system contains more compartments than one plus a reservoir and the efflux of a tracer from the compartments is followed a more complex plot is obtained, which can be analyzed as shown below. The procedure employed is one of graphical analysis and is exemplified in Table 11.1

Let us assume that for times from 1 to 10 min, the values of $c^*/c_{t=0} = y$ were found experimentally (here c stands for concentration of solute, equivalent to s of chapter 3). The decadic logarithm of the values is found in tables and plotted against time (Fig. 11.11). In this semilogarithmic plot, the dependence of y on time for times longer than 5 min is represented by the straight line y_I. It is then possible to express this dependence by

$$y_I = c_I e^{-k_1 t} \tag{3a}$$

or by an equivalent equation, like

$$\ln y_I = \ln c_I - k_1 t \tag{3b}$$

$$\log y_I = \log c_I - k_1 t / 2.303 \tag{3c}$$

Table 11.1. Sample Analysis of a Complex Exponential Curve

(a)	t	0	1	2	3	4	5	6	7	8	9	10
(b)	$c^*/c_{t=0} = y$	1.000	0.583	0.408	0.326	0.279	0.247	0.221	0.199	0.180	0.163	0.147
(c)	$\log y$	0.000	−0.234	−0.389	−0.487	−0.554	−0.607	−0.656	−0.701	−0.745	−0.788	−0.833
(d)	$y_1 = 0.4e^{-0.1t}$	0.400	0.362	0.327	0.296	0.268	0.243	0.220	0.199	0.180	0.163	0.147
(e)	$y - y_1$	0.600	0.221	0.081	0.030	0.011	0.004					
(f)	$\log (y - y_1)$	−0.222	−0.656	−1.091	−1.523	−1.959	−2.398					

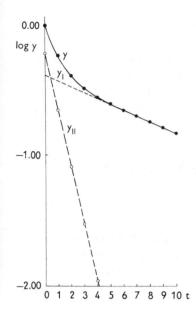

FIG. 11.11. An example of an efflux curve analyzed graphically. For details see the text.

which is the form used for plotting Fig. 11.11. It may be seen that $k_1/2.303 = 0.043$ and hence $k_1 = 0.1$; likewise, $y_{I t=0} = \log c_I = -0.4 = 0.6 - 1$, and hence $c_I = 0.4$. Now the values of the exponential term $y_I = 0.4e^{-0.1t}$ can be calculated also for the first part of the experiment (see (d) in Table 11.1). The values are subtracted from the exponential values of y in (b). The resulting differences are given in (e). Logarithms of these differences are found and plotted against the corresponding times. The straight line y_{II} of the figure is thus obtained. It is clear that the observed dependence can be expressed by the sum of two exponential terms

$$y = c_I e^{-k_1 t} + c_{II} e^{-k_2 t} \tag{4}$$

The value of $k_2/2.303$ from the figure is 0.434, i.e., $k_2 = 1$ and $y_{II t=0} = \log c_{II} = -0.222 = 0.778 - 1$, hence $c_{II} = 0.6$. The time course of the experiment can be described by

$$y = c^*/c_{t=0} = 0.4e^{-0.1t} + 0.6e^{-t} \tag{5}$$

The data in the above hypothetical experiment are idealized; in actual practice, the straight lines will have to be drawn through points with a certain scatter. It is evident that the absolute minimum required for drawing a straight line are three points. From two points the curvature of the experimental line cannot be assessed. Drawing the straight line by the usual method of least squares is not suitable in this case (Robertson, 1957; Solomon, 1960 a) since in semilogarithmic plots one measures concentrations but plots their logarithms, thus obliterating the fact that for short times the values have a greater weight than those for longer times.

By repeating the graphical analysis until a straight line is obtained from the semilogarithmic plots of the difference of the type ($y - y_I$) curves representing a sum of a greater number of exponential terms can be analyzed.

11.6. UPTAKE OF METABOLIZED SUBSTRATES

A somewhat special situation arises if the problem is to analyze the uptake of a substance which is either converted to another metabolite (like sugars) or is incorporated into a polymer or cell component (like amino acids). If the use of inhibitors preventing this subsequent conversion of the transported substance is for any reason impossible, an analysis of the intracellular concentration yields only incomplete information on the uptake parameters since

$$ds_{II}/dt = V_{in} \left(\frac{s_{II}}{s_I + K_{T_I}} \right) - V_{out} \left(\frac{s_{II}}{s_{II} + K_{T_{II}}} \right) - V_M \left(\frac{s_{II}}{s_{II} + K_M} \right) \quad (6)$$

where V_{in} and V_{out} are the maximum rates of flow inward and outward, respectively, K_{T_I} and $K_{T_{II}}$ are the half-saturation constants of uptake and exit, respectively, and the last term expresses the metabolic conversion of the substrate. In a steady state, when $ds_{II}/dt = 0$,

$$V_M \frac{s_{II}}{s_{II} + K_M} = V_{in} \frac{s_I}{s_I + K_{T_I}} - V_{out} \frac{s_{II}}{s_{II} + K_{T_{II}}} \quad (7)$$

If the transport parameters are to be determined it is best to start by measuring the metabolic rate at several concentrations of s_I and hence s_{II}. This being known, we are left with four unknowns which can be determined from four equations, using different concentrations of s_I.

The situation is simpler in an equilibrating system where $V_{in} = V_{out}$ and $K_{T_I} = K_{T_{II}}$ so that only two unknowns are to be determined.

Another possibility of determining the transport parameters of a metabolized substance (R) in equilibrating cells is to make use of its competition with a nonmetabolized analogue (S). Here, at steady state,

$$\frac{s_I'}{s_I' + r_I' + 1} = \frac{s_{II}'}{s_{II}' + r_{II}' + 1} \quad (8)$$

Therefore, $s_I'(r_{II}' + 1) = s_{II}'(r_I' + 1)$ and

★
$$K_{CR} = \frac{s_{II}r_I - s_I r_{II}}{s_I - s_{II}} \qquad (9)$$

The maximum rate of transport is best measured here by tracer labeling.
The same result is obtained even if one assumes $D_C' \neq D_{CS}'$ when (*cf.* Kotyk and Kleinzeller, 1966)

$$K_{CR} = D_{CR}'/D_C' \frac{s_{II}r_I - s_I r_{II}}{s_I - s_{II}} \qquad (10)$$

so that the K_{CR} is unaffected by K_{CS} and D_{CS}'.

12. BIOELECTRICAL
MEASUREMENTS

12.1. INTRODUCTION

Bioelectrical measurements include the determination of potential differences, measurements of the electrical parameters of membranes, and, occasionally, the measurement of ionic fluxes as currents in circuits. Of these, the determination of the resting potential differences is perhaps the most fundamental in transport studies. Data on the potential differences across membranes make it possible to evaluate the total driving force acting on individual ion species so that criteria of the character of the ion transport like the Ussing–Teorell flux ratio may be applied. In a steady state even a simple comparison of the electrical potential difference with the equilibrium potential calculated from the concentrations of the ion on both sides of the membrane may be sufficient for deciding whether the ion is actively transported or not. Although very often performed, the measurements of the potential differences are not quite free of theoretical objections. It was well said by Gibbs that "Again, the consideration of the electrical potential in the electrolyte, and especially the consideration of the difference of potential in electrolyte and electrode, involve the consideration of quantities of which we have no apparent means of physical measurement, while the difference of potential in pieces of metal of the same kind attached to the electrodes is exactly one of the things which we can and do measure," (Gibbs, 1899, quoted by Guggenheim, 1967). It was stressed again by Guggenheim, that "The electrostatic potential difference between two points is admittedly defined in *electrostatics*, the mathematical theory of an imaginary fluid *electricity*, whose equilibrium or motion is determined entirely by the electric field. *Electricity* of this kind does not exist. Only electrons and ions have physical existence and these differ fundamentally from the hypothetical fluid *electricity* in that their equilibrium is *thermodynamic*, not *electrostatic*" (Guggenheim, 1967).

Indeed, the measurement of the potential difference across a membrane, between two electrolyte solutions, cytoplasm being often considered as one of them, often requires an additional assumption. This assumption is that the liquid junction potentials existing between the salt bridges and the electrolyte solution are practically zero. Salt bridges are connections made of a concentrated salt solution, usually 3 *M* or saturated (approximately 4.2 *M* at 25°C) potassium chloride. Salt bridges are employed because, due to the high concentration of KCl in the bridge, the liquid junction will be formed almost entirely by potassium and chloride ions, the concentrations of all other ions being negligibly low. The mobilities of the potassium cation and of the chloride anion are very similar and thus it may be shown, using Henderson's equation for the mixture boundary (section 2.2), that the liquid junction potential will be very low. In electrochemistry, ammonium nitrate is sometimes used instead of potassium chloride when leakage of chloride ions is to be avoided, the mobilities of the cation and of the anion being here practically the same. The same mobilities of ions of opposite signs is a condition *sine qua non* for the salt bridges; as soon as the originally similar mobilities of these ions are modified by a surface charge of the glass inside the glass microelectrode which represents a salt microbridge, or by electrically charged proteins of the cytoplasm, their use may become questionable.

Still, there exist measurements of potential differences in biological material which appear to be free of theoretical objections and do not even require the assumption that salt bridges depress the junction potential. These are measurements of electrical potential differences across complex biological membranes in split chambers, where a surviving layer of cells separates two solutions of the same composition. Let us start the discussion of bioelectrical techniques with these indisputable measurements.

12.2. TRANSEPITHELIAL POTENTIAL DIFFERENCES AND SHORT-CIRCUIT CURRENT TECHNIQUE

In Fig. 12.1 a biological membrane separates two aerated circulating solutions, their composition being either different or, as in many experiments, identical. Epithelial layers of poikilotherms like frog skin are specially suited for studies of this sort, as they may be conveniently experimented

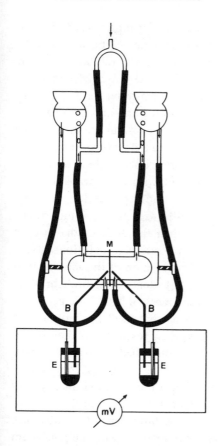

FIG. 12.1. Principle of the electrical potential measurement across epithelial membranes. **M** Membrane, **B** agar bridges, **E** calomel electrodes, **mV** millivoltmeter.

upon at room temperature. Two halves of the split chamber made of polymethacrylate (or of glass if, for instance, oxygen consumption is measured and the ability of polymethacrylate to absorb oxygen would be disturbing) are held firmly together by two screws or by an elastic band, so that no short circuit outside the membrane is present. The bubbling apparatus according to Ussing and Zerahn (1951) shown in Fig. 12.1 is suitable for circulation of aerated media especially when the flux of a labeled substance is measured, the labeled substance accumulating in the fixed volume of the solution from which small samples are taken at suitable intervals. In other cases, continuous perfusion of the chamber halves with fresh aerated solutions from stock bottles or perfusion apparatuses may be preferred. Let us now turn our attention to the circuit employed for measuring the electrical potential. A salt bridge connects each of the half-chambers with a calomel electrode. Bridges may be conveniently made of polyethylene tubing filled with agar gel prepared from the salt solution. Since polyethylene tubing is thermoplastic, gentle heating will aid in forming constrictions at the ends of the individual bridges and a whole series of bridges may be prepared simultaneously by drawing with a water pump the liquefied agar gel (about 1–3% agar in the suitable solution) into the polyethylene tubing. Such agar gel can be repeatedly liquefied on a warm-water bath. The constrictions prevent agar from escaping from the tubing and make the fixation of the bridges to the chamber wall easier. The bridges can be attached to the chamber wall either by pressing the bridge into a hole of suitable diameter in the polymethacrylate directly, or by using rubber tubing as a seal. The ends of the individual bridges may be cut with a razor blade. The position of the bridges in the chamber is not important as long as only the potential difference is measured, but it will be seen to be of importance when the short-circuit current is measured; then the ends of the bridges should be close to the membrane. If the composition of solutions at the two sides of the membrane is identical, the composition of the salt solution in the bridges is not important as long as it is identical in both bridges. 3 M KCl appears to be suitable, but, if even traces of KCl leakage are to be avoided, the same physiological solution as that used for bathing the membranes may be employed in the bridges; their resistance is then somewhat higher so that the measurement is more disturbed by external fields. The calomel electrodes may be prepared as shown in Fig. 12.2; it is very important to keep the platinum contact free of contamination by any solution. A small amount of calomel (Hg_2Cl_2) may be added to the mercury surface, but it seems to be of little importance, a thin layer of this substance originating at any rate from the mere contact of mercury with the potassium

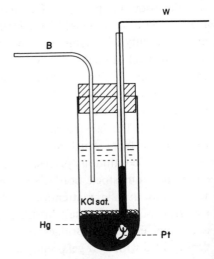

FIG. 12.2. Simple calomel electrode. **Hg** Mercury, **Pt** platinum contact, **B** agar bridge, **W** insulated wire.

chloride solution. However, it is most convenient to use commercial calomel electrodes like those employed for pH measurements as shown in Fig. 12.3. A millivoltmeter or a recording device with a sufficiently high input imped-ance should be used to measure the potential difference across the mem-brane. It is important to check the potential of the measuring electrodes together with their bridges from time to time.

It may be seen in Fig. 12.1 that the electrical circuit is highly sym-metrical. Whatever the jumps of the electrical potential in the circuit and the efficiency of the salt bridges in suppressing the liquid junction potentials may be, they are the same in both halves of the circuit. The electrical po-tential difference measured is, therefore, localized on the biological mem-brane and nowhere else. When different solutions are used at the two sides of the membrane, the efficiency of the salt bridges becomes an appreciable factor. This is not necessarily so, as many important experiments with biological membranes can be carried out with the same solutions at both sides of the membranes, such as measurements of the short-circuit current

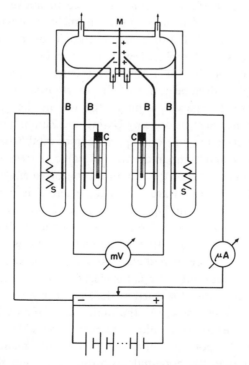

FIG. 12.3. Short-circuit current technique according to Ussing and Zerahn (1951). C Calomel electrodes, S silver–silver chloride electrodes, B agar bridges, M membrane.

described by Ussing and Zerahn (1951). Figure 12.3 shows a simple apparatus for short circuiting biological membranes. Using the electromotive force of a battery an electrical current of the desired intensity is made to flow through the membrane. At a certain current value, equal to that which would flow through the external circuit due to the electromotive force of the membrane itself if the external circuit had zero resistance, the electrical potential difference across the membrane is zero. We can comprehend the situation by realizing that the drop of potential caused by current flow through the membrane resistance just compensates the spontaneous potential difference produced by the membrane under these conditions, or that the flow of ions of the appropriate sign from the bridges is just equal to a similar flow originating in the membrane. The fact remains that under the conditions of such total short circuiting there is no electrical potential difference and no concentration gradient of any ion between the two solutions. There is no simple physicochemical force which would explain the flow of current from one solution to the other and thus the flux of ions corresponding to this current must be due to a direct coupling with a metabolic reaction in the membrane, i.e., due to active transport. There are several possibilities of identifying the ion, the flow of which corresponds to the short-circuit current. For example, individual ions may be omitted from the solutions (and replaced with another inert or unphysiological ones) and the change in the short-circuit current may be recorded. A change or complete disappearance of the short-circuit current upon such omission indicates the participation of the ion in the generation of the short-circuit current. Best, as done by Ussing and Zerahn (1951) for sodium, the net fluxes of the individual ionic species can be established by the double-labeling technique and compared with the value corresponding to short-circuit current. (The ratio of the flux of an ion of valency z to the current density across the same unit area is $1 : zF$, where F is one Faraday.)

In the device shown in Fig. 12.3 silver–silver chloride electrodes are used as the current electrodes, connected with the media with the same agar bridges as those used with the calomel electrodes. The ends of the bridges leading from these current electrodes should be as distant as possible from the membrane for only then is the electrical field across the membrane nearly homogeneous and the current density nearly the same across the whole area of the membrane. On the other hand, the ends of the bridges from the calomel electrodes should be as near as possible to the membrane, since then only the potential difference across the membrane proper is measured rather than the potential drop due to current flow through the resistance of any appreciable thickness of the media. It might be found convenient to place the silver–silver chloride electrodes directly in the solutions bathing the surfaces of the membrane as this results in no harmful effects. This is understandable since the solubility product of silver chloride is low (1.73×10^{-10} at 25°C) and the concentration of silver cations in the media usually containing chlorides is therefore extremely low. The silver–silver chloride electrodes are prepared simply by connecting silver wires with the positive pole of the D.C. source (small battery with a potentiometer) and thus made to function as an anode in practically any chloride solution (3 M or saturated KCl, physiological saline and the like). A piece of graphite or platinum wire can serve as cathode. When silver wires are covered with a layer of silver chloride they are ready for use as current electrodes. Silver–silver chloride electrodes prepared under special precautions are sometimes used in combination with bridges for potential difference measurements but calomel electrodes appear to be more stable.

A number of technical improvements of the short-circuit current technique were reported, the most obvious of which are an automatic maintenance of the membrane potential at zero level and multichannel chambers. Both these improvements are combined in the apparatus described by Morel and co-workers (1961). The apparatus makes it possible to record simultaneously the short-circuit currents through six different 0.8-cm² areas of a single surviving frog skin, arranged in pairs symmetrical with respect to the longitudinal axis of the animal, one member of each pair serving as a control. There is a common chamber on one side of the skin and six individual chambers, each containing 3 ml of the medium, on the other side. The potential difference across each area of skin is kept at zero by means of a servomechanism and the currents are separately recorded by a six-channel microammeter. Another automatic apparatus with which short-circuit current across ten different pieces of frog skin or bladder can be registered simultaneously was recently described by Ivanov and Natochin (1968).

Apparatuses in which measurements of the short-circuit current may be combined with microelectrode measurements of intracellular potentials will be described later in this chapter and the simultaneous measurement of short-circuit current and transport of water will be discussed in chapter 13. An interesting device in which the effects of hydrostatic pressure, as well as of mechanical deformations, on the short-circuit current across frog skin are registered was recently used by Nutbourne (1968).

12.3. INTRACELLULAR POTENTIAL MEASUREMENTS USING MICROELECTRODES

Unlike the measurements of electrical potential differences between two controlled media as discussed above, the measurements of intracellular potentials, using microelectrodes introduced by Ling and Gerard (1949) (they are in fact salt microbridges) rest heavily on the assumption that salt microbridges are very efficient in supressing liquid junction potentials. It is thus assumed that the jump of the electrical potential by which the microelectrode proper contributes to the potential differences encountered in the circuit (the so-called tip potential) is either negligibly small or that it has practically the same value in the experimental medium as in the cytoplasm. This assumption is often not justified; the cell must not be irreversibly damaged by the microelectrode entry so that the dimensions of the microelectrode tip must be kept as small as possible. The opening in the tip is thus very narrow so that the surface charge of the glass becomes important and the tip potential may reach considerable values.

Adrian (1956) showed that tip potentials are correlated with electrode resistance, that they vary logarithmically with the concentration of alkali metal ions in the external medium, and explained their origin as being due to contamination of the tip by which the mobility of chloride anions in the tip is reduced, so that microelectrodes filled with 3 M KCl become negative inside. Agin and Holtzman (1966) suggest that tip potentials are due to the difference between the interfacial potential at the inner and the outer surface of the glass, as seems to be the case with the asymmetry potential of glass pH electrodes. They demonstrated that tip potentials can be eliminated in the presence of trace concentrations of thorium chloride (and even reversed at some concentrations) and that physiological concentrations of calcium chloride also reduce tip potentials appreciably. The authors found that at the required concentrations of thorium chloride neither the resting nor the action potential of frog sartorius muscle fiber is affected.

The possibility of tip contamination, especially by cytoplasmic components, and the sudden change in the ionic environment of the tip when the cell membrane is pierced remain serious problems of the microelectrode technique. For a truly pessimistic view on the use of microelectrodes Tasaki and Singer (1968) should be consulted. Nevertheless, in many cases practically the same values of intracellular potentials are found with a number of electrodes. As not all of them are likely to be contaminated in the same way, the data obtained may be fairly reliable. For safer control, it may be useful to measure tip potentials not only in the external physiological media but also in solutions approximating the ionic composition of the cell interior.

Detailed descriptions of the microelectrode technique are found in Donaldson (1958) and Nastuk (1964). Here only a few suggestions concerning the fabrication and use of glass microelectrodes will be briefly mentioned. The most convenient way to prepare glass microelectrodes is to use pullers, most often solenoid-operated microelectrode pullers, e.g., those produced by Industrial Science Associates, Ridgewood, N.Y. Borosilicate (e.g., Pyrex) tubes 1–2 mm in diameter are usually employed as the starting material and micropipettes with a tip diameter of about 1 μ or less are pulled out from them. It is also possible to prepare such microelectrodes by hand with the aid of a gas burner, and one who is skilled in the technique will maintain that hand-pulled microelectrodes are superior to others. Micropipettes of convenient length (some 5 cm) are filled by using a bottle as shown in Fig. 12.4. If methanol is used for the first filling, no heating is necessary, the reduction of air pressure produced by an ordinary water pump being sufficient to produce vigorous boiling. Boiling should be repeated several times; when suction by the water pump is released air bubbles in the microelectrodes are seen to shrink and depart from the microelectrodes. After several repetitions no air bubbles are visible in the microelectrode tips even when pressure is substantially reduced and methanol boils on the surface rather than in the micropipettes. The holder with micropipettes may now be transferred to distilled water. Due to difference in the density of methanol and water the replacement proceeds rapidly and

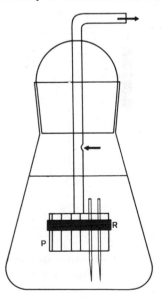

FIG. 12.4. Bottle for filling glass microelectrodes under reduced pressure. **P** Polymethacrylate holder with grooves, **R** rubber band.

after about one hour distilled water may be replaced with $3 M$ potassium chloride. After several hours the microelectrodes are ready for use. It is of advantage to filter all the liquids used (methanol, water, $3 M$ KCl) before filling to remove all particles which could subsequently contaminate the microelectrode tips. Sometimes $3 M$ KCl is used for filling directly but then it must be hot and suction applied and released several times as in the methanol procedure.

The first check of microelectrode quality is usually done before filling by microscopy. The second check consists in measuring the resistance of the filled microelectrodes. Microelectrodes of the same tip resistance are likely to have similar tips and the higher the resistance, the smaller usually the tip diameter with electrodes of otherwise analogous shape. A simple apparatus for measurement of microelectrode resistance is shown schematically in Fig. 12.5. The microelectrode is immersed in a salt solution, a physiological saline being used most often. Contacts between the medium and the $3 M$ KCl inside the microelectrode may be made by using silver–silver chloride electrodes, i.e., silver wires covered with a thin layer of chloride. Voltage E of some 100 mV is applied from a battery using a low-resistance potentiometer and measured by a high input impedance millivoltmeter. When switch S is open, practically no current flows through the circuit so that the total voltage E is measured. When, at the second stage, the switch is on, current flows through the microelectrode and $(E - R_x i)$ instead of E is measured, where R_x is the microelectrode resistance. Current i being

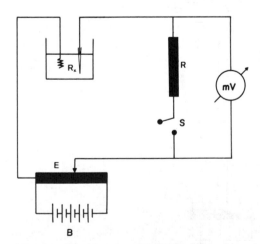

FIG. 12.5. A simple apparatus for measurement of the microelectrode resistance. R_x Microelectrode resistance, **R** standard resistance, **B** battery, **S** switch, **E** potential applied from the voltage divider. For explanation see the text.

equal to $E/(R_x + R)$ (all other resistances in the circuit are negligibly small), the second measurement represents

$$E - R_x i = E - E \frac{R_x}{R_x + R} = \frac{R}{R_x + R} E$$

The ratio of the difference between the first and the second reading $ER_x/(R_x + R)$ to the value of the second reading $ER/(R_x + R)$ is then the ratio of the microelectrode resistance to the known standard resistance R. The usual resistance of glass microelectrodes being several $M\Omega$ or several tens of $M\Omega$, a resistance of some 10 $M\Omega$ is suitable as a standard.

Measurement of the microelectrode resistance during or after measurement of the membrane potential and measurement of the tip potential after such experiment will give information on any changes of the microelectrode during its penetration into the cell.

During use the microelectrodes are held most conveniently with a polyamide screw in a plastic holder, attached to a suitable micromanipulator. Electrical connection may be effected with a silver–silver chloride electrode or, better still, it may be connected through a thin agar bridge, prepared as described above, to a calomel electrode which is more stable.

Potential differences can be measured in several ways. After suitable preamplification, the deflected oscilloscope ray can be returned to zero level using a suitable calibrator. Alternatively, a sensitive commercial millivoltmeter with a high input resistance can be used directly.

The tissue on which the measurement is to be done should be fixed during the experiment in a suitable medium, such as by pinning it in a simple polymethacrylate chamber. Somewhat more complicated apparatuses were described for microelectrode studies of epithelial membranes, where the media circulate and the overall potential difference across the preparation as well as the short-circuit current are recorded simultaneously (Whittembury, 1964; Cereijido and Curran, 1965; Janáček et al., 1968).

12.4. CATION-SENSITIVE GLASS MICROELECTRODES

Knowing the difference of the electrical potential across a membrane and the concentration of an ion at the two membrane sides we still may not be able to assess the simple physicochemical driving forces acting on the ion. The overall concentration is determined by chemical analysis but this, in view of possible binding of the ion by cytoplasmic constituents or of compartmentation of the intracellular content, may not reflect the actual concentration in the space of interest. Strictly speaking, the ion activities in the proximity of the membrane should be considered instead of the overall concentrations, this being the rationale for applying the technique of cation-sensitive glass microelectrodes.

The theory of cation-sensitive glasses, the dependence of ionic selectivity on the composition of glass, etc., has been reviewed by Eisenman (1962) who himself contributed considerably to its development. Here it may do to recall that there exist glasses from which microelectrodes can be prepared which are preferentially either potassium- or sodium-sensitive and the behavior of which, moreover, can be satisfactorily described by the simple Nikolsky equation, derived on the basis of ion-exchange properties of glasses (see Nikolsky and Shults, 1963, or Kleinzeller $et\ al.$, 1968, for references). The potential of a preferentially potassium-sensitive microelectrode can be described by the formula

$$\psi = \psi_0 + S \log[a_{K^+} + k_{K-Na}\, a_{Na^+}] \tag{1}$$

where k_{K-Na} is a positive number smaller than one and a_{K^+} and a_{Na^+} are activities of the two alkali ions. The potential of a preferentially sodium-sensitive microelectrode can be expressed analogously by

$$\psi = \psi_0 + S \log[a_{Na^+} + k_{Na-K}\, a_{K^+}] \tag{2}$$

The symbol S stands for the experimentally determined slope which is usually not very far from the theoretical value for the equilibrium potentials. Later in this section it will be shown how the values of S and of k are determined graphically.

It is perhaps needless to emphasize that upon introduction into a cell the cation-sensitive microelectrode registers not only the change in the ionic environment but also the change of the electrical potential. The membrane potential must be hence measured separately with a conventional Ling–Gerard microelectrode and the value subtracted from the reading with the cation-sensitive microelectrode. If the measurement is carried out with two electrodes of different selectivities toward sodium and potassium, the activities of the two ions can be calculated from the two equations obtained (Lev, 1964b). Occasionally, the effect of the low intracellular activity of sodium may be neglected and the activity of potassium can be calculated directly from the measurement with the potassium-selective electrode. The types of glass used for intracellular pH measurement are usually so selective that the effect of other ions may be neglected and calibration may be carried out simply by using several buffer solutions.

The cation-sensitive microelectrodes are usually prepared on a micro-electrode puller in the same way as conventional electrodes and their tips are then sealed, usually in the proximity of or by a direct contact with an electrically heated platinum loop under microscopic control. With the exception of the tip, which is introduced into the cell, the electrode must be insulated. According to Lev and Buzhinsky (1961) (see also Kleinzeller *et al.*, 1968, for a detailed description) an ordinary Pyrex micropipette of a large tip diameter (2–4 μ) is used for the insulation. (Tips of this size may be obtained conveniently by reducing slightly the heating in the micro-electrode puller at the moment when the tip is formed.) Some picein is introduced into the tip of this Pyrex jacket by dipping the tip briefly into melted picein and by washing away the superfluous picein from the outer surface with chloroform. The cation-sensitive microelectrode, filled with a suitable salt solution by the usual procedure (boiling in methanol, described in section 12.3) is then introduced by micromanipulation into the jacket, the picein being melted by an electrically heated platinum loop. An electrode with a gradual transition between the sensitive tip and the jacket, in which both the tip and part of the insulating jacket are introduced into the cell (e.g., muscle fibre; Lev and Buzhinsky, 1961), as well as a plain electrode, only the tip of which is introduced into the cell (e.g., flat epithelial cell; Janáček *et al.* 1968), can be prepared in this way. Other types of insulation were described, such as by coating with shellac or by simultaneously pulling two concentric capillaries, the inner one sensitive, the outer one of soft insulating glass (Kostyuk and Sorokina, 1961).

One of the glasses that can be used with advantage for making potassium-selective electrodes is NAS 27-5 (Na_2O, 27 mol%; Al_2O_3, 5 mol%; SiO_2, 68 mol%). The tips made of this glass deteriorate quickly in aqueous solutions or in moist air and, therefore, the electrodes should be prepared shortly before each experiment. The glass designated as KABS 20-9-5 (K_2O, 20 mol%; Al_2O_3, 9 mol%; B_2O_3, 5 mol%; SiO_2, 66 mol%) made at the Faculty of Physical Chemistry, University of Leningrad (Lev and Buzhinsky, 1961) is somewhat more stable but less selective. Glasses used for sodium-selective microelectrodes are NABS 25-5-9 (Na_2O, 25 mol%; Al_2O_3, 5 mol%; B_2O_3, 9 mol%; SiO_2, 61 mol%) and NAS 11-18 (Na_2O, 11 mol%; Al_2O_3, 18 mol%; SiO_2, 71 mol%). The latter glass is more selective than the former but it softens only at a high temperature and the preparation of microelectrodes from it is more difficult.

The prepared microelectrodes are generally calibrated in pure solutions of KCl and of NaCl. Some solutions suitable for calibration, their activity coefficients, and the logarithms of their activities (against which the mea-

Table 12.1. Solutions Suitable for Calibration of Potassium- and Sodium-Sensitive Electrodes

	c, mM	Activity coefficient $\bar{f}$	$\log a = \log c + \log \bar{f}$
NaCl	1.00	0.993 [1]	−3.016
	3.16	0.936 [1]	−2.529
	10.0	0.89 [1]	−2.051
	31.6	0.81 [1]	−1.592
	100.0	0.78 [2]	−1.108
	316.2	0.71 [2]	−0.649
	1000	0.66 [2]	−0.180
KCl	1.00	0.993 [1]	−3.016
	3.16	0.936 [1]	−2.529
	10.0	0.89 [1]	−2.051
	31.6	0.81 [1]	−1.592
	100.0	0.77 [2]	−1.114
	316.2	0.69 [2]	−0.661
	1000	0.60 [2]	−0.222

[1] Values calculated according to the formula $\log \bar{f} = -0.51 \, (10^{-3}c)^{1/2}$.
[2] Values from the *Handbook of Chemistry and Physics*, The Chemical Rubber Co., 47th ed., p. D-89.

sured potentials should be plotted) are shown in Table 12.1. An example of calibration curves of a potassium-selective microelectrode filled with 1 M KCl is shown in Fig. 12.6 (microelectrodes made of the NAS 27-5 glass give curves which are similar to those plotted in the figure). Since two parallel straight lines are obtained the behavior of the microelectrode may be apparently well described by the Nikolsky equation (*1*). The necessary constants may be obtained directly from the graph. ψ_0 is obtained by extrapolation of the calibration line for potassium as the intercept on the ψ axis. The value of ψ_0 is the result of two factors, the first of which is that the inside electrolyte has a concentration, rather than an activity of 1, the second is the usual existence of an asymmetry potential due to the fact that the inner and outer glass surfaces are not identical. The value of k_{K-Na} can be calculated from the vertical distance of the two lines, as shown in the figure, and the slope S may be read directly as the segment on the ψ axis corresponding to a unit segment on the $\log a$ axis. Sometimes, especially with sodium-selective glasses, the behavior of the electrodes cannot be satisfactorily described by the Nikolsky equation for a wide

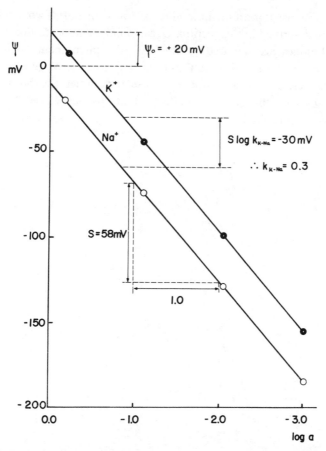

FIG. 12.6. Example of the calibration curves of a potassium-selective microelectrode. For explanation see the text.

enough range of concentrations. Therefore, a microelectrode of this type should be calibrated in a very narrow range of concentrations for the measurement of which it will be actually used as, otherwise, according to Lev (see Kleinzeller *et al.*, 1968), more complicated equations would have to be used to describe the behavior of the microelectrode.

Since the resistance of the cation-sensitive microelectrodes is usually of the order of 10^{10}–$10^{12}\,\Omega$, a millivoltmeter with a very high input resistance must be used (like the Cary vibrating reed electrometer or the Vibron electrometer). It should be stressed that at these high resistance values in the measuring circuit care must be taken that sufficient screening is provided.

Only a few of the bioelectrical measuring techniques which were considered to be most important in studies of ion transport across cell membranes, were described in this section. Occasionally, elaborate electrophysiological techniques are required to elucidate the phenomena of ion flows across cell membranes and their rapid changes as take place especially in excitable tissues. For information on these techniques the book by Donaldson (1958) or the excellent collection of articles edited by Nastuk (1964) may be consulted.

13. VOLUME FLOW
MEASUREMENTS

The experimental approaches applied in volume flow measurements (the theoretical implications of these are described in section 4.2.) are summarized in the following scheme:

13.1. Volume flow between cell and its surroundings

13.1.1. Determination of volume flow from cell weight changes

13.1.2. Estimation of cell volume
  a) direct measurement of cell diameter or thickness
  b) measurement of the volume of cells packed by centrifugation
  c) optical methods

13.2. Transcellular volume flow and flow across cell layers

13.2.1. Determination based on the change of concentration of an impermeant substance

13.2.2. Determination based on weighing

13.2.3. Determination based on direct volume measurement
  a) apparatuses with calibrated vessels and capillaries
  b) automatic maintenance of volume in one compartment

A few words will now be said about each of the methods listed, together with some practical examples from the literature. The results may be expressed either in units of volume flow or in units of water flow; the precision of the biological measurement not warranting a distinction between the two ways.

13.1. VOLUME FLOW BETWEEN CELL AND ITS SURROUNDINGS

13.1.1. Determination of Volume Flow from Cell Weight Changes

Weighing of incubated tissue slices, after their blotting with filter or tissue paper, at given time intervals, may give a sufficiently precise measure of the volume flow between the cells and their surroundings. The method was used by Whittembury and co-workers (1960) for determining the equivalent pore radius (see section 3.1 of this book) in slices of *Necturus* kidney tissue. At intervals of 3–4 minutes a slice was taken out of the solution, blotted, weighed, and returned to the solution. To obviate any doubt as to whether the weight changes measured correspond to intracellular or extracellular water content changes, the measurement should be combined with an extracellular space evaluation done at suitable time intervals, e.g., by the inulin technique, described in chapter 10 of this book.

13.1.2. Estimation of Cell Volume

(a) Direct Measurement of Cell Diameter or Thickness. The volume of regularly shaped cells (spherical or cylindrical) and its changes may be evaluated from the microscopic diameter measurement with a calibrated eyepiece using simple geometrical formulae. The thickness (and, at the same time, the volume) of the epithelial cell layer of the frog skin can be measured by focusing sharply with a water immersion lens on the cornified layer and then on a suitably chosen melanin granule in the melanophore layer immediately beneath the epithelium and evaluating the difference from the calibration of the fine screw of the microscope (Ussing, 1961; MacRobbie and Ussing, 1961).

(b) Measurement of the Volume of Cells Packed by Centrifugation. The volume of cells in suspension may be determined and its changes followed by measuring cells packed by a standard centrifugation in calibrated tubes (haematocrit, see chapter 10). The extracellular space in the packed layer may be evaluated by the inulin method. To measure the volume of the isolated cells of the renal cortex special polymethacrylate tubes were used (Bosáčková, 1963). The bore of their upper part was 16 mm in diameter, that of the lower part 3 mm, and it was graduated. Centrifugation of 20 min at $2000\,g$ was used to obtain the final cell volume.

(c) Estimation of Cell Volume by Optical Methods. Correlations may be found between the volume of cells in suspension and some of the optical properties (absorbance, light scattering) of these suspensions. Under standard conditions, carefully preserved, functional dependences of the optical properties on the cell volume may be determined, permitting an uninterrupted monitoring of the cell volume. The techniques are described in chapter 10.

13.2. TRANSCELLULAR VOLUME FLOW AND FLOW ACROSS CELL LAYERS

Important conclusions have been drawn from transcellular flows observed with algae (Dainty and Ginzburg, 1964a) and from studies of the volume flows across a number of epithelial membranes. (As will be seen in chapter 22, the pathway of the volume flow in this case appears to be, at least partly, intercellular rather than transcellular.) Flows of this kind proceed between two compartments which are usually much larger than the cell volume and hence neither hydrostatic pressure nor concentration differences are likely to influence substantially the flow rates which can be observed for longer periods than flows between the cell and its surroundings.

13.2.1. Determination Based on the Change of Concentration of an Impermeant Substance

Volume flow increases the concentration of an impermeating substance in the compartment where it originates or decreases its concentration in the compartment to which it flows, the product of the concentration and of the volume remaining constant. The substance used must be not only impermeating but also sufficiently inert chemically and determinable with sufficient accuracy. Inulin-^{14}C used, for example, by Skadhauge (1967) studying the cloacal water and electrolyte resorption in fowl by means of *in vivo* perfusion, appears to be suitable for the purpose.

13.2.2. Determination Based on Weighing

It may be occasionally found useful to prepare vesicles from epithelial membranes, changes of the vesicle giving the volume flows per exposed membrane area. Gravimetry representing a highly exact technique, the precision of this method is given chiefly by the degree of standardization of the blotting procedure used which must be done gently so as to do no

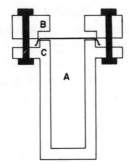

FIG. 13.1. House's chamber for determining very small volume flows across the frog skin. The exposed skin area is 1.25 cm². Volume of the compartment (**A**) is generally several cm³. The ring (**B**) is clamped to the chamber (**C**) by four screws with nuts and washers. The original apparatus contains electrodes for spontaneous potential measurement and short-circuit current recording.

damage to the membrane. Contact with dry glass or paper surface will do in most cases. Vesicles or bags made of the skin of frog legs were used for such measurements (Huf *et al.*, 1951), one serving as the control for the other. Similar paired vesicles can be prepared from the bilobed frog bladders (e.g., Natochin *et al.*, 1965). Everted sacs from rat ileum were used to evaluate volume flow (Green *et al.*, 1962). The weight of the emptied sacs should be determined at least at the beginning and the end of the experiment to evaluate which part of the weight change is due to swelling and shrinking of the tissue itself.

An accurate method of determining very small fluxes across membranes which can be clamped in split chambers was developed by House (1964, 1968) and is shown in Fig. 13.1. The chamber is made of polymethacrylate (lucite) or polytetrafluoroethylene (teflon). Volume flow across the frog skin was determined from changes in chamber weight after the fluid in the upper compartment had been sucked out through a glass tube provided with a nozzle (1.5 mm in diameter) by a vacuum pump. The chamber was handled in rubber gloves and at each experimental time interval (usually 2-hour periods were used), the removal of the fluid and weighing were repeated several times and evaluated statistically. The method permits flows greater than $0.5 \text{ mg} \cdot \text{cm}^{-2} \cdot \text{hr}^{-1}$ to be detected.

13.2.3. Determination Based on Direct Volume Measurement

(a) Apparatuses with Calibrated Vessels and Capillaries. There exist a number of devices in which volume changes in one compartment result in a measurable movements of the meniscus (or an air bubble) in a calibrated tube or capillary. The basic features of a very satisfactory one (Dainty and Ginzburg, 1964a) are shown schematically in Fig. 13.2. The difference in osmotic pressure between the two compartments results

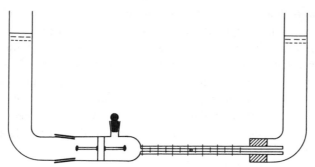

Fig. 13.2. Apparatus for measuring transcellular osmosis according to Dainty and Ginzburg (1964 *a*). For explanation see the text.

in transcellular osmosis across a long algal cell sealed within the small hole in the dividing wall with vaseline. The volume changes in the right-hand side compartment can be measured by the movement of the air bubble in the capillary tube. The apparatus is immersed in a water bath in which the temperature is controlled accurately within $\pm0.01°C$. The left-hand side compartment is open to the exterior and the liquid in it can be readily replaced.

A different approach using calibrated vessels consists in measuring the volume flow from the medium on one side of an epithelial vesicle to its other side which is in contact with moist air only. An epithelial vesicle either continually perfused, like the intestinal preparation developed by Smyth and Taylor (1957), or one filled with a suitable solution, like the gall bladder preparation of Diamond (1964), is suspended in a hermetically closed vessel containing moist air or oxygen. The transported fluid is collected in a graduated stem which forms the bottom part of the vessel. The whole apparatus is kept at constant temperature.

(b) Automatic Maintenance of Volume in One Compartment.

Finally, an example of an automatic apparatus, registrating the volume flow, will be briefly described. The apparatus developed by Bourguet and Jard (1964) permits an automatic registration of volume flows of 0.2–8 μl min^{-1} per 2 cm^2 of an amphibian membrane (skin or bladder). An isolated amphibian membrane separates two solutions of different osmolarity in a split chamber. Only the compartment with higher osmolarity is perfused continually with an aerated solution and held at negative hydrostatic pressure of some 20 cm H$_2$O, so that the membrane is pushed against and firmly supported by a nylon mesh. Volume flow across the membrane would thus result in a volume decrease in the other compartment, if this change were not prevented by an automatic addition of liquid. Each volume decrease interrupts the electrical contact between a platinum wire and the liquid in a capillary, separated from the compartment by a rubber membrane. The interruption operates a servomechanism of low inertia, by which liquid is injected into the compartment from a microsyringe. The microsyringe axis holds a potentiometer which records the voltage proportional to the injected volume. The injection is interrupted every minute and the servomechanism returns the potentiometer through a differential to zero position.

14. USE OF ARTIFICIAL MEMBRANES

14.1. INTRODUCTION

The intriguing properties of cell membranes have stimulated attempts at modelling a functional membrane proceeding from synthetic components or greatly simplified naturally occurring constituents. A number of laboratories are now engaged in experiments with the interfaces between bulk aqueous solutions and lipid, micellar, and colloidal suspensions, with monolayers of lipid, protein, and synthetic polymers at the air–water interface, with filter-supported lipid in an aqueous medium, and with aqueous soap films in air.

By far the greatest attention has been devoted to artificial bilayer membranes formed in an aqueous medium from amphiphilic lipids (Mueller *et al.*, 1962). The principal component of these membranes are phospholipids; in the hydrated form these lipids give rise to lamellar forms in which the molecules are arranged in layers with regularly arranged polar groups. These membranes thus represent a direct analogy of cell membranes, both as to composition and as to structure (*cf.* chapter 1). Likewise, the thickness of these membranes resembles that of the plasma membranes (50–100 Å) and they are relatively easy to study under optimal observation conditions.

14.2. MEMBRANE MATERIALS

The lipids used for the preparation of artificial membranes are either purified chemical individuals or else multicomponent mixtures of lipids isolated from animal or plant materials. The purified lipids, in their turn, may be of natural origin or fully synthetic. Most frequently used are membranes containing lecithin isolated from natural sources but membranes formed from diglycerides, sphingomyelins, phosphatidyl serine, phosphatidyl inositol, and dioleyl-phosphatidyl choline have been used one time or another. The animal materials used for the extraction are either whole organs or their parts (beef or rat brains) or else isolated membrane fractions (erythrocyte membranes, synaptic vesicles, mitochondria, etc.). Of the plant materials used mention may be made of spinach chloroplasts (Ti Tien *et al.*, 1968) or of micellar suspensions of *Neurospora crassa* (Howard and Burton, 1968).

The lipids are dissolved in organic solvents, most often in a mixture of chloroform and methanol, or in hydrocarbons, such as *n*-decane. In spite of the fact that the composition of the solutions used for membrane preparation is chemically defined, one must be wary of possible differences in the composition of the membrane formed with respect to the original solution. Thus, using labeled phosphatidyl ethanolamine and cholesterol in decane, it has been found that the ratio of the two lipids in the membrane is the same as in the decane solution but the molar ratio of the solvent to the dissolved substances is 10:1 while in the solution it is 700:1 (Henn and Thompson, 1968).

Prior to membrane formation, it is usual to add to the solution an excess of hydrocarbons or fatty acids, or else cholesterol and α-tocopherol. In the first case, (when adding squalene, mineral oil, caprylic acid) the compounds added may serve at the same time as solvent for the principal components. In the second case, cholesterol enhances the stability of the membrane and increases the dielectric strength while α-tocopherol accelerates the formation of the secondary black membrane and also acts as an antioxidant.

The thick newly formed lipid membrane and the thinning membrane contain a continuously decreasing amount of lipid. The amount may be determined from the volume of the membrane if the surface area and the thickness in the individual phases are measured. These parameters can be determined by optical methods but these require more time than is usually available if the measurement is to be of significance. Once the bilayer membrane is formed such measurements are more reliable. Thus, a membrane of surface area of 1 mm^2 and a thickness of 50 Å has a volume of 5×10^{-12} liters. If the area of a lipid molecule situated in the bilayer membranes is assumed to be 50 Å^2 the membrane will contain 4×10^{12} molecules (6.7×10^{-12} mol) lipid (see Howard and Burton, 1968).

Lipids are present in cell membranes as well as in other structures derived from natural sources mostly as lipoproteins. Water molecules play an important role in the combination of the lipid chains with proteins. For this reason, dehydration agents, such as ethanol, methanol, acetone, etc. are used for the extraction. Most lipids are not readily soluble in these solvents and hence some less polar solvents had to be added, such as chloroform, ether, petroleum ether. Some lipolytic enzymes are activated by the commonly used solvents, in particular at a higher temperature, and hence it is usual to work at temperatures only slightly above zero (see Hanahan *et al.*, 1960). The extraction mixture is protected from direct sunlight (induction of peroxidative changes) and most operations are carried out in an inert gas in the presence of antioxidation agents, such as α-tocopherol, mercaptoethanol, or dithiothreitol (see Howard and Burton, 1968).

The mixture of lipids in the organic solvent is washed with water so as to remove some of the impurities such as amino acids, urea, and inorganic ions. The classical extraction procedure is the one by Mueller and co-workers (1963) modified from Folch and Lees (1951).

The white matter of fresh beef brain is homogenized in a nitrogen atmosphere for 2 min at 4°C in a mixture of chloroform and methanol (2:1) with an addition of 0.01% α-tocopherol. The mixture is centrifuged and filtered, the filtrate emulsified by adding water, and the emulsion evaporated to dryness under nitrogen at reduced pressure. The residue is dissolved in chloroform, filtered, and methanol is added. The filtrate is again emulsified with water and evaporated and the procedure repeated several times to remove the last traces of protein. The residue after one cycle contains still 0.6% protein, after three cycles only 0.15%. The solution is then washed with water, the layer of organic solvents is filtered and the solution mixed with 1/3 its volume of methanol. In this form, the solution may be kept under nitrogen for several months, in sealed ampoules for a year. It is also possible to store the freeze-dried lipids in small ampoules in amounts required for a single experiment. The lipid content of the residue is usually determined by analysis of phosphate.

14.3. FORMATION OF MEMBRANES

Artificial membranes are usually prepared in a rather simple apparatus consisting of two half-chambers filled with an aqueous medium and separated by a partition with a small opening. The most common procedure is again that due to Mueller and co-workers (1963) and is called the *painting method*. The opening (usually 0.5–5 mm²) over which the film is to be formed is smeared with the lipid solution (plus the additive agents) with a fine brush. The formation of the membrane is observed in reflected light at a 10- to 15-fold magnification. The membrane first formed is thick and displays interference colors. After a while, the so-called secondary black begins to form, usually at several places simultaneously. The membrane then thins out spontaneously to a bimolecular leaflet (Fig. 14.1) and the edge becomes annular. The use of the brush is not particularly recommended as it may easily contaminate the material. A plastic spatula, on the other hand, is readily cleaned and may be used for years (see Howard and Burton, 1968). It is also possible to use a syringe filled with the lipid solution if a short piece of plastic tubing is attached to the needle tip (Läuger *et al.*, 1967).

Other techniques described include the *marginal suction method* wherein the membrane is formed in a Teflon ring with radial boring and provided with a syringe needle filled with the lipid solution. The whole device is immersed in the aqueous solution. By injecting the lipid solution to fill the ring a biconcave disc is formed inside. This disk is then flattened by sucking off the excess material, whereby a bimolecular membrane is formed spontaneously (see Howard and Burton, 1968).

A simpler method is the *dipping* one which is used mostly for testing the efficacy of membrane. A loop made of wire or a glass or plastic fiber is immersed in the lipid solution and then transferred to a vessel with the aqueous solution (see Howard and Burton, 1968).

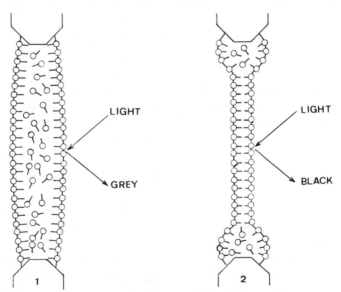

FIG. 14.1. Two stages of thinning of a lipid membrane. **1** Thick lipid membrane, **2** bilayer lipid membrane.

Finally, mention should be made of the *bubble-blowing method* wherein the syringe with a blunt needle filled with the aqueous medium is introduced into the lipid solution so as to trap a surface layer of lipid material. Then it is immersed into a vessel with the aqueous solution and a bubble is blown (Mueller *et al.*, 1964).

14.4. DESIGN OF EXPERIMENTAL CHAMBERS

The principle of artificial membrane chambers is very simple. The design most often used is a split chamber divided by a suitable partition with an opening into two compartments filled with the aqueous medium (Fig. 14.2). The membrane is formed at the partition opening, thus separating the two aqueous solutions. The chamber of this type is easily made from a polyethylene cup by drilling into its wall an opening about 2 mm in diameter. The cup is then placed in a dish filled with the aqueous solution. The membrane is smeared over the opening and observed in reflected light. Appropriate electrodes for electrical measurements can also be introduced into the cup.

To facilitate work with artificial membranes more complicated chambers are frequently constructed. These can be taken apart; the partition separating the two compartments is replaceable and its opening is heat-polished. The chambers contain special wells for magnetic stirrers, thermometer, electrodes, tubing for perfusion, removal of samples, or addition of reagents. The whole chamber may be enclosed in a constant-temperature jacket. A good chamber must permit easy access to all the components and simple washing to avoid chemical contamination; each component should be readily replaceable. A good chamber should permit optical measure-

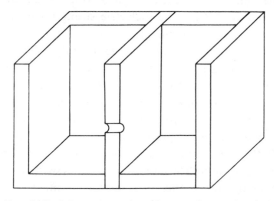

Fig. 14.2. Schematic section of a membrane chamber.

ments, it should be screened from side light and from electric fields, it should have built-in temperature control and be shock-resistant. It is essential that the two compartments be fully insulated with respect to passage of ions and electrons save across the membrane in the partition opening. This requires that the device be constructed from nonconductive material and that all the accessories which come into contact with the solutions (electrodes, thermometers, thermistors, stirrers) must be perfectly insulated from earth and from each other.

The choice of materials for chamber construction is restricted by the requirement of chemical inertness toward both the aqueous and the lipid phase, sufficiently low conductivity, high transparency, and, last but not least, good workability. Although glass would seem to meet all these requirements, difficulties due to its poor workability prevent it from being used more widely for chamber construction. Moreover, for some reason, the membrane material does not adhere to glass as readily as it does to other materials (e.g., teflon). For the partitions between chambers it is more common to employ polyethylene, polypropylene, and teflon which possess the desired chemical and electrical properties, are easily shaped but, on the other hand, are not transparent, are difficult to cement, and actually are only semisolid. The material of choice for the chamber itself is polymethacrylate (lucite and the like) which is a nonconductor, is transparent, workable, cementable, and fairly resistant to organic solvents.

14.5. ELECTRICAL MEASUREMENTS

The principal parameters measured on simple phospholipid membranes are electrical resistance R_m and capacity C_m which form here a parallel combination, and the time constant τ which is related to the former two according to $\tau = R_m C_m$. The dielectric strength can be estimated as the potential V/cm at which the membrane bursts.

The electrical methods used for membrane analysis include transitional direct current or alternating current determinations either at constant current or at constant voltage conditions. The D.C. measurements under constant-current conditions are most often used for their simplicity and low susceptibility to error. In the technique used, the D.C. pulses pass through a current-limiting resistor into electrodes placed at both sides of the membrane and further through the membrane. The potential drop at the membrane resistance is detected by another pair of electrodes with a D.C. amplifier with a high input resistance and, after amplification, is recorded photographically on an oscilloscope. The arrangement is shown schematically

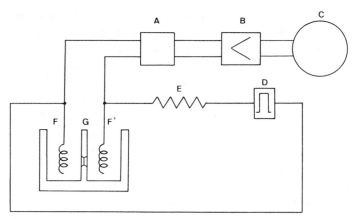

FIG. 14.3. Electrical measurement circuit. A Cathode follower, B amplifier, C oscilloscope, D pulse generator, E resistor, F, F' electrodes, G membrane partition.

in Fig. 14.3. The current passing through the circuit and thus through the
membrane can be recorded either as voltage drop on a resistance of known
magnitude or directly with the aid of a differential amplifier and hence
conducted to the second oscilloscope trace. The membrane resistance R_m
can then be computed from the simultaneously measured current I and the
potential drop across the membrane V_m from Ohm's law ($R_m = V_m/I$) and
normalized by multiplication with membrane area A so that it is expressed
in $\Omega \cdot cm^2$. The specific resistance can be calculated from the value of the
resistance R_m by using the relationship $\varrho = R_m A/\delta$ where δ is the assumed
thickness of the membrane, i.e., 50 Å. The membrane resistance and the
current passing through it can be estimated without directly measuring
the current if the output potential of the source and the value of the resistance in series with the membrane are known. The voltage drop at this
resistance represented by the difference between the voltage of the source
and the voltage drop across the membrane determines the current which
passes through the resistance and hence also the current passing through
the membrane $I = (V_S - V_m)/R_s$ and, since the membrane potential is
being measured, its resistance is $R_m = V_m/I$.

The time constant τ is found from the course of the transmembrane
potential V_m as time required for achieving $0.632\, V_{m,\max}$ with a pulse lasting at least 5τ. The membrane capacity can be calculated from $C_m = \tau/R_m$.

Methods of measurement with a sinusoidally alternating current are
suitable for a continuous registration of changes of electrical parameters
during membrane formation, thinning out, bilayer plateau, and membrane

breakdown. In this type of measurement, an alternating current passes through the membrane and brings about an alternating potential decrease which is amplified by an amplifier responding only to A.C. voltage, rectified, and registered as a value proportional to membrane impedance $Z = R/(1 + j\omega C_m)$ where $\omega = 2\pi f$ and f is the frequency of the voltage used. It is an advantage of the method that the measurement is not affected by any D.C. potential differences applied simultaneously across the membrane.

By adsorbing suitable compounds on the bimolecular lipid membrane, one can alter its electrical properties and one can observe electrokinetic phenomena resembling a cell action potential. These phenomena are most easily studied by means of the above-mentioned D.C. pulse analysis. At constant current conditions, the oscilloscope shows the time dependence of the transmembrane potential which corresponds to the change of membrane resistance occurring above a certain threshold potential when the membrane resistance jumps to a different stationary state. If both the current and the voltage across the membrane are measured one can compute the current–voltage characteristic, i.e., dependence of the passing current on the applied voltage. This characteristic may also be measured directlyunder' constant-voltage conditions when the transmembrane potential is maintained constant by eliminating the limiting resistance so that one then measures only the current passing through the membrane.

14.6. ELECTRICAL MEASUREMENT APPARATUS

The instruments used for electrical measurements on artificial membranes can be divided into two groups. The first are those supplying the current to be applied to the membrane, the other are those recording the magnitude and shape of the voltage drop on the membrane. The voltage sources may be either A.C. or D.C. Commercial generators of sinusoida-frequency are mostly used as A.C. sources, as they permit regulation ol both the voltage and the frequency over a wide range. As D.C. sources, use is made of batteries with proper resistors and switches as well as of rather complex electronic square-wave generators which produce pulses of D.C. voltage of an exactly known shape and duration at exactly preset intervals. It is important that the source has a high output resistance (which may be achieved even in a low-resistance instrument by inserting in the circuit a resistance, the magnitude of which exceeds substantially that of the measured fobject). In this case the current passing through the circuit will be deter mined by the additional resistor so that changes in resistance of the measured object have but a negligible effect on this current. This effect will be the lower the smaller the resistance of the measured object and the greater the resistance in series. Then, if the resistance of the measured object changes, it will show a change of potential proportional to its change in resistance, since the current remains effectively constant.

For recording the changes of the transmembrane potential one must amplify the voltage, its original value being too low for common recorders (about 150 mV). The amplifier with the input directly connected with electrodes on each side of the membrane must also have a high input resistance to reduce as much as possible the shunting current flowing through the input resistance of the amplifier and to prevent distortion of the results. If the membrane circuit contains a current-reducing resistor it will be considerably greater under constant-current conditions than the membrane resistance and will form, together with the parallel combination of resistances of the membrane and of the instrument input, a resistance divider. A low input resistance of the amplifier will thus depress the transmembrane potential and much of the current will pass through the input circuit of the amplifier rather than through the membrane. Consequently, the effect will be minimized by low input resistance in the amplifier. The input circuit of the amplifier must also have a low capacity, in combination with the high resistance of the circuit; this capacity will give rise to a large time constant. The requirements are best met by amplifiers with their input circuit arranged as a cathode follower where R_{input} may attain 10^{15} Ω and C_{input} 1–2 pF.

The measuring instrument is connected with the solutions by salt bridges and electrodes which convert the electron current to an ion one and vice versa. In most cases, use is made of silver–silver chloride or calomel electrodes as are described in the chapter on bioelectric measurements.

14.7. PROPERTIES OF ARTIFICIAL MEMBRANES

14.7.1. Electrical Properties

The electrical parameters of artificial membranes resemble those of natural ones only if the lipid membranes are modified by other substances usually adsorbed on their surface. Nonmodified membranes represent a parallel combination of resistance and capacity and their resistance in media containing NaCl or KCl may attain 10^6–10^9 $\Omega \cdot cm^2$ but its value is difficult to reproduce and depends probably on the presence of trace impurities. If, however, the chloride anion in the medium is replaced by iodide, the resistance may drop by three orders of magnitude. On restoration of the original chloride medium the resistance will rise to approximately the initial value. The resistance drop is independent of the cation present (Läuger *et al.*, 1967 *a*). The potential differences produced by salt gradients indicate, on the other hand, that the membranes are more permeable to cations than to anions (Andreoli *et al.*, 1967 *a*).

The current–voltage characteristic of nonmodified membranes is linear over a range of ± 50–150 mV. The capacity of the membrane corresponds to the biological ones, lying between 0.4 and 1.4 $\mu F/cm^2$, indicating an identical thickness of the membrane. The capacity shows a frequency independence over a wide range, is fairly reproducible and rises with D.C. polarization (by 10% for every 100 mV). This induced potential difference compresses the membrane and its capacity, being a function of thickness, will rise. The membrane capacity depends linearly on the membrane area (Hanai *et al.*, 1965).

The properties described so far can be modified by adding various proteins or polypeptides which are adsorbed to the membrane surface and alter its properties as was observed by Mueller and Rudin. These materials decrease substantially the membrane resistance, alter the current–voltage characteristic, and give rise to electrokinetic phenomena on the membrane. Thus, for example, an extract of *Aerobacter cloacae*, called the excitability-inducing material, when added to one of the membrane sides, decreases the electrical resistance of the membrane from $10^8 \, \Omega \cdot cm^2$ to $10^3 \, \Omega \cdot cm^2$ (Mueller and Rudin, 1967), i̇.e., to a value comparable with that of the plasma membrane. With respect to jump changes of the current and voltage (as long as they are subthreshold) the membrane will behave as a simple parallel combination of resistance with capacity, but above certain threshold values of these jumps a 2- to 10-fold restorative rises of resistance may be observed. This is the so-called half-action potential. When the current is interrupted the membrane resistance will drop reversibly to its original value. The membrane resistance thus fluctuates by jumps between two values depending on the applied voltage (current). In the presence of the excitability-inducing material the membranes are cation-selective but do not distinguish between Na^+ and K^+. With membranes containing sphingomyelin the material will give rise to a fully developed action potential which has the same magnitude and time course as the action potential of biological systems, and possesses all the electrokinetic features observed in cell action potentials (Mueller and Rudin, 1967). Repetitive firing and rhythmic firing may be observed at a suitable composition of the membrane material, proper setting of the salt concentrations and the amount of the excitability-inducing material. These action potentials can be blocked reversibly by local anaesthetics at physiological concentrations (e.g., with cocaine).

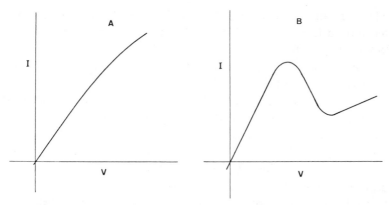

Fig. 14.4. Current–voltage characteristics of artificial membranes. **A** Nonmodified membrane, **B** characteristic of modified membrane with section of negative resistance.

An important feature of the artificial membranes is the change of their current–voltage characteristic (Fig. 14.4), in particular the appearance of a section of this characteristic displaying the so-called negative resistance (Mueller and Rudin, 1963, 1968). This particular section is found with all excitable biological membranes. Similar results can be obtained with membranes composed of nonanimal materials and modified in various ways (*cf.* Monnier *et al.*, 1965; Monnier, 1968). Cyclic peptides (like valinomycin) can bring about different degrees of ionic selectivity, the membranes then being able to distinguish between individual cations, favoring the permeability to potassium over that to sodium by a ratio of 300:1 (Andreoli *et al.*, 1967b; Lev and Buzhinsky, 1967). For a treatment of the conductivity mechanisms of artificial membranes the reader is referred also to Eisenman (1968), Tosteson (1968) and Liberman and Topaly (1968).

14.7.2. Permeability

The water permeability of artificial membranes has been measured by two methods. The first is based on estimating water flow along an osmotic gradient. Using a Tris buffer medium, one side of the membrane was gradually fed with a concentraced solution of urea through a micrometer syringe. To maintain the planar state of the membrane the other side received Tris to replace the water loss. The micrometer syringe volumes were read as a function of time and the volume flow of water was determined for each urea concentration. The values of the water permeability constant P_{H_2O} obtained by this method vary between 3 and 24 $\times$ 10^{-4} cm/sec. Other modifications of the osmotic method yielded values of 104 $\times$ 10^{-4} cm/sec (see Howard and Burton, 1968). A method based on the diffusion of tritiated water applied to membranes from beef brain and human erythrocytes gave values of 3–4 $\times$ 10^{-4} cm/sec.

Permeability to labeled urea of membranes from human erythrocyte lipids was found to be 5.9 $\times$ 10^{-7} cm/sec (Wood *et al.*, 1968).

The permeability constant of D-glucose was tested by using 0.7 *M* ^{14}C-labeled sugar and human erythrocyte lipids and found to be 5.1 $\times$ 10^{-8} cm/sec which contrasts with the value found in intact human erythrocytes, 1 $\times$ 10^{-4} cm/sec. Moreover, the artificial membranes showed no depression of glucose permeability by phlorizin. Using lipids from rabbit erythrocytes, the permeability constant for glucose was similar to the *in vivo* value (5 $\times$ 10^{-8} as compared with 3 $\times$ 10^{-8} cm/sec) (Wood *et al.*, 1968).

Permeability to acetylcholine, salicylamide, and some other compounds was also tested on artificial membranes but the results are not very promising.

At any rate, there is evidence that lipids alone cannot fulfil the function of natural membranes and that proteins are required for the process. At present, the problems of attaching suitable proteins to suitable membranes at suitable concentrations appear formidable but there is no doubt that this is the only way to modeling functional transporting systems of living cells.

15. ASSAY OF TRANSPORT PROTEINS

Because of the lack of any chemical reaction catalyzed by the proteins involved in membrane transport (with the possible exception of some of the HPr system components described in chapter 8) the identification of such proteins *in vitro* is based solely on their binding affinity for their natural or modified substrate.

A number of methods for the estimation of such binding exist, such as *equilibrium dialysis* and *ultrafiltration, ultracentrifugation, conductometry, "column chromatography"* with bound protein or with bound substrate, and some special techniques.

For all these techniques, with the exception of conductometry and the special ones, the use of labeled binding substrates is a necessity because the lower the concentration of the substrate the greater the relative binding will be.

15.1. EQUILIBRIUM DIALYSIS AND ULTRAFILTRATION

In *equilibrium dialysis*, the solution of protein is separated from that of the ligand (the small molecule or ion) by a semipermeable membrane, such as the 8/32 Visking tubing or another similar brand of cellophane, with pores not greater than about 50 Å, preventing the passage of compounds above 10,000 daltons. The solutions are then agitated for several hours until equilibrium is established. Then, using the symbols of the theoretical section, the concentration of free substrate s will be the same at the protein side of the membrane (I) and at the nonprotein side (II).

However, the estimation of total S will include at side I a certain amount
of S bound to the protein C, cs, as follows from the dissociation equation
of a protein–substrate complex with a single binding site

$$K_{\text{diss}} = (c_t - cs)s/cs \qquad (1)$$

Then $cs = c_t s/(K_{\text{diss}} + s)$ which is identical with the classical Michaelis–
Menten formula and, therefore, can be plotted in any of the reciprocal
versions described on p. 234, e.g., $1/cs$ against $1/s$, yielding the important
parameters K_{diss} and c_t. It has become a convention in studying binding

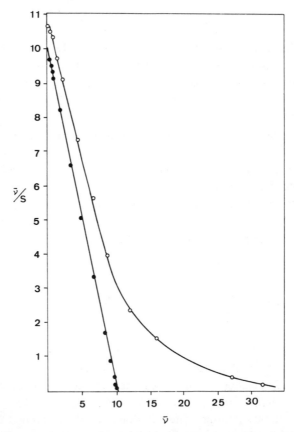

Fig. 15.1. A Scatchard plot showing the relationship between the fraction of occupied
binding sites and the substrate concentration. The straight line with solid circles corresponds to binding to a single set of binding sites, the curve with open circles to two sets
of binding sites with different affinities. (From Edsall and Wyman, 1958.)

interactions to designate the fraction of the protein molecules which bind substrate, i.e., $cs/(cs + c) \equiv s/(s + K_{diss})$, as $\bar{v}$. If a protein molecule contains n independent identical binding sites

$$s/(s + K_{diss}) = \bar{v}/n \qquad (2)$$

Scatchard (1949) plotted $\bar{v}/s$ against $\bar{v}$ (Fig. 15.1) to obtain a straight line with the abscissa intercept equal to n, the maximum number of binding sites, the slope of the line being equal to $-1/K_{diss}$. This type of plot is useful even if the protein molecule contains more than one independent but unequal binding site. As s increases, $\bar{v}/s$ tends to zero but $\bar{v}$ approaches the limiting value of $n = \Sigma n_i$ (where n_i is the number of one type of binding sites). Due to the formal identity of the expressions with those of enzyme kinetics, plots describing positive or negative interactions in transport may be applicable to the examination of static binding.

Ultrafiltration is based on the same principle as equilibrium dialysis but here the ligand is mixed with the protein solution directly and pressure is applied to separate the protein-containing solution from the filtrate again across an ultrafiltration cellophane membrane. With advanced ultrafiltration techniques including protein-retentive centrifuge cones individual separations can be achieved in a matter of minutes, thus obviating the danger of contamination or breakdown during the lengthy equilibrium dialysis. The only complicating factor as compared with equilibrium dialysis is the necessity of calculating the affinity constant and the total number of binding sites from a set of quadratic equations since here the volume of solution without protein is not larger than that with protein (as is generally the case in equilibrium dialysis) so that the decrease in ligand concentration due to binding cannot be neglected. Then

$$K_{diss} = \frac{(c_t - cs)(s_t - cs)}{cs} \qquad (3)$$

The equation is easily solved for two or more values of s_t but the reciprocal plot of $1/cs$ against $1/s$ does not yield a straight line.

In both the equilibrium dialysis and the ultrafiltration techniques one should observe the effect of the electric charge of the protein on the distribution of ions about the membrane which would result in a Gibbs–Donnan equilibrium (see p. 95). This effect (if the binding of ions is tested) can be minimized in ultrafiltration by keeping the filtrate amount as small as practicable.

15.2. ULTRACENTRIFUGATION

If either a crude preparation or a purified transport protein is available one can examine the sedimentation profile of the ligand in the presence of protein in an (ultra)centrifuge tube and assess the amount of binding. Obviously, if the ligand is of greater size, its own sedimentation has to be taken into account but then only simple corrections are needed.

15.3. COLUMN CHROMATOGRAPHY

The technique described by Bobinski and Stein (1966) consists of adsorbing the protein examined to a firm support in a chromatographic column (e.g., DEAE–cellulose) and running through it a solution of the ligand, together with a similar but differently labeled substance, such as ^{3}H-D-glucose and ^{14}C-L-glucose. If the material shows an affinity for the ligand but not for the differently labeled analogue the former will emerge from the column somewhat later than the analogue. Knowing the concentration of the ligand and the amount of the bound protein, the retardation can serve as a semiquantitative measure of the affinity.

Another technique using a solid support for one of the interacting substances is that described by Hummel and Dreyer (1962) and developed by Fairclough and Fruton (1966). Unlike in Bobinski's technique, the small labeled ligand molecule is fixed here in a Sephadex or Bio-gel column by running its solution through the column. The protein is then placed in the same solution of the ligand and applied to the column. A peak of radioactivity superposed on the background level emerges at the exclusion volume of the column, corresponding to the CS complex. A trough of equal size appears at some distance after the peak as corresponds to the amount of S bound to C (Fig. 15.2). This elegant method is applicable to a wide range of binding interactions but attention should be called to a correction required when drawing quantitative data from the area of the trough mentioned above. The number of μmoles bound is obtained as the difference between the apparent amount of μmoles (calculated from the radioactivity corresponding to the trough area) and the expression $AC(w + \bar{V} + \bar{V}w)/b$ where A is the baseline radioactivity in (counts/min)/ml, C the amount of protein used in g, w the water content of the protein sample expressed as a decimal fraction, $\bar{V}$ the partial specific volume of the protein which may lie between 0.7 and 0.8 ml/g and b the counts/min per μmole of ligand under standard conditions of measurement.

15.4. CONDUCTOMETRY

If the binding of an ion, particularly one of a high transference number, to a protein is investigated, one can measure the decrease in conductance of the solution subsequent to the binding. The equations derived for a quantitative assessment of the number of ions bound to protein are still partly empirical and can be used if only one species of ion is bound to the protein (e.g., Na^+ or Cl^-) (Scatchard et al., 1950; Karpenko et al., 1968).

The specific conductance of a solution containing the salt A^+B^- will be

$$\varkappa_1 = \varkappa_{A+} + \varkappa_{B-} \tag{4a}$$

If a protein with cation-binding properties is added to the solution its specific conductance will change to

$$\varkappa_2 = \varkappa'_{A+} + \varkappa_{B-} + \varkappa_{C+} \tag{4b}$$

where the contribution of the slow large protein molecule $\varkappa_{C+}$ can be neglected. Without doing injustice to the sophisticated derivation we may note the useful formula emerging from it for the number of cations $\bar{\nu}$ bound to a protein molecule:

$$\bar{\nu} = \frac{(\varkappa_1 - \varkappa_2)(V_o + V_s)}{nl_{A+}} \tag{5}$$

where V_o is the original volume of solution and V_s the volume of salt solution added, n is the number of equivalents of independent binding sites (possibly identical with the number of moles of protein), and l_{A+} is the equivalent conductivity of the cation ($mho \cdot cm^2 \cdot mol^{-1}$).

15.5. OTHER METHODS

Rather specific techniques can be occasionally used for the determination of binding of a small molecule to a large one, among them spectroscopy—if a dye changes its spectrum upon binding to a protein (e.g., Klotz, 1953), and nuclear magnetic resonance—if an ion is attached to a large molecule (e.g., Cope, 1965).

Comparative Aspects

16. BACTERIA

16.1. INTRODUCTION

The number of microorganisms classed under bacteria in the broad sense is of the order of 10^3 and it would be presumptuous to make attempts at generalizing their membrane transport mechanisms. In fact, only about a dozen bacteria have been examined in greater detail in this connection and the transport properties of literally no single one of them are understood in their entirety. The selection of bacterial objects for study has been largely fortuitous and there is no certainty that the existing transport data are typical or representative for the whole group of bacteria. Thus, for instance, of the relatively sizable group of strict anaerobes none has been examined for its transport parameters although some novel mechanisms might be operative there. Still, the few better known species display some characteristics in common, in particular (1) the distinctly active character of all the transports studied, including those of monosaccharides, disaccharides, amino acids, organic acids, and inorganic ions, and (2) the definite association of the transport proteins with a genetic locus on the bacterial chromosome.

It was on the basis of studying the bacterium *Escherichia coli* that the concept of permeases was advanced which by now has proved its worth both in terms of concrete observations and as a stimulant of vigorous disputations.

Genetics played a significant role in the elucidation of the permeability of *E. coli* in that mutants could be prepared which possessed no active transport of a given substrate although the substrate could be metabolized by metabolic enzymes. This phenomenon of *crypticity* was striking with lactose in a mutant deficient in lactose "permease" and has been since repeatedly observed. Now that the existence of either constitutive or, very often, inducible, proteins catalyzing active transport of solutes into bacterial cells has been established there is no need to trace the history of the permease theory but, in spite of its general acceptance, the molecular mechanism by which permeases function is not understood and no unanimity exists in the probable interpretations. However, whatever the mechanisms of permease action may be, it seems highly probable that, along with the permease or as a mechanistic component of it, the transport of the various substances in bacteria proceeds by carrier diffusion, the parameters of which have been established in many individual cases.

In the following, we shall discuss briefly the transports of the various classes of compounds as long as they have been studied, the greatest amount of information being available from *E. coli, Staphylococcus aureus, Salmonella typhimurium,* and *Streptococcus faecalis* but also from *Aerobacter aerogenes, Streptomyces hydrogenans,* and several others. It is an unfortunate circumstance that numerous strains of the same species, this being particularly true of *E. coli,* have been used by different authors and the results rather indiscriminately compared. It is a general experience that there may be not only quantitative but also qualitative differences between individual strains and one must be wary when examining data reported by different authors.

16.2. SUGARS

16.2.1. *Escherichia coli*

The transport of mono- and disaccharides in *E. coli* is associated with a whole array of genetically defined permeases so that one almost wonders about the usefulness of such a variety of systems, some of which do not even seem to be able to transport any substrate of physiological importance. Table 16.1 gives an insight into the inducer and transport specificities of four such permeases (Rotman *et al.,* 1968).

The *methylthiogalactoside* permease I is apparently identical with the original β-galactoside permease treated in detail by Rickenberg and co-workers (1956) and by Kepes (1960). Kinetically, the transporting system is one of mediated diffusion with superimposed active transport which can be suppressed by metabolic inhibitors such as azide or dinitrophenol. The half-saturation constants of uptake by *E. coli ML 308* at 26°C are 5×10^{-4} M for methylthio-β-galactoside, 2×10^{-5} M for β-thiodigalactoside, and 2.5×10^{-4} M for phenylthio-β-galactoside. In a galactosidase-less mutant, lactose had a K_T of 9×10^{-4} M. The maximum accumulation ratios may

Table 16.1. Inducer (I) and Substrate (S) Specificities of Some Sugar Permeases of *Escherichia coli K12* and Derivative Strains (in Relative Units of Efficiency)

Sugar	Galactose permease [a]		Methylgalactoside permease		Methylthiogalactoside permease I		Methylthiogalactoside permease II	
	I	S	I	S	I	S	I	S
D-Galactose	100	100	2	?	+	?	18	?
Methylgalactoside	32	1	10	100	—	3	0	84
Methylthiogalactoside	27	0	13	4	+	20	0	100
Lactose	—	0	—	3	—	100	—	3
o-Nitrophenylgalactoside	—	0	—	0	—	—	0	0
D-Fucose	60	0	100	12	?	3	0	0
Melibiose	85	—	18	—	+	—	100	—

[a] Can be induced also by D-arabinose, L-arabinose, L-galactose, and maltose.

attain values of 400:1. A somewhat peculiar aspect of accumulation of methylthiogalactosides is that they are transported inward up to a certain limiting concentration (e.g., methylthio-β-galactoside up to 160 μmol/g dry weight). This behavior is reminiscent of that observed in uphill-transporting yeasts (*Rhodotorula gracilis, Candida beverwijkii*, etc.) and its explanation is by no means simple. It would seem as if the cell were invested with a certain adsorption capacity for the sugar in question but this explanation must be ruled out because the sugars are intracellularly in an osmotically free state and, moreover, empty bacterial membranes show the same phenomenon. Hence one is compelled to resort to a complicated feedback-type regulation of the uptake rates such as suggested by Ring and co-workers (1967) or to an excess-substrate inhibition as described in the theoretical section (p. 143).

The uptake of methylthiogalactosides (and probably of other permease-transported sugars) appears to be associated with a splitting of ATP in a 1:1 stoichiometric ratio with concomitant increase in O_2 uptake by the bacteria. The coupling of the transport process with metabolic energy (in *E. coli ML 308-225*) appears to consist of increasing the half-saturation constant of the exit reaction by as much as two orders of magnitude ($K_{ZS} > K_{CS}$, cf. p. 139) (Wilkler and Wilson, 1966). The half-saturation constant for the entry process is unaffected by metabolic coupling and/or addition of metabolic inhibitors (Table 16.2). This type of metabolic cou-

Table 16.2. K_T Constants for β-Galactoside Transport in *Escherichia coli*

Sugar and inhibitors	Entry, mM	Exit, mM
Lactose	0.8	16
Lactose + 30 mM NaN₃ + 1 mM iodoacetate	1.0	0.7
o-Nitrophenylgalactoside	0.9	20
o-Nitrophenylgalactoside + 30 mM NaN₃ + 1 mM iodoacetate	1.5	1.3

pling is sometimes called the *pull* mechanism, in contrast with the *push* mechanism where metabolic energy is used to reduce the half-saturation constant of the entry, leaving the exit unaffected.

On the other hand, an approach based on the estimation of the proportion of carrier molecules associated with galactoside in *E. coli ML 30* (using the fact that galactoside binding protects the carrier from the inhibition by N-ethylmaleimide) indicates that the role of energy coupling lies in increasing the affinity of the carrier for entry rather than decreasing the affinity for exit (Schachter *et al.*, 1966). The problem is difficult to resolve at present but, if analogies with other systems (amino acids, sugars in yeast) are permitted, the pull mechanism appears to be perhaps more likely.

It has been suggested by Pavlasová and Harold (1969) that the membrane of *E. coli* must be proton-impermeable to effect active transport of galactosides.

A galactoside-binding protein has been identified and isolated from *E. coli* membranes (p. 196).

The *galactose* permease is an entity distinct from the galactoside permeases, showing a K_T for D-galactose of $7 \times 10^{-6} M$ and a maximum accumulation ratio s_{II}/s_I of about 10,000:1 (Horecker *et al.*, 1960). Actually, at very low D-galactose concentrations, the function of another transport mechanism is indicated (in *E. coli ML 32400*) with a K_T of $5 \times 10^{-7} M$ and maximum accumulation ratios of 20,000 (Rotman and Radojkovic, 1964). Overlapping specificities of several carriers (or permeases) in microorganisms are a general phenomenon so that it is likely that only one of the above K_T's belongs to the galactose permease proper (on the basis of affinity of an isolated galactose-binding membrane protein from *E. coli* the value of $5 \times 10^{-7} M$ is the more acceptable one). The transport of

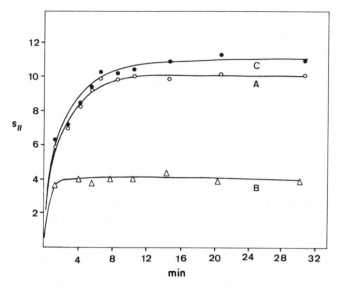

FIG. 16.1. Uptake of thiomethylgalactoside by *Escherichia coli W 2244*. s_{II} is expressed in 10^3 counts/min. **A** Untreated cells, **B** cold-shocked cells, **C** cold-shocked cells incubated with the heat-stable protein HPr (Kundig *et al.*, 1966).

galactose is one of the systems sensitive to the presence of phospho-enol-pyruvate (Kaback, 1968) as are described in chapter 8. The uptake of 10^{-5} *M* galactose by isolated *E. coli* membranes is stimulated by about 20-fold by an addition of 10^{-1} *M* phospho-enol-pyruvate.

Another important transport system of *E. coli* is that for *α-glucosides*. It is a constitutive system, capable of transporting glucose, α-methylglucoside, and probably a number of related sugars, such as mannose, 2-deoxy-glucose, etc. The transport of α-glucosides is not inhibited by fructose or galactose but is inhibited by metabolic poisons like iodoacetate. Glucose was found to compete with α-methylglucoside uptake four times more efficiently than with its exit (Halpern and Lupo, 1966). Here again, an effect of the phospho-enol-pyruvate hexose-phosphotransferase system is apparent, the uptake being stimulated by phospho-enol-pyruvate in isolated bacterial membranes and restored to cold-shocked cells by the shock fluid (Kundig *et al.*, 1966, working with *E. coli W 2244*) (Fig. 16.1).

In spite of the interaction of glucose with α-methylglucosides, there seems to be another distinct *glucose* permease in *E. coli* which transports glucose, galactose, fructose, 2-deoxyglucose, D-fucose, but not 3-O-methyl-D-glucose (Rogers and Yu, 1962).

Other sugar permeases have been reported to operate in *E. coli* (Kepes and Cohen, 1962). A *glucuronide* permease, specific for compounds with a free carboxyl at C_6 can be induced by methylglucuronide. A *maltose* permease can be induced by maltose and is inhibited by sodium azide. *Ribose* uptake can be induced while ribose utilization is constitutive, this suggesting the existence of a ribose permease. Novotny and Englesberg (1966) described in *E. coli* B/r an inducible, energy-dependent system for the uptake of L-arabinose which was apparently capable of transporting also D-xylose, D-fucose, and D-galactose, judging from the inhibition of L-arabinose uptake. The K_T for L-arabinose was 1.3×10^{-4} M, V was 1171 μmol/min per mg dry weight. Here again, in common with the transport of galactosides, the intracellular level of L-arabinose tends to a constant value, irrespective of the extracellular concentration although the system has not been tested at concentrations where s_{II}/s_I ratios of less than unity might be predictable.

Winkler (1966) described an uphill transport system for hexose phosphates, inducible by mannose-6-phosphate, fructose-6-phosphate, and glucose-6-phosphate, all the sugar phosphates serving also as substrates for the system, together with 2-deoxyglucose-6-phosphate. The permease does not transport galactose-6-phosphate and α-methylglucoside phosphate.

The potential effect of phospho-enol-pyruvate on the permeases listed here has not been tested explicitly but at least the pentoses do not seem to require the system for uphill transport. On the whole, however, the role of phosphorylation in sugar transport in *E. coli* is incontestable and begins to show signs of ubiquity. Incidentally, the phosphorylation accompanying the membrane transport of many sugars brings up a semantic question. Is the transport of sugars in bacterial cells actually a primary active transport? If the sugar occurs at one face of the membrane in the free form while at the other face it is only phosphorylated, the subsequent hydrolysis occurring possibly in a different kinetic (and morphological?) compartment, the actual movement of free sugar into cells is not against its concentration gradient and hence not primarily active although it is directly coupled with metabolic energy.

16.2.2. Other Bacteria

The discovery of a phospho-enol-pyruvate-requiring mechanism for sugar transport in *E. coli* (see Kundig *et al.*, 1964, 1966) has prompted intensive examination of other bacterial species in this respect. The most conclusive evidence has been obtained from *S. aureus*, which has been investigated in a series of studies by Egan and Morse (1965a,b, 1966) and found to lose its sugar-transporting activity (for lactose, maltose, sucrose, trehalose, galactose, fructose, ribose; but not for glucose) by a single-gene mutation. They observed that lactose ($K_T = 5 \times 10^{-6}$ M), α-methylgluco-side (2.5×10^{-5} M), sucrose (K_T very high), and maltose (10^{-4} M) are accumulated as derivatives rather than as free sugars. Hengstenberg and co-workers (1967, 1968) showed them to be phosphates, the phosphorylation being effected by the HPr system of Kundig and co-workers (1966).

The phosphorylation of sugars during membrane passage is not restricted to the above-named but includes also thiomethyl-β-galactoside, the classical substrate of galactoside permease, the resulting product being a 6-phosphate (Laue and MacDonald, 1968). The further fate of these phosphates is a hydrolysis which, in the case of lactose, yields glucose and galactose-6-phosphate.

Other microorganisms have been shown to possess the HPr system, thus *S. typhimurium* (Simoni *et al.*, 1967), *Bacillus subtilis* (Kundig *et al.*, 1965), *A. aerogenes* (Tanaka *et al.*, 1967; Hanson and Anderson, 1968), and *Pseudomonas aeruginosa* (Phibbs and Eagon, 1969). In *A. aerogenes*, the transport system includes a two-componental enzyme II, one of the components being a high-molecular, the other a specifier protein which acts to increase the affinity of the enzyme complex for sugar, in this case for fructose. It is likely that this specifier protein is an essential component of all the HPr systems.

A rather unusual observation was made in *S. typhimurium* (Hoffee *et al.*, 1964) where the level of α-methylglucoside is increased by the presence of azide or dinitrophenol or under anaerobic conditions. The effect was found to be caused by a lowered exit from cells associated with the decreased level of ATP which acts as a modifier of the exit process. At the ATP concentrations which are optimal for metabolism, on the other hand, the uptake is distinctly depressed (*cf.* Kepes, 1964). Moreover, the uptake of α-methylglucoside by *S. typhimurium* is influenced not only negatively but also positively by other sugars (Table 16.3). It would be of interest to examine this phenomenon in the light of the recently described HPr systems.

Uptake of sugars has been demonstrated in a number of other bacteria, including some Gram-negative marine species (Hamilton *et al.*, 1966) and *Serratia*, *Azotobacter*, and *Pseudomonas* (Nicolle and Walle, 1963) but no attempts were made to analyze the process in detail. Wilkins and O'Kane (1964) described uphill transport of glucose and galactose by *S. faecalis* which was inhibited by uranyl ions and by *p*-chloromercuribenzoate but not by iodoacetate. A labile intracellular binding is invoked to account for the results obtained.

Table 16.3. Relative Effect of Various Sugars on the Intracellular Level of α-Methyl-glucoside in *Salmonella typhimurium* *

Sugar added	% level of MG	Sugar added	% level of MG
None	100	Galactose	92
Glucose	1.8	Lactose	100
α-Methylglucoside	6.8	L-Glucose	105
β-Methylglucoside	8.3	Trehalose	120
Maltose	68	Cellobiose	125
2-Deoxyglucose	82	L-Arabinose	130
Mannose	85	Fructose	130
Sucrose	86		

* Outside concentration of α-methylglucoside (MG) was 4×10^{-5} M, that of the other sugars 4×10^{-4} M. (After Hoffee *et al.*, 1964.)

16.3. AMINO ACIDS

16.3.1. *Escherichia coli*

The uptake of amino acids by *E. coli* is brought about by a set of constitutive proteins into an intracellular pool, an ill-defined term, employed to suggest that the intracellular amino acids are available for transport as well as for enzyme reactions and incorporation into proteins. Unlike the case of yeast species, all the amino acids pass through the pool before being incorporated into proteins (Britten and McClure, 1962). The distribution ratios (s_{II}/s_I) attained by various amino acids are very high (28,000 for valine, 14,000 for glutamic acid, 7300 for proline, and 2300 for aspartic acid).

At least nine different permeases (Kepes and Cohen, 1962; Halpern and Even-Shoshan, 1967; Piperno and Oxender, 1968) have been identified in *E. coli* cells, the substrates preferred by the individual transport systems being listed below. I: valine, leucine, isoleucine; II: alanine, glycine, serine, threonine; III: phenylalanine, tyrosine, tryptophan; IV: methionine; V: proline; VI: arginine; VII: histidine; VIII: lysine; IX: glutamic acid. There may still be other permeases for amino acids not listed here. The transport constants of some of the amino acids (for *E. coli K 12*) are shown in Table 16.4.

A remarkable feature of amino acid uptake by *E. coli K 10* (apparently shared by other microorganisms) is the fact that the uptake is repressible

16.4. Transport Parameters of Amino Acids in *Escherichia coli K12* *

Amino acid	K_T, μM	V, $(\mu mol/g)/min$
Alanine	9.2	2.34
Glycine	3.8	2.16
Isoleucine	1.22	1.92
Leucine	1.07	3.16
Valine	8.0	2.82
Methionine	2.3	0.78
Phenylalanine	0.7	1.50
Tryptophan	0.9	1.18
D-Alanine	8.3	0.72

* From Piperno and Oxender (1968).

by precultivation of the bacterium on the respective amino acid (Inui and Akedo, 1965). The uptake is shown to persist even in isolated *E. coli W* membranes and has been studied there in some detail (Kaback and Stadtman, 1968). It was found to depend on pH, aerobic conditions, and temperature and it is blocked by a number of inhibitors: 2,4-dinitrophenol, sodium azide, carbonyl cyanide *p*-trifluoromethoxyphenylhydrazone, potassium arsenate, potassium cyanide, iodoacetamide, and *p*-chloromercuriphenylsulfonate. However, the transport (of glycine, at least) is not affected by fluoride, ouabain, antimycin A, and chloramphenicol. It is noteworthy that a great majority of glycine carbon was soon converted to phospholipids when incubated at 37°C but not at 0°C. This observation should sound a note of warning with respect to interpretations of transport parameters in a simple manner, based on a plausible membrane model. Apparently, even in such simple systems as isolated cell membranes, the transported molecule may be subjected to a variety of interactions which can distort the results quite substantially. It is advisable in such cases to use nonmetabolized analogues, such as α-aminoisobutyric acid, azaserine, cycloleucine, and the like.

The complex nature of the amino acid transport systems is emphasized in the work of Halpern and Even-Shoshan (1967) who assume that the uptake (of L-glutamic acid) is effected by an "allosteric permease" which, in succinate-grown cells, is stimulated by γ-aminobutyrate (a type of allotopic increase of binding site affinity), noncompetitively decreased by aspartate and α-ketoglutarate (an allotopic decrease of binding site affinity) and competitively decreased by D-glutamate and derivatives (competition for binding sites).

A membrane protein involved in the transport of leucine and related amino acids has been isolated in two laboratories from *E. coli* in a crystalline state (p. 196).

The cells of *E. coli W* apparently possess specific transport mechanisms even for peptides (Payne and Gilvarg, 1968). It is now assumed that the oligopeptide system does not differentiate strictly between the amino acid components of the oligopeptide since homogeneous tripeptides of glycine, lysine, ornithine, and tyrosine compete for a common system. However, the cells probably contain specific dipeptide-transporting systems which do not mediate the translocation of higher peptides.

16.3.2. Other Bacteria

A variety of amino acid transports have been observed in other microorganisms: *Mycobacterium avium* (Yabu, 1967), *Mycobacterium fermentans* and *Mycobacterium hominis* (Razin *et al.*, 1968), *S. typhimurium* (Ames, 1964), *S. faecalis* (Mora and Snell, 1963; Bibb and Straughn, 1964; Holden and Utech, 1967), *Lactobacillus arabinosus 17-5* (Holden and Holman, 1959), and *S. hydrogenans* (Ring and Heinz, 1966; Ring *et al.*, 1967).

Only a few of the kinetically interesting observations will be mentioned here. *S. faecalis* cells transport glycine, L-alanine, and D-alanine with a high Q_{10} quotient against a considerable concentration difference. Protoplasts show a striking stimulation of uptake by pyridoxal and by K^+ and an inhibition by Na^+ while intact cells are completely insensitive to these agents.

S. typhimurium contains specific permeases for histidine, phenylalanine, tyrosine, and tryptophan and a general permease for aromatic amino acids. Thus the K_T of histidine by the specific permease is $1.7 \times 10^{-7} M$, by the general permease $1.1 \times 10^{-4} M$. It is remarkable that glucose decreases the pool size of these amino acids (*cf.* yeast cells, p. 314).

Working with α-aminoisobutyric acid (AIB) in *S. hydrogenans*, Ring and co-workers found an accumulative transport with a K_T of $3.8 \times 10^{-5} M$ and a V of 29 (μmol/min)/g. It has a Q_{10} of 3.7 and is inhibited by a variety of inhibitors. While the efflux of AIB is stimulated by extracellular concentrations of AIB, the influx decreases on preloading cells with AIB, the value of V being apparently decreased (Fig. 16.2). The inhibition is effected also by other neutral amino acids but not by glycine, serine, histidine, and tryptophan. The explanation suggested by the authors is one of negative

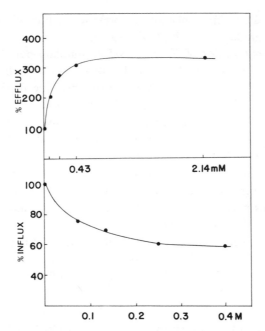

Fig. 16.2. Relative efflux (upper curve) and influx (lower curve) of labeled α-amino-isobutyric acid (AIB) against different concentrations of nonlabeled AIB on the *trans* side of the cell membrane of *Streptomyces hydrogenans*. (According to Ring *et al.*, 1967).

feedback affecting one of the intramembrane reactions, whereby carrier molecules are hypothetically converted from an inactive to an active form. Since the observation may not be restricted to this particular microorganism it would be worthwhile to develop the model kinetically.

16.4. CATIONS

In common with the transport of organic compounds, the uptake and/or loss of inorganic ions are associated with a genetically controlled membrane protein or proteins so that it is possible to prepare mutants defective in the transport of one or another ion.

The best-known case is the transport system for potassium ions in *E. coli*. The transmembrane movements of Na^+ and K^+ were found to be such as to create a high intracellular concentration of K^+ and a low concentration of Na^+ which is a feature common to most cell types (Schultz and Solomon, 1961; Schultz *et al.*, 1964; Epstein and Schultz, 1966). While the fluxes of both K^+ and Na^+ are suppressed by the absence of glucose or by an addition of iodoacetamide or dinitrophenol, the presence of fluoride affects only the K⁺ uptake. Cyanide has no effect whatsoever.

While parent strains of *E. coli CBH* accumulate potassium from low concentrations the *RD-2* mutant can do so only from high K^+ concentrations (Table 16.5). The phenomenon is probably due to the loss of concentrative ability for potassium, passive facilitated diffusion (or rather distribution according to a Donnan ratio) remaining in operation.

The defect of potassium transport is accompanied by decreased uptake of phosphate. If, however, K^+ is added in excess (raising the concentration from 0.02 mM to 50 mM) the uptake of phosphate is increased more in the mutant than in the parent strain (10-fold as contrasted with 1.75-fold). This anomalous behavior is probably linked with the function of an intracellular enzyme (Damadian, 1967).

A different phenotype of K^+-deficient mutant was described by Lubin and Kessel (1960) where the mutation affected the outward movement of potassium rather than its uptake. The exit flux of potassium was clearly raised (Günther and Dorn, 1966).

The uptake of K^+ in *E. coli*, including its genetic associations, is resembled by the cation uptake by *S. faecalis*, except that there the energy supply depends solely on the metabolism of exogeneous substrate. The microorganism possesses no endogenous reserves (Harold and Baarda, 1967). K^+ and Rb^+ are taken up against a concentration gradient, electroneutrality being preserved by simultaneous extrusion of Na^+ and H^+. In a mutant described by Harold and co-workers (1967) the accumulative capacity is lost, the defect residing in the outward rather than in the inward movement of K^+.

A practically linear proportionality between intracellular *potassium* content and the concentration of *sodium* in the medium was found in *A. aerogenes* (Tempest and Meers, 1968).

Table 16.5. Steady-State Potassium Fluxes in *Escherichia coli* *

Strain	K_o^+, mM	Flux of K^+, (pmol/cm²)/sec
Parent	0.026	1.60
Mutant	0.026	0.50
Parent	1.0	1.58
Mutant	1.0	1.39

* From Damadian (1966).

16.5. ANIONS

The uptake of phosphate by bacteria seems to follow a pattern similar to that in yeast cells, there being competition between phosphate and arsenate (Harold and Baarda, 1966; in *S. faecalis*).

The chloride concentration inside logarithmic phase *E. coli* cells is roughly 3 times greater than outside, corresponding to a membrane potential of 29 mV (Schultz *et al.*, 1962).

17. YEASTS AND FUNGI

17.1. INTRODUCTION

Of the hundred or more yeast species recognized by modern taxonomists only about a dozen have been examined in more detail from the point of view of their transport properties. Among the several hundred fungal species the situation is even less comforting, transport data being available in a rather happenstance manner and by no means permitting of a more general picture of the membrane transport mechanisms to be made.

The yeast species studied (but by no means understood) include *Saccharomyces cerevisiae*, *Saccharomyces carlsbergensis*, *Saccharomyces ellipsoideus*, *Saccharomyces fragilis*, *Saccharomyces globosus*, *Candida utilis*, *Candida beverwijkii*, *Candida albicans*, *Candida lipolytica*, *Torulopsis candida*, and *Rhodotorula gracilis*. A number of other species have been used for the purposes of taxonomy where, implicitly, transport was of crucial importance but the results are of little significance for the study of transport mechanisms, being of the yes-or-no character and restricted to a few arbitrarily chosen compounds.

The fungus species investigated more closely are *Neurospora crassa* and *Aspergillus niger* and, again, as with the yeasts, a few others, where the transport step was of economic or taxonomic importance.

It goes without saying that the information gathered from the studies is not of equal quality and, perhaps with the exception of baker's yeast, *S. cerevisiae*, in none of the species have more than two groups of substances been investigated. Hence, it is impossible to draw any general conclusions about the types of transport existing in the various species. In fact, the scant data that are available indicate a wide variety of transport mechanisms ranging from simple diffusion to pinocytosis, their quantitative properties being so varied that it is hardly an exaggeration to claim that within the group of yeasts alone there are more dissimilarities in the transport of substances than there are in all vertebrates.

For laboratory experiments, yeasts have distinct advantages over other cell types. They are easily grown even in rather acid media so that they are not greatly endangered by bacterial contamination; they can be cultivated under a variety of conditions so as to produce different physiological states of cells and, what with the progress of continuous-flow cultivation, the suspension samples can contain a practically homogeneous cell population; they are rigid and can withstand any practicable osmolarity of the medium, from redistilled water to 5 osmolar sodium chloride; unlike animal cells, they are readily separated from the medium when necessary (p. 213). Unlike bacteria, they are clearly visible under the light microscope.

As will be seen in the following paragraphs, most of our information on membrane transport in yeasts originates from work done on *S. cerevisiae* which, for apparent practical reasons, has been in the center of attention of investigators since Pasteur's time a century ago. Still, in many respects, baker's yeast is not necessarily typical of the whole group as it does not possess some transport mechanisms operative in other "wilder" yeasts which, unlike baker's yeast, have not been cultured by man for literally millions of yeast generations. It will be attempted here to include as much information obtained from work with other yeast species as is technically possible.

17.2. SUGARS

17.2.1. Monosaccharides

17.2.1.1. Baker's Yeast

The more important natural monosaccharides and some of their derivatives that have been investigated in baker's yeast are shown in Table 17.1.

The maximum rates of uptake range from 2 to 50 (μmol/ml)/min. The affinities of the monosaccharides for transport are the lower, the more changes of orientation of hydroxyl groups the sugar molecule has undergone with respect to glucose. The relative importance of the positions decreases from C_1 to C_3 to C_4 to C_5 to C_2.

All these sugars were found to enter the cell by facilitated diffusion but their space of distribution depends on the structureal characteristics of the monosaccharide. The essential feature of the molecule viewed in the *Cl*

Table 17.1. Half-Saturation Constants of Monosaccharide Transport in Baker's Yeast

Sugar	K_T, mM	Sugar	K_T, mM
D-Glucose	5	L-Sorbose	21
D-Mannose	27		1100 [a]
2-Deoxy-D-glucose	4.5	D-Xylose	130
D-Allose	225 [a]	D-Arabinose	115
D-Galactose	35 [b]	D-Lyxose	95
1,5-Anhydroglucitol	50 [a]	D-Ribose	600
3-O-Methyl-D-glucose	250 [a]		2000 [a]
L-Glucose	100	2-Deoxy-D-ribose	2000 [a]
L-Rhamnose	1000	L-Arabinose	1000
D-Fructose	17	L-Xylose	400

[a] Values from Cirillo (1968); other values from Kotyk (1967).
[b] Prior to induction; after induction $K_T = 4$ mM.

chair conformation (most stable for D-glucose and related sugars) is an equatorial hydroxyl group at carbon 4 and/or, when the *1C* conformation is more stable (as with most of the L-sugars), an equatorial group at carbon 2. It was then suggested that two constitutive carriers operate in the yeast cell. One, with a low specificity, permits the sugars to reach only a part of the cell water volume (about 40–50%), the other (showing the above-described structural requirements) enabling them to cross a further intracellular barrier and to distribute themselves in the entire water space of the cell (roughly 2/3 of the entire cell volume). There is another inducible carrier which shows a preferential specificity for sugars with an axial hydroxyl group at carbon 4 of the *Cl* chair conformation (like D-galactose). This carrier begins to function after an hour or more of incubation in the presence of D-galactose (with some strains also D-fucose and L-arabinose) (Fig. 17.1).

The transport proceeds only up to a diffusion equilibrium with no requirement for metabolic energy or ionic cofactors although there is indirect evidence that metabolizable sugars D-glucose, D-mannose, D-fructose (and D-galactose after induction of galactokinase) can potentially be transported by an energy-dependent process (van Steveninck and Rothstein, 1965; van Steveninck and Dawson, 1968) but their asymmetric steady-state distribution cannot be established because of rapid intracellular utilization. This active transport may involve phosphorylation of some membrane components (Deierkauf and Booij, 1968) but it does not appear to involve phosphorylation of the sugars as some earlier theories suggested (recently,

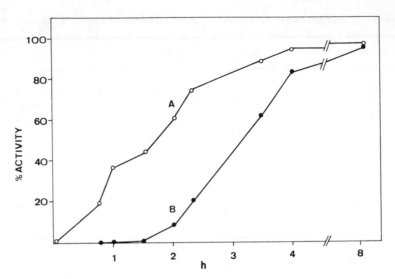

Fig. 17.1. Induction of galactose carrier (A) and galactokinase (B) during incubation of *Saccharomyces cerevisiae* with D-galactose in a non-growth medium. (From Kotyk and Haškovec, 1968.)

however, van Steveninck, 1968, described a primary phosphorylation of 2-deoxy-D-glucose during transport into baker's yeast). The phosphorylated carrier hypothesis can be formulated as follows:

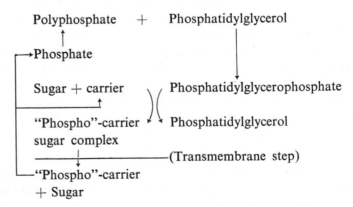

All the carriers appear to be mobile in the sense that the phenomenon of countertransport can be readily demonstrated. The specific carrier (for glucose-type sugars) was found to move 2–3 times faster when carrying a sugar molecule than when in the free form.

The specific carrier (but not the other one) is markedly inhibited by

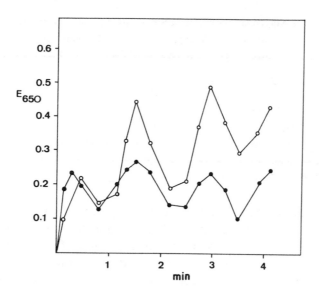

F$_{IG}$. 17.2. Uptake of 0.1 M D-glucose by *Saccharomyces cerevisiae* aerobically without (open circles) and with (black circles) 5×10^{-4} M uranyl nitrate. The values of extinction E_{650} are directly proportional to intracellular free glucose.

uranyl ions (from 10^{-6} M upward) which is incidentally the only known distinct inhibitor of sugar transport in baker's yeast.

The time curve of uptake of monosaccharides, both metabolized (like D-glucose) and nonmetabolized (like D-xylose and L-xylose), when analyzed in detail shows oscillations that are practically temperature-independent with a period of 70–75 sec (Fig. 17.2).

The phenomenon might be associated with volume changes of (a part of) the cell.

All the salient features of monosaccharide transport are preserved in protoplasts prepared from yeast by treatment with snail gut enzymes.

17.2.1.2. Other Yeast Species

A carrier-mediated transport of monosaccharides up to a diffusion equilibrium appears to operate also in other *Saccharomyces* species. On the other hand, in *Torulopsis*, *Candida*, and *Rhodotorula*, monosaccharides can be transported actively uphill, by a typical primary transport process requiring no known cofactors or activators. The steady-state distribution of some sugars in *R. gracilis* is shown in Table 17.2.

A peculiar property of most of these transports is that, unlike the case of *Escherichia coli* and other bacteria, the transport is so tightly coupled

Table 17.2. The Maximum Accumulation Ratio s_{II}/s_{I} of Monosaccharides in *Rhodotorula*
gracilis (Aerobically, Using 0.1 mM Solutions) *

D-Xylose	1000
D-Arabinose	40
L-Xylose	28
L-Rhamnose	90
D-Glucose	2 (glucose is rapidly utilized)

* From Kotyk and Höfer (1965).

with metabolism that there is no simple way of applying inhibitors to differentiate between the facilitated diffusion and the active components of
the process. Another striking feature of the transport is that the s_{II}/s_{I} ratio
attained at steady state is greater than 1 at low outside concentrations but
less than 1 (s_{II} tending to be constant) at very high outside concentrations
(Fig. 17.3); the deviations from an even distribution according to a diffusion
equilibrium both are linked with energy metabolism, i.e., the pump or
pumps operating in these species can reverse their direction according to
the prevalent substrate concentration (Deák and Kotyk, 1968).

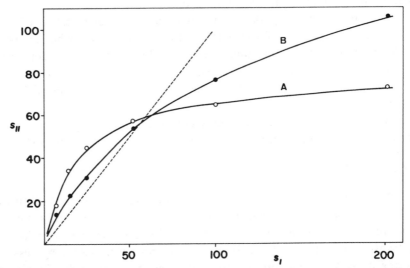

FIG. 17.3. Distribution of D-xylose between the intracellular (s_{II}) and the extracellular
(s_{I}) compartment of the yeast *Candida beverwijkii*. The concentrations shown are in
mg/ml cell water. A Untreated yeast, B yeast treated with 50 mM potassium sorbate.
(Adapted from Deák and Kotyk, 1968.)

Besides its primary uphill transport of many nonmetabolized mono-saccharides, *R. gracilis* has the apparently rather rare distinction of taking up at least one of its potential growth substrates, D-ribose, by simple dif-fusion (no temperature effect, no saturation dependence of rate, no effect of inhibitors or other potentially competing sugars) (Horák and Kotyk, 1969).

The temperature characteristics of monosaccharide transport in yeasts at about 30°C correspond to an enzyme-mediated reaction, the Q_{10} ranging from 1.8 to 3.2, depending on the "active" character of the transport. At temperatures between 0 and 10°C, however, Cirillo and co-workers (1963) found a Q_{10} of 16, suggesting very drastic changes in the properties of the plasma membrane constituents at those temperatures.

17.2.1.3. Fungi

The uptake of L-sorbose by wild-type *Neurospora crassa R1 38A* was found to be active in the sense of Kedem (p. 132) but did not proceed uphill, with $K_T = 116$ mM and $V = 3$ (μmol/ml)/min (Crocken and Tatum, 1967). In various mutants, as well as in the wild-type strain *74-OR 23-1A*, Klingmüller (1967) found a K_T for sorbose uptake of 6 mM and a V of 220 (μmol/ml)/min. His *Neurospora* transported sorbose (as well as glucose and fructose) against a gradient of 600:1. This shows the extreme variability of results one is faced with when working with microorganisms, in particular with those that are not sufficiently defined over many generations of main-tenance in standard collections.

Glucose uptake by conidia of *Neurospora sitophila* is sensitive to ura-nium ions, much like the uptake in yeast (Cochrane and Tull, 1958).

17.2.2. Oligosaccharides

In contrast with the uptake of monosaccharides, the disaccharides (including 1-alkyl hexoses) can be taken up against concentration gradients even in baker's yeast. The role of enzymic (and frequently inducible) trans-porters is quite evident (de la Fuente and Sols, 1962). Work done on maltose penetration has contributed to the elucidation of the phenomenon of cryp-ticity observed in microbial mutants lacking the specific transport system for a given substrate while the metabolic enzymes are present in the cell. With cryptic cells, a cell homogenate will display higher rates of utilization of the substrate than the intact cell.

The specificity pattern for the uptake of disaccharides varies from

Table 17.3. Uptake of Disaccharides by *Saccharomyces* Species *

Disaccharide	*Saccharomyces*			
	japonicus	*fragilis*	*cerevisiae*	*italicus*
Maltose	−	−	+	+
Sucrose	−	+	+	−
Lactose	−	+	−	−
Melibiose	−	+	−	−
Trehalose	−	+	+	−
Cellobiose	−	−	−	−

* After Kocková-Kratochvílová, personal communication.

species within the genus *Saccharomyces* (Table 17.3) and has been used, in conjunction with other criteria, for taxonomic classification of the species.

At least three specific permeation systems for disaccharides have been identified in baker's yeast. The first is induced by isolmaltose, α-methyl-glucoside, α-methylthioglucoside (and probably by other α-alkylglucosides) and transports actively all the inducing substrates, but, in addition, is competitively inhibited by some other sugars, such as trehalose, maltose, and glucose. A number of sugars, for instance galactose, melibiose, cellobiose, and lactose, are inactive. It is an interesting feature of the transport system that it can be readily differentiated into a facilitated diffusion and a superimposed active mechanism (e.g., by using 10 mM sodium azide or 0.4 mM 2,4-dinitrophenol) but both these transport modes are under the control of a single (MG$_2$) gene, the facilitated diffusion being constitutive, the active uptake inducible. In mg$_2$ mutants no transport occurs (Okada and Halvorson, 1964).

The second inducible disaccharide permease is that for maltose which transports only maltose and α-methylglucoside but the inducer spectrum of which is fully identical with that of the isomaltose permease, indicating a close linkage of the coupling enzymes for both reactions.

The third inducible system is that described by Kaplan (1965) to transport β-glucosides, in particular cellobiose. It appears, however, that the system is of the cell surface type, where cellobiose is split prior to entering the cell membrane so that the sugar species transported (apparently actively) is glucose (Kaplan and Tacreiter, 1966).

17.3. POLYOLS

Polyol penetration was studied rather cursorily in *T. candida*, *Pichia delftensis*, and *C. utilis*. The uptake patterns of the yeast species vary but it can be said in general that polyols, if taken up at all, are transported against a concentration gradient (Barnett, 1968). In *T. candida*, both sugars (L-sorbose) and polyols (eryhthritol, ribitol, D-arabinitol) are transported uphill and unusual two-site interactions between the two classes of compounds can be observed (Haškovec *et al.*, in preparation), the role of the growth substrate being apparently of paramount importance for the uptake specificity.

17.4. AMINO ACIDS

Transport of amino acids in baker's yeast can proceed against a substantial concentration gradient such that, even when protein synthesis and metabolic conversions of amino acids are not blocked, intracellular concentrations of at least 200 times higher than those in the medium can be found (e.g., with $10^{-5} M$ cysteine). There is apparently no single universal carrier for all amino acids in baker's yeast but rather a number of more or less specific carriers are involved (Grenson *et al.*, 1966; Grenson, 1966; Gits and Grenson, 1967). Their specificities may overlap so that some amino acids can be transported by several such carriers with the consequence that the mutual competition for transport is not symmetrical. The apparent K_T and V for transport determined in the presence of actidione which blocks

Table 17.4. Kinetic Parameters of Amino Acid Uptake by Baker's Yeast *

Amino acid	K_T, mM		V, (μmol/ml cell)/min	
	I	II	I	II
Glycine	0.70	—	63	—
Cysteine	0.25	—	14	—
Lysine	0.13	0.8	1.5	2.5
Methionine	0.06	1.1	6	37
Aspartic acid	0.28	3.3	7	17
Leucine	0.13	4.0	15	48

* Values obtained by Poncová and Kotyk.

protein synthesis are shown for some amino acids in Table 17.4. The two different values found for some of the amino acids reflect the multiplicity of carriers involved in the transport of each amino acid, it being rather likely that the values reported may still represent hybrids of pure transport constants of the various carriers which cannot be distinguished by a simple reciprocal plot of $1/s$ against $1/v$.

The situation is not quite clear as to the space of distribution of amino acids in yeast, there having been suggestions of at least two pools, an expandable one and a metabolic one (Halvorson and Cowie, 1961) but no definitive evidence is available on this point.

The outflow of amino acids from yeast is rather peculiar (Eddy, 1963), being very slow in general and practically unaffected by the presence of the same or of another amino acid in the medium (Fig. 17.4). This might suggest that the intracellular amino acid pool is not in direct contact with the plasma membrane. The lack of outflow makes it impossible to demonstrate phenomena like countertransport or to study cooperative effects such as described on p. 240.

Unlike the transport of sugars, the transmembrane movement of amino acids is affected by a number of agents, first of all by metabolic inhibitors which depress the steady-state level very substantially. It is perhaps signifi-

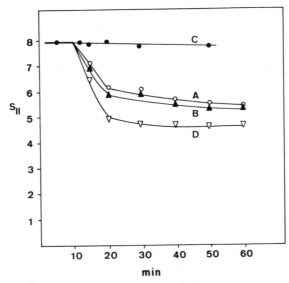

FIG. 17.4. Efflux of ^{14}C-glycine from preloaded *Saccharomyces cerevisiae* into distilled water (**A**), 10 mM nonlabeled glycine (**B**), 5×10^{-4} M uranyl nitrate (**C**) and 5×10^{-4} M 2,4-dinitrophenol (**D**). s_{II} is expressed in 10^3 counts/min.

Table 17.5. Effect of Alkali Metal Ions on the Rate of Uptake of Some Labeled Amino Acids by Baker's Yeast *

Cation in the medium	Glycine	L-Aspartic acid	L-Lysine
None	100	100	100
10 mM Na$^+$	115	106	110
50 mM Na$^+$	132	112	120
100 mM Na$^+$	138	115	118
10 mM K$^+$	82	100	85
50 mM K$^+$	115	120	102
100 mM K$^+$	117	118	98

* Values after 30 min of incubation in the presence of actidione, in % of the control (Poncová, Říhová and Kotyk, unpublished).

cant that uranyl ions show a specific effect on the transport of some amino acids, in common with their influence on monosaccharide transport.

The uptake of some amino acids as studied in this laboratory is affected by the presence of alkali metal ions, Na$^+$ having a stimulating effect on the uptake of most amino acids, K$^+$ having a rather varied effect (Table 17.5).

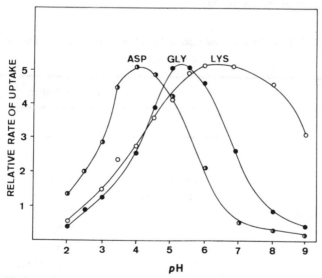

FIG. 17.5. pH dependence of amino acid uptake by *Saccharomyces cerevisiae* treated with actidione.

The uptake of some amino acids, in its turn, elicits a loss of Na^+ from sodium-rich yeast (Conway and Duggan, 1958) and uptake of potassium by ordinary yeast (Eddy and Indge, 1962).

The pH of the medium (using nonmetal buffers) has a pronounced influence on the amino acid uptake (Fig. 17.5), the shift of the optimum following qualitatively the isoelectric point of the amino acid.

The amino acid transport systems possess some features that are found in the transport across the intestinal wall. One of them is the distinct sensitivity to the presence of monosaccharides. Some sugars depress the intracellular steady-state level of amino acids (glucose, mannose, fructose), others raise it (D-xylose, L-sorbose), and others still stimulate the uptake at low and inhibit it at high concentrations (2-deoxy-D-glucose) (Poncová and Kotyk, 1967). The effect is different with different amino acids and is not reciprocal, suggesting the operation of a rather complex mechanism in the transport of solutes into baker's yeast.

In *N. crassa*, amino acids are taken up against a gradient, several genetic loci being involved in the production of different amino acid permeases (Wiley and Matchett, 1966; Jacobson and Metzenberg, 1968; Roess, 1968; Pall, 1969) (Table 17.6).

Similarly, in *Penicillium chrysogenum* (Benko and Segel, 1968) amino acid permeases for methionine, phenylalanine, and leucine have been described. It is of interest that nitrogen starvation increases the V for amino acid transport as much as 100 times.

Table 17.6. K_T Values of Amino Acid Uptake by Different Transport Systems of *Neurospora crassa* (mM)

Amino acid	Transport system		
	I	II	III
L-Tryptophan	0.05–0.06	0.04–0.05	—
L-Leucine	0.11–0.12	0.004	—
L-Phenylalanine	0.04–0.05	0.002	—
D-Phenylalanine	—	0.02–0.03	—
L-Asparagine	—	0.008	—
Glycine	—	0.005	—
L-Arginine	—	0.0002 (K_i)	0.002
L-Aspartic acid	—	1.2	—

17.5. OTHER ORGANIC COMPOUNDS

A great variety of substances other than the above-mentioned are known to penetrate into yeast cells, including low-molecular organic acids, higher fatty acids, nitrogenous bases, dyes, water-soluble as well as lipid-soluble vitamins, etc. Very little is known about the mechanism of their penetration but it is rather unlikely that they should each use a special carrier system. It is not impossible that some cross the plasma membrane by virtue of their solubility in lipids, others by sharing sites on rather non-specific membrane carriers.

It was shown by Malm (1950) that the ability of monocarboxylic acids to penetrate the baker's yeast cell is rather high while di- and tricarboxylic acids penetrate poorly. There is a certain amount of potassium ions exchanged for the carboxylic acid, in particular for formate. Suomalainen (1968) demonstrated that keto acids are taken up by yeast cells at a rate more or less proportional to their chain length, this suggesting penetration through the lipid parts of the membrane. However, some organic acids, particularly intermediates of carbohydrate metabolism (pyruvate, tricar-boxylic cycle acids) probably employ special permeases for entering into yeast cells.

Äyräpää (1950) showed that nitrogenous bases (ranging from methyl-amine to atropine) seem to penetrate by virtue of their lipid solubility but that small molecules like NH_4^+ and hydrazine can enter by special mechanisms.

The uptake of purines by *T. candida* proceeds either uphill (guanine, xanthine, uric acid) or only up to a diffusion equilibrium (adenine, hypo-xanthine), the differentiation being apparently associated with subsequent metabolism of the first-named group (Roush and Shieh, 1962).

Alkanes are a group of substances that have come to the attention of research workers through extensive attempts at deparaffinating some mineral oils. Some yeast species, notably *C. lipolytica*, thrive on an emulsion of medium-sized (like hexadecane) alkanes with buffer. Electron microscopy suggested that pinocytosis might be involved in the uptake (Ludvík *et al.*, 1968) but there are apparently no specific carriers for the alkanes present and the alkane is probably simply dissolved in the cell lipids.

No systematic effort seems to have been expended to study the transport mechanisms of these various compounds either in other yeast species or in fungi and the piecemeal evidence available indicates a great variability, both quantitative and qualitative, as one goes from species to species.

17.6. CATIONS

In the field of transport of ions the great majority of data again come from work on *S. cerevisiae*. All the inorganic cations investigated cross the membrane by the mediation of one or more carriers, the uptake and/or loss being coupled with the metabolic reactions of the yeast cells.

The alkali metal ions Na^+ and K^+ are present in baker's yeast in the following concentrations (using values after a 60-min incubation of an approximately 2% suspension, with the indicated medium components, expressed in m-equiv/g dry weight):

Medium content	Na^+	K^+
H_2O	0.010	0.58
0.1 M NaCl	0.047	0.54
0.1 M KCl	0.008	0.60
0.1 M KCl + 0.1 M glucose	0.009	0.73

There is clearly present a mechanism pumping sodium outward and potassium inward (reminding one of the situation in human and some other erythrocytes). The movement of K ions is of particular interest as it proceeds apparently by two different mechanisms. In the absence of glucose, K^+ is taken up alone by a carrier while in the presence of a fermentable sugar there occurs a practically stoichiometric exchange of K^+ for H^+ so that external values as low as pH 1.6 can be reached in a thick suspension (Conway and Duggan, 1958). The active nature of the K^+ uptake and H^+ extrusion provoked Conway to advancing a redox-pump hypothesis to account for the coupling of uptake with energy sources (e.g., Conway, 1953). The hypothesis proceeds from a consideration of an enzyme redox system

$$EH_2 \rightleftharpoons E + 2H \rightleftharpoons E + 2H^+ + 2e^- \tag{1}$$

whose potential (E_E) depends on pH while that (E_M) of a metal enzyme system

$$M^- \rightleftharpoons M + e^- \tag{2}$$

is pH-independent.

If the concentration of the reductant is the same as that of the oxidant in both systems and if the systems are electrically connected, the free energy change associated with the passage of electron equivalents from M to E is given by

$$nF(E_M - E_{E_a}) + nRT \ln(H)_a = \text{const.} \tag{3}$$

If (H) is raised so that the potential of the E-system is equal to that of the M-system, then

$$nF(E_M - E_{E_b}) + nRT \ln(H)_b = \text{const.} \qquad (4)$$

Since the first term is zero

$$RT \ln(H)_b = \text{const.} \qquad (5)$$

Subtraction of (4) from (3) yields an expression

★ $$F(E_M - E_{E_a}) = RT \ln(H)_b/(H)_a \qquad (6)$$

reflecting the fact that the entire available free energy associated with the passage of one electron equivalent from M to E at pH_a can be used for raising one hydrogen ion equivalent from pH_a to pH_b. If then the movement of H^+ is coupled with that of K^+ or of other ions, the translocation of these ions, too, can be effected at the expense of the free energy of the redox system.

Although the role of ATP in K^+ and H^+ transport in the yeast seems rather likely, the redox theory need not be refuted altogether as it offers a plausible explanation (at least thermodynamically speaking) of such processes as hydrogen ion secretion by the gastric mucosa and of ion transport in plants. At any rate, one can envisage ATP-driven redox carriers and it has been calculated that for 1 mole ATP split a concentration ratio of hydrogen ion carrier across the membrane equal to $10^8:1$ can be attained (Netter, 1961).

Whatever the energy coupling of the hydrogen–potassium transport

Table 17.7. Kinetic Constants of Cation Uptake by Baker's Yeast *

Cation	K_M of uptake, mM	K_i of modifier, mM
H^+	0.2	0.02
Li^+	27	19
Na^+	16	14.4
K^+	0.5	1.6
Rb^+	1.0	—
Cs^+	7.0	1.3

* From Armstrong and Rothstein (1964, 1967).

may be, the carrier shows affinities for other alkali metal ions, even if lower than for either hydrogen or potassium ions (Table 17.7).

The cations are seen not to compete simply for the carrier site but to act also as modifiers in a noncompetitive manner, without necessarily being transported. This follows nicely from a Hunter and Downs plot (*cf.* p. 238) of, say, the interaction between H^+ and K^+ (Fig. 17.6).

Likewise, bivalent cations Mg^{2+} and Ca^{2+} can modify the transport of alkaline ions but themselves show a very low affinity for the hydrogen–potassium carrier (K_M of 0.5 M and 0.6 M, respectively). Alkaline earth metals as well as other bivalent cations can be bound to the cell surface with different specificities by fixed anionic sites but also can be transported into a nonexchangeable intracellular pool by a special carrier. The uptake K_M's range from 0.01 mM to about 1 mM in the series Mg^{2+}, Co^{2+}, Zn^{2+} < Mn^{2+} < Ni^{2+} < Ca^{2+} < Sr^{2+}, the uptake being greater in the presence of glucose (either anaerobically or aerobically) and much enhanced by pre-treatment with phosphate, suggesting an essential role of phosphate in the function of the bivalent-metal carrier (*cf.* the situation in the kidney). The uptake of bivalent cations is generally compensated by an efflux of K^+ (or Na^+ from sodium-rich cells) (Fuhrman and Rothstein, 1968).

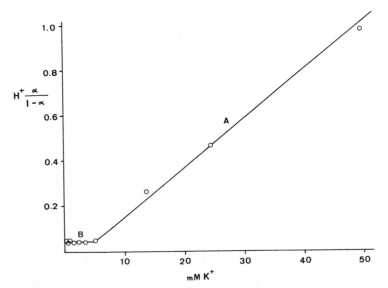

FIG. 17.6. A Hunter–Downs plot of the uptake of potassium by baker's yeast at *p*H 4.5 (using the value at *p*H 8 as control). Hydrogen ions are seen to act as a noncompetitive inhibitor at low concentrations **B** and as a competitive one at higher concentrations **A**. (Adapted from Armstrong and Rothstein, 1964.)

A special mention should be made of the interaction of uranyl ions with the yeast cell, resulting in a blockade of monosaccharide transport. Uranyl ions do not penetrate into the yeast cells (Rothstein *et al.*, 1948) but seem to be attached on the cell surface to polyphosphate ($K_{\text{diss}} = 3 \times 10^{-7}\,M$) and (at higher concentrations) to carboxyl groups (Demis *et al.*, 1954), which are involved in the transport of sugars but also apparently of amino acids and possibly other compounds. Ba^{2+}, Ca^{2+}, Be^{2+}, Mg^{2+}, and Zn^{2+} compete with the binding of uranyl at lower affinities.

In *N. crassa*, the uptake of K^+ and Na^+ resembles that in baker's yeast. The steady-state concentrations of these ions in a growing cells are $180 \pm 16\,\text{m}M$ K^+ and $14 \pm 2\,\text{m}M$ Na^+, K ions being lost from cells only at outside concentrations below 0.3 mM K^+ or in the presence of nystatin. Rb^+ can replace K^+ in the transport system while Na^+ cannot (Slayman and Tatum, 1964). The discrete character of the potassium carrier in this fungus was demonstrated by the isolation of a potassium-negative transport mutant (Slayman and Tatum, 1965) which required a three times greater concentration of K^+ for optimum growth than the parent strain.

17.7. ANIONS

Phosphate and arsenate compete for the same uptake sites in baker's yeast, the affinities being practically the same ($K_\text{M} = 4 \times 10^{-4}\,M$ according to Rothstein and Donovan, 1963; or $10^{-5}\,M$ according to Borst Pauwels *et al.*, 1965) but the V being 0.3 (μmol/ml)/min for phosphate and only about one-half that for arsenate. The transport, however, takes place practically solely in the presence of glucose, indicating a dependence on metabolic energy, and the difference in K_M's observed by different authors can be attributed to this very fact.

The uptake of phosphate is influenced by the presence of cations but it is not clear whether this is a direct molecular interaction or an indirect effect via metabolic processes of electrochemical equilibria.

Chloride ions appear to be taken up by the yeast cell passively.

The transport of sulfate into baker's yeast is an active process requiring metabolic energy and stimulated by the presence of nitrogen-containing compounds (Kotyk, 1959; Maw, 1963) (Table 17.8). The uptake is inhibited by 2,4-dinitrophenol, iodoacetate, sodium azide, sodium arsenate, and (probably at a later stage of sulfate metabolism) by sulfite and thiosulfate, as well as by methionine, cysteine, homocysteine, and ethionine. The effect of the amino acids is limited to 30–37% of the sulfate uptake.

Table 17.8. Uptake of $^{35}SO_4^{2-}$ by Baker's Yeast

Preincubation	Gaseous phase during incubation	Glucose present in incubation medium	Counts / min
In complete medium	Air	+	3320
In complete medium	Air	−	1170
In complete medium	Nitrogen	+	2950
In complete medium	Nitrogen	−	41
Without nitrogen	Air	+	260
Without nitrogen	Air	−	61
Without nitrogen	Nitrogen	+	190
Without nitrogen	Nitrogen	−	57

* Values taken after a 20-min incubation in a synthetic medium with 0.4 mM potassium sulfate.

18. ALGAE AND HIGHER PLANTS

18.1. INTRODUCTION

The complicated pathway followed by water and mineral compounds which enter the plant roots from the soil and are therefrom transported through xylem vessels into the overground parts of plants has attracted the attention of plant physiologists for many years. Nevertheless, much of the mechanism of this efficient system as well as of that of the organic substances supply by the phloem is not yet fully understood, especially the nature of the driving forces involved. On the other hand, a number of useful data characterizing the water and ion uptake at the membrane level have been obtained as can be seen from several reviews and monographs (Blinks, 1955; Arisz, 1956; Epstein, 1956; Laties, 1959a; Russel and Barber, 1960; Fried and Shapiro, 1961; Dainty, 1962; Brower, 1965; Legett, 1968; Schilde, 1968a, Briggs et al., 1961; Steward and Sutcliffe, 1961; Jennings, 1963). The whole of the nutritional requirements of autotrophic organisms like plants is met by the uptake of the inorganic substances, this uptake being subject to various regulations. The aim of this chapter is to describe the present ideas about the transfer of ions and water across the cell membranes, the related bioelectric phenomena, and cell activities behind these processes.

18.2. MEMBRANE POTENTIALS

Although the biophysical aspects of transport in higher plants have been dealt with by some authors the investigation of these phenomena is most advanced with algae. Both for the kinetic studies of membrane transport and for the conclusions concerning the character of the transport well-defined media on both sides of the transporting barrier are essential. In this respect, difficulties are encountered with the tissues of higher plants containing a complicated network of conductive vessels which cannot be easily separated from the metabolically active cells, in contrast with the comparatively homogeneous extracellular space of the animal tissue. It is also more difficult in the higher plants than in algae to evaluate the contribution of the ions bound to cell walls to the chemical analysis of the plant tissue.

Among algae, suspensions of cells like *Scenedesmus* and *Chlorella* have been employed by plant biophysicists, or else single cells of large dimensions like those of the *Characeae* (*Chara, Nitella*) have been investigated. The cell wall may be analyzed separately in these giant plant cells but even here, concomitant with the advantages offered by the giant size of the cells, some complications not common with animal tissue are encountered when the parameters determining the character of the transport are measured. Apart from the membrane between the rigid cell wall and the cytoplasm, the so-called plasmalemma, there is still another membrane surrounding the large central vacuole of plant cells, the so-called tonoplast. When measuring the intracellular potentials of algae the tip of the microelectrode often penetrates the thin layer of protoplasm to enter the vacuole so that two potential steps in series are measured rather than a single membrane potential: one of them across the tonoplast, the other across the plasmalemma plus the cell wall. Actually, some recent attempts succeeded in measuring separately the two potential steps in some large algae as well as the value of the potential difference between the wall and the medium.

The rigid cell wall is composed of polysaccharides like cellulose, hemicelluloses, pectins, etc. (Probine and Preston, 1960) and behaves as a weakly acid ion-exchange resin. The cell wall thus contains nondiffusible negatively charged particles so that the amounts of the diffusible cations and anions in the wall are different, the wall forming the so-called Donnan free space (D.F.S.), "free space" generally designating a compartment which is freely accessible to solutes from the external medium. In addition to the Donnan free space, the so-called water free space (W.F.S.) was defined for plant tissues and cells in which there are equal amounts of positive and negative charges of diffusible particles in the solution. The meaning of D.F.S. and W.F.S. was discussed in considerable detail by Briggs and collaborators (1961). A number of authors came to the conclusion that the two spaces are limited to the cell wall (e.g., Dainty and Hope (1959a)), the intracellular ions exchanging only very slowly due to permeability barriers of which the plasmalemma appears to be the most significant. Dainty and Hope

(1961) called the attention to the fact that a Donnan equilibrium need not always be the most suitable explanation of the distribution of ions between the external medium and the D.F.S. When the charged system is not homogeneous but is composed of charged surfaces spaced at a certain distance, the relationship between the equivalent width of the D.F.S. and the concentrations of ions in the medium is better expressed using the Guy–Chapman theory of the diffuse electrical double layer at the charged surface. The charge density in the cell wall derived from this theory is approximately 4×10^{-5} C·cm^{-2}. The concentration of the nondiffusible anion in the D.F.S. was determined to lie between 0.6 and 0.8 equiv/liter (Dainty and co-workers, 1960; Spanswick and Williams, 1965). Nagai and Kishimoto (1964) measured the electrical potential in the wall of the alga *Nitella flexilis* against the external medium and studied the dependence of this potential difference (E_w) on the concentration of electrolytes in the medium. Whereas in 0.1 M KCl E_w is practically zero, in 10^{-4} M it reaches -90 mV. Analogous results were obtained by Vorobyev and Kurella (1965) and Vorobyev and co-workers (1967) with the alga *Nitella mucronata*. Another potential jump which can be registered when introducing a microelectrode carefully into an alga is localized across the plasmalemma. This membrane is visible on electron micrographs as a single unit membrane about 75 Å thick and appears to lie very close to the cell wall.

Most often, the potential difference across the plasmalemma represents the most substantial potential step; the values measured with microelectrodes across the tonoplast are usually considerably lower. The protoplasm is electrically negative with respect to both the external solution and to the vacuolar sap, i.e., when proceeding with a microelectrode from the outside the negative potential registered in the protoplasm layer drops by a few millivolts when the tip of the microelectrode penetrates into the vacuole. Spanswick and Williams (1964, 1965) found in *Nitella translucens* a potential difference of -134 to -138 mV across the plasmalemma and $+18$ to $+24$ mV across the tonoplast. There are, however, some algae with an exceptionally high potential difference across the tonoplast, which exceeds the plasmalemma potential difference; as a result of this, the overall potential difference between the external medium and the sap in the vacuole is low and positive. In the marine alga *Valonia ventricosa* the overall potential difference is $+10$ to $+20$ mV, even if the potential difference across the plasmalemma amounts to -70 mV (Gutknecht, 1966). In *Chaetomorpha darwinii*, potential differences across the plasmalemma and the tonoplast of -35 and $+45$ mV, respectively, were found (Dodd *et al.*, 1966). Some authors assume that in most algae the potential difference across the tono-

Table 18.1. Potential Differences in Various Algae (mV)

Species	Total p.d.	Plasmalemma p.d.	Tonoplast p.d.	Reference
Chara australis	−159			Hope and Walker (1960)
Chlorella pyrenoidosa	−40			Barber (1968a)
Hydrodictyon africanum	−90	−116	+26	Raven (1967a)
Hydrodictyon reticulatum	−79			Janáček and Rybová (1966)
Nitella clavata	−120			Barr and Broyer (1964)
Nitella translucens	−120	−138	+18	Spanswick and Williams (1964)
Nitellopsis obtusa	−150			MacRobbie and Dainty (1958a)
Chaetomorpha darwinii	+10	−35	+45	Dodd et al. (1966)
Gracilaria foliifera	−81			Gutknecht (1965)
Halicystis ovalis	−80			Blount and Levedahl (1960)
Rhodymenia palmata	−65			MacRobbie and Dainty (1958b)
Valonia ventricosa	+11	−71	+88	Gutknecht (1966)

plast is negligible, others fear that the contamination of the microelectrode tip by protoplasm prevents one from establishing its value. Values of the overall potential difference, called the vacuolar potential, for several algae are summarized in Table 18.1.

18.3. ION CONTENTS AND ACTIVITIES

As in practically all living cells, the prevailing intracellular cation of plants is potassium, sodium being present in considerably lower amounts. In large marine algae or in fresh-water *Characeae* one can analyze the ionic content of the cell sap separately from the protoplasmic layer and the cell wall by using methods like micropuncture or centrifugation. The ions do not seem to be equally distributed in the thin protoplasmic layer. Kishimoto

Table 18.2. Intracellular Concentrations of Ions in Algae *

Species	Na, mM	K, mM	Cl, mM	Reference
Chara	48	80	106	Hope and
australis	(2.2)	(0.06)	(2.4)	Walker (1960)
Chlorella	1.1	114	1.3	Barber (1968a)
pyrenoidosa	(1.0)	(6.5)	(1.0)	
Hydrodictyon	17 vac.	40 vac.	38 vac.	
africanum	51 cyt.	93 cyt.	58 cyt.	Raven (1967a)
	(1.0)	(0.1)	(1.3)	
Hydrodictyon	4	139	45	Janáček and
reticulatum	(0.5)	(1.5)	(0.9)	Rybová (1966)
Nitella	39	73	127	Barr and
clavata	(3.0)	(0.1)	(3.3)	Broyer (1964)
Nitella	65 vac.	75 vac.	160 vac.	Spanswick and
translucens	14 cyt.	119 cyt.	65 cyt.	Williams (1964)
	(1.0)	(0.1)	(1.3)	
Nitellopsis	54	113	206	MacRobbie and
obtusa	(30)	(0.65)	(35)	Dainty (1958a)
Chaetomorpha	25	540	600	Dodd *et al.* (1966)
darwinii	(500)	(13)	(523)	
Gracilaria	66	680	462	Gutknecht (1965)
foliifera	(471)	(11)	(532)	
Halicystis	257	337	543	Blount and
ovalis	(498)	(12)	(523)	Levedahl (1960)
Rhodymenia	25	560		MacRobbie and
palmata	(467)	(11)	—	Dainty (1958b)
Valonia	44 vac.	625 vac.	643 vac.	Gutknecht (1966)
ventricosa	40 cyt.	434 cyt.	138 cyt.	
	(508)	(12)	(596)	

* The values in parentheses represent external concentrations.

(1965), using the combined techniques of centrifugation and perfusion, separated the flowing cytoplasm from the chloroplast layer, situated in the proximity of the plasmalemma and he obtained the following values: Flowing protoplasm 125 mM K$^+$, 4.9 mM Na$^+$, and 35.9 mM Cl$^-$; chloroplast layer 110 mM K$^+$, 26 mM Na$^+$, and 136 mM Cl$^-$. Analogous results were obtained by MacRobbie (1964). With smaller algae only the overall cell concentrations are available so far. Table 18.2 summarizes the data on sodium and potassium concentrations from various algae; chlorides as the principal accumulated anions in most algae are also included. There are some rare algal species, the growth of which is independent of the presence of chlorides in the saline or which appear to extrude them actively, e.g., the marine red alga *Porphyra perforata* (Eppley, 1958). Other anions are often accumulated against their concentration gradient, such as nitrates, sulfates, or phosphates.

The regulated uptake of salts in plant cells is considerable. Algae, whether from sea, brackish or fresh water, maintain a higher osmotic pressure than that of the surroundings (Collander, 1936; Blinks, 1951). In freshwater algae the osmotic value of the internal solution is usually about 5 atm. With marine algae, the scatter of values is substantial; whereas in some there is little excess osmotic pressure over that of sea water, in others it is almost equal to the osmotic pressure of sea water (about 23 atm for 3.5% salinity), i.e., their osmotic value is nearly twice that of sea water. Green and Stanton (1967) developed a simple method for measuring cell turgor, i.e., the difference between the inner and the outer osmotic pressure, using a micromanometer. Commonly the turgor pressure can be measured by changing the osmolarity of the outer saline and by determining its value at the moment of starting plasmolysis.

The chemical analysis of the ionic cellular composition itself does not provide sufficient basis for an accurate evaluation of the electrochemical ionic equilibria between external and internal media. What is easily accessible to chemical determination are concentrations rather than activities. Recently, the activity of potassium ions in the vacuole and the cytoplasm of *Chara australis* was estimated by means of glass cation-selective electrodes (Vorobyev, 1967). The value of 115 mM given for cytoplasmic activity appears to be so high as to suggest that in this plant cell there is no bound potassium. In the same algal species Coster (1966) measured the activity of chloride ions by the potentiometric method. The values obtained for vacuolar and cytoplasmic activity were approximately 100 mM and 10 mM, respectively. The low level of cytoplasmic activity as compared with the considerably higher concentrations found in this layer (Coster and Hope, 1968) suggests that cytoplasmic chlorides are mostly accumulated in the chloroplast gel layer, which is in agreement with the already mentioned findings of Kishimoto (1965) and MacRobbie (1964).

18.4. ION FLUXES

The existence of two membranes in plant cells, the plasmalemma and the tonoplast, causes some complications when the fluxes of ions are measured by the usual isotope technique, as these are to be determined separately across the two barriers. MacRobbie and Dainty (1958a) performed a kinetic analysis of the efflux and influx curves of the isotopically labeled ions for *Nitellopsis obtusa*, where the plasmalemma fluxes considerably exceed the fluxes across the tonoplast. The authors obtained the following figures for the individual tonoplast fluxes: 0.4 pmoles $Na^+/cm^2 \cdot sec$, 0.25 pmoles $K^+/cm^2 \cdot sec$, and 0.5 pmoles $Cl^-/cm^2 \cdot sec$. The plasmalemma fluxes were about 8 pmoles $Na^+/cm^2 \cdot sec$, and 4 pmoles $K^+/cm^2 \cdot sec$. Also in *Nitella axillaris* (Diamond and Solomon, 1959) the tonoplast fluxes are much lower than those across the plasmalemma and hence the compartmental analysis much easier than with *N. translucens* (MacRobbie, 1962) where the situation is reversed.

Perfusion methods (Tazawa, 1964), including sampling of the vacuolar content, were employed to determine the individual fluxes across tonoplast and plasmalemma by Coster and Hope (1968). An influx of 1–4 pmoles $Cl^-/cm^2 \cdot sec$ across the plasmalemma and a large influx across the tonoplast of about 100 pmoles $Cl^-/cm^2 \cdot sec$ were observed in *C. australis* using the last-named technique.

In general, experimental results based on fluxes are in accordance with the concept that algae behave like a three-compartment system, the compartments in series being the cell wall, the protoplasmic layer, and the vacuole. The ionic fluxes in plants are smaller than in animal cells, being of the order of 10^{-12} moles$\cdot cm^{-2} \cdot sec^{-1}$, for this reason, a long duration of the flux experiments is necessary. In another part of the present chapter, the effect of light conditions and metabolic inhibitors on influxes and effluxes of the principal ions is discussed in more detail (see section 18.7).

18.5. ACTIVE TRANSPORT

In the following the problems concerning the relation between the membrane potential and the distribution of ions in the outer and inner medium will be considered. Once the ionic concentrations (or ideally activities) are estimated and the potential differences across the plasmalemma and tonoplast are determined, the Nernst potentials of individual ions can be compared with the measured potential to establish which ions are in electrochemical equilibrium and which are not. Obviously, even in plant cells the activity of ionic pumps is the cause of the assymetrical distribution of ions. Even if a much greater diversity is found among plant species it appears that in analogy with animal cells, a sodium pump, actively extruding Na^+ from the cells, is operative in most algae. The equilibrium concentration of potassium as a highly permeant cation is then many times higher in the cell than outside. It appears, moreover, that in algal cells the passive distribution of K^+ is not the general case; evidence has accrued in favor of an active potassium pump which takes part in accumulating this ion within the cells. Both these mechanisms are situated at the plasmalemma membrane. Nevertheless, actively transporting systems, presumably for sodium if only cations are considered, are also probably located at the tonoplast. In addition to carrier systems transporting the cations there are apparently widespread anionic pumps in algae, as well as in other plant cells; of these, the chloride pump seems to be of most general occurrence. Thus for *N. translucens* (Spanswick and Williams, 1964) an active sodium extrusion and an active potassium and chloride uptake were identified at the plasmalemma, while at the tonoplast Na^+ is extruded into the vacuole, potassium and chloride being near electrochemical equilibrium. In *Valonia ventricosa* (Gutknecht, 1966; Blei, 1967) sodium seems to be pumped from the protoplasmic layer into both the vacuole and the outer medium. Potassium in this alga is also actively transported into the vacuole across the tonoplast and probably also from the surroundings into the cytoplasm across the plasmalemma. In contrast with most other algae, the chloride influx from sea water into the vacuole does not seem to be due to an active process.

Using a fine microelectrode technique, Barber (1968a) succeeded in measuring the membrane potential in the cells of *Chlorella pyrenoidosa*, the diameter of which hardly exceeds 5μ. The procedure involved fixation of the cells in the tip of a glass capillary, to which suction was applied by a syringe; then the microelectrode was advanced by micromanipulation and inserted into the cell. The average potential was estimated to be -40 mV. Comparing this value with the Nernst potentials which were -3 mV for E_{Na^+}, -71 mV for E_{K^+}, and $+8$ mV for E_{Cl^-}, the author concluded that K^+ and Cl^- are actively accumulated and sodium pumped in the outward direction.

Sodium and potassium are not the only cations subjected to active transfer. Spanswick and Williams (1965) drew attention to the existence of a calcium pump, located probably at the tonoplast, which favors an active uptake of Ca into the vacuolar sap. A striking discrepancy is found in *Nitella clavata* between the currents carried by Na^+, K^+, and Cl^-, and the total membrane current. Evidence was provided by Kitasato (1968) that the disagreement might be due to a high H^+ conductance. The resting membrane potential is more sensitive to $[H^+]_o$ than to $[K^+]_o$, giving practically no response to changes of $[Na^+]_o$. The author illustrates further that the disagreement between the E_M and the Nernst E_{H^+} indicates an active H^+ extrusion from the cells.

Considering the transport of anions, active uptake of phosphate (Mac-

Robbie, 1966a,b; Smith, 1966), nitrate (Pražmová and Rybová, unpublished results) and sulfate (Vallée and Jeanjean, 1968) were observed in different algal species.

To demonstrate more directly which ionic species are actively transported, Blount and Levedahl (1960) applied the short-circuit current technique of Ussing to *Halicystis ovalis*. To achieve conditions when no electrochemical forces act on passive movements of ions, the vacuolar sap was exchanged for sea water by perfusion and the potential difference between medium and the vacuole was short-circuited. In such an arrangement, only the actively transported ions move across the membrane. Using isotopes, an active efflux of Na^+ and an active influx of Cl^- were identified in *Halicystis*.

Much effort was devoted to the problem of describing mathematically the transmembrane potential difference but it appears that the plant material is in this respect even more refractory than animal tissues. Nevertheless, it was shown that in *C. australis* (Hope and Walker, 1961) and *N. translucens* (Spanswick et al., 1967) the membrane potential in calcium-free media is satisfactorily determined by a modified Goldman equation, containing only expressions for Na and K:

$$E_M = \frac{RT}{F} \log \frac{[K^+]_o + \alpha[Na^+]_o}{[K^+]_i + \alpha[Na^+]_i} \qquad (1)$$

where $\alpha = P_{Na}/P_K$.

The value of α equals 0.06 for *C. australis* and 0.27 for *N. translucens*. This difference in the values of the α coefficient implies that the sensitivity of the potential difference of various algae toward changes in the external sodium and potassium concentrations will be different. As a matter of interest it is possible to mention that the E_M of the mold *Neurospora crassa* (Slayman, 1965a) gives over a considerable range of concentrations a similar response to changes in K_o^+ and in Na_o^+, the slope of the linear relationship being 45 mV/log unit in the case of potassium and 33 mV/log unit in the case of sodium.

It could be perhaps useful to recall at this place the role of the cell wall potential. On transferring the cell from one saline to another an undesirable potential difference between the cell wall and the medium may be formed which does not cease until a new Donnan equilibrium is established. To avoid this, Hope and Walker (1961) and Spanswick and associates (1967) when deriving equation (1) kept the sum of $Na_o + K_o$ constant. Even then some time is required to reach the diffusion equilibrium due to the presence of unstirred layers (Dainty, 1962).

In the presence of calcium ions the effect of the external concentration of sodium and potassium ions on the plasmalemma potential is markedly decreased (Spanswick et al., 1967; Kishimoto et al., 1965). Calcium itself affects the E_M level and the membrane resistance increases with raised Ca_o. Spanswick et al. demonstrated that addition to the above Goldman-type equation of terms for Cl^- or Ca^{2+} is of no help. The fluxes of calcium are substantially lower than those of Na^+ and K^+, P_{Ca} being thus negligibly small when compared with P_K or P_{Na}, which means that calcium can hardly participate directly in the potential formation. It is not possible in this case to express the E_M value by a Goldman equation and it should be concluded that, in addition to the suggested ionic fluxes, there must exist still another charge-transfer mechanism, such as an electrogenic pump. Then the sum of the active and passive fluxes will be zero, whereas the sum of the individual passive fluxes will differ from zero and the Goldman equation cannot be applied.

Equation (1) was used (Hogg et al., 1968) to explain satisfactorily the temperature dependence of the transmembrane potential difference as being due to differential changes in the permeabilities of sodium and potassium.

In plant cells, mainly anionic electrogenic pumps are invoked and, of these, most often the chloride pump (e.g., Barr, 1965). Hope (1965) assumed the operation of a bicarbonate electrogenic pump in C. australis. On the other hand, it is postulated that the active transport of sodium and potassium cations might be carried out by an electrically neutral Na^+–K^+-coupled pump (Dainty, 1962; MacRobbie, 1962; Raven, 1967a; Barber, 1968b). Kitasato (1968) came to the conclusion that an electrogenic H^+ pump comes into play in Nitella clavata. Slayman (1965b) presumes that in Neurospora crassa both an active component, inhibited by azide and 2,4-DNP, and a passive component, sensitive to K_o, take part in the production of the membrane potential. At high concentrations of calcium, the resistance to the passive component is increased, the active component becoming dominant.

In experiments using vacuolar perfusion only minute changes in the potential difference were noted when replacing the vacuolar sap with artificial solutions containing Li^+, Na^+, or Rb^+ or when substituting chlorides with other anions (Tazawa and Kishimoto, 1964). This may be due to the fact that the changes in the vacuolar solution cause only very slow changes in the composition of the protoplasmic layer. Kishimoto (1965) observed that the potential difference drops almost linearly (with a slope of approximately 50 mV) with log K^+ in the vacuole if the potassium concentration is varied from 1 to 50 mM. On increasing further the vacuolar K_o (up to 200 mM) the change of the potential difference across the tonoplast is diminished, the total potential difference approaching zero.

18.6. SOURCES OF ENERGY FOR TRANSPORT

The aim of many experiments was to obtain a clearer insight into the mechanism of ion transport across plant membranes and to establish how the active transfer is coupled with energy supply. In plants the following mechanisms may be involved: (1) utilization of ATP, produced in respiratory or photosynthetic processes, (2) direct coupling with reduction–oxidation processes in either the respiration or the photosynthetic chain reactions. According to these possibilities several hypotheses were formulated. Lundegårdh worked out a detailed theory on this point (1955, 1961). Anions are supposed to move stepwise into the cell interior while electrons flow simultaneously in the opposite direction along the enzymatic system of cytochromes. At the same time, the cations, released by the same enzyme system during oxidation of substrate, are accumulating in the cells in exchange for H^+. The author suggests that the proper absorption process takes place deep in the protoplasmic layer near the vacuole. This idea of "anion respiration" is basically similar to the redox-pump theory developed by Conway (cf. section 20.3) and it stresses the importance of anion pumps in plants. However, a selective step in the anion transport appears to be required, the affinity of the transport mechanism being not the same for different anions.

Likewise, Robertson (see Briggs et al., 1961) conceives of anion transport as the regulatory principle in the accumulation of salts, the cations entering only passively. He suggests a possible coupling of the electron transfer during the early photosynthetic reactions with the uptake of chlorides. Lundegårdh (1961) points also to the significance of chloride transport in connection with their specific role of indispensable cofactors in the photolytic water breakdown.

The carrier hypothesis, as summarized by Epstein and Hagen (1956) among others, prefers the idea of ion transport as being due to specific carriers. The rate of ion uptake is then kinetically described in analogy to an enzymatic reaction by an equation of Michaelis–Menten type, and the dissociation constant of the substrate–carrier complex can be evaluated (cf. chapter 3). ATP formed during respiration was defined as the energy donor by Laties (1959b) among others.

It has been known for a long time that light is necessary for accumulation of ions in plants; hence, the role of photosynthetic processes as energy donors in ion transport is proposed. MacRobbie (1964, 1965, 1966a,b) directed her investigation to these aspects trying to solve the question whether the ion transport is connected with the consumption of energy-rich compounds built up during photosynthetis or whether there exists some direct coupling with the photosynthetic electron transfer. In the chain of photosynthetic reactions two basic steps combined with the absorption of two light quanta may be differentiated. They are the so-called first photo-synthetic system, represented by the reduction of ferredoxin with the sub-sequent transfer of electrons to NADP (the electron donor of CO_2 reduction in the Calvin cycle) and second photosynthetic system, connected with photolysis of water and O_2 liberation and leading to the reduction of plasto-quinone. A "noncyclic" ATP formation takes place during the spontaneous electron flow from plastoquinone to chlorophyll$_1$. A cyclic photophospho-rylation (i.e., ATP formation coupled to a light-stimulated electron flow in the cycle) comes into play when the second photosynthetic system is blocked. MacRobbie took the advantage of the possibility that the two photosynthetic systems can be separated by using either a light source of a suitable wavelength or specific inhibitors. On the basis of the finding that in *N. translucens* DCMU (dichlorophenyldimethylurea) in low con-centrations brings about a substantial decrease of chloride uptake, whereas the accumulation of potassium is relatively insensitive to this inhibitor of water photooxidation, and that the inhibitors of photophosphorylation (imidazol, CCCP—carbonyl cyanide *m*-chlorophenylhydrazone) greatly impair potassium uptake, the author concluded (1) that the active entry of chlorides is directly coupled to the second photosynthetic system, with-out consuming ATP and (2) that the active accumulation of potassium is dependent on ATP synthesized during photophosphorylation. Analogous conclusions were made by Raven (1967b) who demonstrated that in the alga *Hydrodictyon africanum* not only the light-stimulated uptake of po-tassium but also the light-stimulated extrusion of sodium is blocked by inhibitors of photosynthesis.

An active DCMU-inhibited chloride pump was observed to work in *Hydrodictyon reticulatum* (Pražmová and Rybová, to be published). Sodium and chloride effluxes were enhanced by light in the latter alga whereas no action of light could be ascertained on potassium efflux (Rybová and Janáček, 1968; Fig. 18.1). In *N. translucens* and *C. australis* the chloride effluxes are actually greater in darkness (Hope *et al.*, 1966).

A neutrally coupled Na–K pump was envisaged to function in *N. translucens* (MacRobbie, 1962) and in *H. africanum* (Raven, 1967a) on

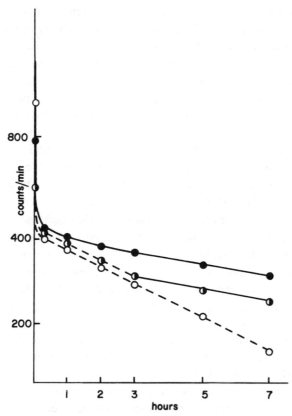

FIG. 18.1. Efflux of ^{22}Na from the alga *Hydrodictyon reticulatum* in the light (dashed line) and in darkness (solid line).

the basis of tracer flux measurements and the observation that ouabain, a known specific inhibitor of sodium pump in animal tissue, brings about a decrease in active Na$^+$ efflux and K$^+$ influx. However, no effect was exhibited by ouabain on the membrane potentials of *N. translucens* in experiments performed by Spanswick and Williams (1964). In *H. reticulatum*, a rather unusual effect of ouabain, swelling with simultaneous uptake of KCl, was observed (Janáček and Rybová, 1966). Nevertheless, the isolated membrane Na,K-sensitive ATPase fraction from the latter alga was influenced by ouabain (Wins and Kleinzeller, personal communication). This does not mean that the Na,K–ATPase system is not operative in the transport process but that, probably due to steric hindrance, ouabain cannot exert its expected effect in most plant cells (see also Dodd *et al.*, 1966; Gutknecht, 1967; Barber, 1968*b*). In *H. reticulatum* the inhibitor appears, moreover, to act on a process different from the pump mechanism, causing rather a relaxation in the cell wall structure and thereby in the turgor pressure, water being consequently osmotically driven into the cells.

Returning once more to the occurrence of a cation-coupled pump, the findings of Barber (1968*b*) may be cited. In *Chlorella pyrenoidosa*, he observed the sensitivity of sodium efflux to the presence of external potassium, this being generally considered as an indication in favor of the coupling mechanism. In another paper by the same author (Barber, 1968*c*) results are reported showing that a light-stimulated and carrier-mediated K^+–K^+ exchange is also likely to exist in *Chlorella*; the author suggests that this exchange may be in some stages of development temporarily converted into a K^+–H^+ exchange mechanism.

All the reported differences found in various algae point again to the diversity of phenomena in plants, which might be due either to species variability or to the adaptation of the cell to living conditions.

The active uptake of potassium and chloride, as well as the active sodium extrusion are supposed to be located at the plasmalemma membrane (or in the adhering chloroplast layer) even if definite information especially about the direct coupling of electron flows in chloroplasts and chloride accumulation is lacking. As the plasmalemma active transport explains only the protoplasmic accumulation of minerals, an additional step was proposed (MacRobbie, 1964) to be involved in vacuolar absorption; an energy-dependent formation of small vesicles in the cytoplasm loaded with salt solution, the content of which is subsequently ejected into the vacuole. A similar idea was suggested earlier by Sutcliffe (1960) and supported by Gutknecht to be probably valid for *Valonia* cells (1967). The observation by Arisz (1953) that the cytoplasmic uptake of chlorides in the leaves of *Vallisneria* is insensitive to dinitrophenol, whereas the vacuolar uptake is affected, is consistent with this two-step system of salt accumulation.

18.7. EFFECTS OF LIGHT ON MEMBRANE POTENTIAL DIFFERENCES

Already at the beginnings of the bioelectrical investigations of algae much attention was devoted to the problems of the changes in the potential difference and in other electrical parameters by the action of light (Blinks, 1955). The usual response of the electrical potential in dependence on the light conditions may be characterized as an increase in the potential dif-

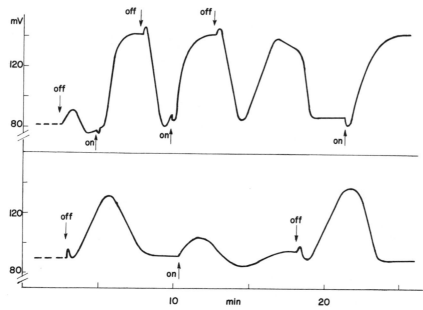

Fig. 18.2. Changes of the potential difference with light conditions in *Hydrodictyon reticulatum*. (Top curve) normal response, (Bottom curve) "hybrid" response.

ference during the light period and a decrease in darkness, the rate and extent of the changes varying much according to the species under study (Blinks, 1955; Nagai and Tazawa, 1962). A "reversed" effect of the light and dark periods on the potential difference, i.e., a drop in the light and rise in the darkness, has also been described. The role of previous light conditions on the response was discovered by Nishizaki (1963). In the alga *H. reticulatum* a "hybrid" response may be obtained in some cases (Fig. 18.2), i.e., a transient peak of the hyperpolarizing potential appears in the darkness, the steady state of the potential remaining unchanged (Metlička and Rybová, unpublished results). As was shown a couple of years ago by Blinks (1955), the complete reversal of the normal light response in *Halicystis ovalis* can be induced by addition of NH_4^+ ions to the saline. This is also true for *Hydrodictyon* (Fig. 18.3); within one hour after addition of 10^{-4} M NH_4Cl to the medium the potential difference is substantially decreased and a slow further decrease is brought about on switching on the light while an increase results on switching it off. Ammonia and several amines, functioning as uncouplers of photophosphorylation, are known to cause a reversal of the chloroplast volume changes induced by light (Izawa, 1965). Here again a relation between the potential changes and the processes

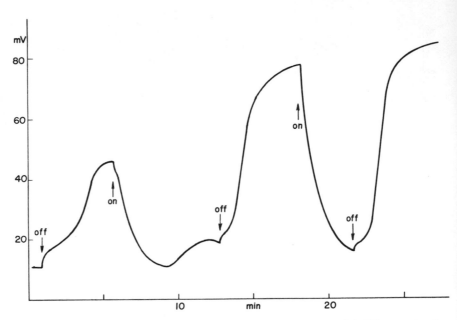

Fig. 18.3. Effect of light and darkness on the membrane potential difference in the presence of 10^{-4} M ammonium chloride in *Hydrodictyon reticulatum*.

in the chloroplast may be envisaged, as will be discussed below in more detail. Any response in *H. reticulatum* to light in normal artificial pond water is completely inhibited within a few minutes by DCMU (Metlička, Rybová, unpublished results) which might indicate that a chloride conductance change is involved in these effects. It may be visualized that the photoelectric phenomena are dependent on the photosynthetic reactions in the chloroplasts but, due to the complexity of these processes, it is very difficult to assess the underlying mechanisms. For the light-triggered change in the potential difference at least the following mechanisms can be responsible in principle: (1) A permeability change caused, for example, by the changed ionic composition at the plasmalemma, (2) a change in the pump activity (for analogous reasons), (3) a change in the local concentration of an ion which takes part in the formation of the potential difference. pH changes might be of great importance here. In *Ulva lobata* and *Ulva expansa* a rapid uptake of H^+ in exchange for Na^+ upon onset of photosynthesis was reported. The presence of HCO_3^- was found to be of importance for these short-term ion translocations (Cummins *et al.*, 1969). A light-triggered DCMU-inhibited uptake of protons has been recently reported by Ben-Amotz and Ginzburg (1969) in *Dunaliella parva*.

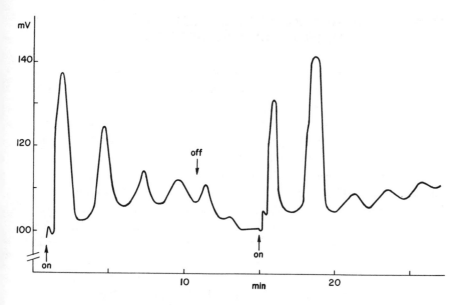

Fig. 18.4. Light-induced oscillations of the intracellular potential in *Hydrodictyon reticulatum* in artificial pond water containing 2 μg valinomycin/ml.

In addition to the bioelectric light-influenced changes which are suspected to be associated with photosynthetic phenomena, very rapid photoelectric effects in the alga *Acetabularia* have been recently reported by Schilde (1968b), caused probably by a direct effect of light on the plasmalemma membrane.

Light-induced oscillations of potential and increased numbers of peaks caused by the addition of valinomycin (a cyclic oligopeptide known to act as a K^+-specific carrier in artificial membranes and mitochondria) were obtained with *Hydrodictyon* (Metlička and Rybová, 1967; Fig. 18.4). The oscillation may arise spontaneously wherever there are more nonlinear processes occurring in a system. With valinomycin, either a direct action on the plasmalemma or a chloroplast-mediated influence can be assumed. Karlish and Avron (1968a) demonstrated that valinomycin has an uncoupler-like effect on the chloroplasts if the membrane permeability of these particles to H^+ ions is simultaneously enhanced. The complexity of the processes involved in oscillatory behavior was shown by Ried (1968) who divided oscillations of oxygen evolution in *Chlorella fusca* into several components differing in their dependence on the wavelength of the light source and sensitivity to inhibitors.

18.8. ION TRANSLOCATIONS ACROSS THE CHLOROPLAST MEMBRANE

Much attention has been devoted during the last years to the correlation between conformational changes, ion translocations, and energy processes in chloroplasts during the light and the dark period. The outer chloroplast membrane is believed to be permeable to salts and sugars but not to proteins. The double-layer membrane of the thylakoid represents a true permeability barrier to salts. Figure 1.2 shows the chloroplast structure of *Scenedesmus quadricauda*, this alga of about 5 μ in diameter having only one chloroplast. There appears to be much in common between the light-induced changes in chloroplasts (Nobel and Packer, 1965) and the processes in mitochondria during different stages of activity (chapter 25). The aim of Dilley and Vernon (1965) was to clarify how volume changes are connected with the movement of ions. They observed that the light-induced consumption of H^+ ions is accompanied by an efflux of K^+ and Mg^{2+}, electroneutrality being thus maintained. The authors ascribe the characteristic light-dependent shrinkage of the chloroplasts to the subsequent passive efflux of water which relieves the change in osmotic pressure. Packer and co-workers (1966) have drawn the attention to the action of light on the chloroplast volume in dependence on the suspension media. If a NaCl solution is used for the suspension, a specific swelling is brought about in the light and the total process is not fully reversible. On the other hand, if salts of weak organic acids are used for suspension, the light-triggered shrinkage is fully reversed in the dark stage. The mechanism of shrinking in these favorable conditions is explained as follows. The light-induced active uptake of hydrogen ion results in the formation of a new equilibrium inside the chloroplasts according to the dissociation constant of the weak acid. The equilibrium of the undissociated acid across the chloroplast membrane is thus displaced and, consequently, organic anions leave the cell together with water to preserve the osmotic balance. The authors claim that such a mechanism can better explain the conformational changes as the movements of inorganic cations are too small to be responsible for the volume changes.

A detailed flash-light analysis of the photosynthetic processes was performed in Witt's laboratory. On the basis of absorption changes of both the pigments and added pH indicators, the authors (Schliephake *et al.*, 1968) concluded that as photosynthesis sets in a proton is taken up concomitantly with the electron transfer initiated by the light activation of chlorophyll molecules. Moreover, an electric field across the thylakoid membrane is formed. In an elementary process on the molecular level, the two light reactions cause a translocation of two hydrogen ions and a potential difference of 50 mV is established. The light-induced ATP synthesis in the chloroplasts is due to the efflux of H^+, which is driven by the potential difference. The results of Witt and co-workers appear to be consistent with Mitchell's hypothesis (see chapter 25). The finding by Jagendorf and Uribe (1966) that synthesis of ATP may be evoked by an artificially produced pH gradient is also considered to favor the chemiosmotic concept. As in the case of mitochondria, Mitchell's hypothesis has not been generally accepted to be valid for chloroplasts and objections have been raised against it. Karlish and Avron (1968b) suggest an alternative explanation for the role of the proton pump in a facilitation of the co-transport of the negatively charged complex of phosphate, ADP, and Mg^{2+} into the chloroplasts.

18.9. ELECTRICAL PROPERTIES

To complete the description of ion-transporting plant membranes some interesting electrical properties and basic electrical parameters may be mentioned. The electrical behavior and parameters of the plasmalemma and the tonoplast are most often investigated by studying the electrical potential responses to square pulses of current applied through a microelectrode. The most important part of the resistance between the medium and the vacuole appears to be localized in the plasmalemma and, moreover some rectification of the current has been described, the resistance to the current flowing from the vacuole into the external medium being somewhat lower than that in the opposite direction. In *C. australis* 12.1 k$\Omega \cdot$cm^2 was found for the plasmalemma and 1 k$\Omega \cdot$cm^2 for the tonoplast (Findlay and Hope, 1964a), this being in agreement with previous measurements of Walker (1960). In *N. translucens*, resistances between 21.4–24.8 k$\Omega \cdot$cm^2 were reported together with the space constant λ of 2.6–3.0 cm (Williams *et al.*, 1964; Bradley and Williams, 1967a). The tonoplast resistance was found to increase markedly when the vacuole was perfused with dilute saline, this being assumed to represent a protection against the loss of electrolytes from the cytoplasm (Kishimoto, 1965). The capacitance of the plant membrane, like that of most animal cell membranes, is of the order of 1 μF$\cdot$cm^{-2}.

It may be seen from the current–voltage characteristics that the membrane behaves as an ohmic resistance only in limited segments. The phenomenon of negative differential resistance was observed in *Valonia* (Blinks, 1955) and more recently also in cells of *N. translucens* (Bradley and Williams, 1967b). An analogous effect was analyzed in artificial membranes (see section 14.7.1). If a hyperpolarizing current of low density of about 2.0 $\mu A \cdot cm^{-2}$ is applied to *Nitella*, then in the region where the membrane potential is by about 40 mV more negative a sudden increase in the membrane resistance may be observed with some cells. The hyperpolarizing response to current pulses below this region was capacitative while with higher current densities an inductive response was obtained. A similar hyperpolarizing response to square-wave pulses, i.e., a hyperpolarizing voltage jump with a simultaneous resistance increase, was described by Kishimoto (1966) for *Nitella flexilis*, incubated in concentrated salt solutions of monovalent but in the absence of divalent cations. It seems that in these effects two different structural states of the membranes with different conductance properties may be involved, the intensity of the applied current determining the dominance of one or another configuration.

With higher intensities of the increasing hyperpolarizing current, the response of measured voltage in normal saline' becomes smaller and smaller until the differential resistance attains a zero value and the membrane is "punched through" (Coster, 1965; Williams and Bradley, 1968a).

Along with the hyperpolarizing phenomena, responses to the depolarizing current are of interest. As a common manifestation of excitability, an action potential may be observed in algae. In contrast with the nerve and the muscle, where the action potential is of functional significance (*cf.* chapter 20 and 21), the action potential in algae is somewhat different. The duration of the spike in algae is considerably longer, of the order of seconds. Whereas in the excitable animal tissue a sudden increase of permeability toward sodium ions is the underlying mechanism, in *Characeae* chloride permeability is increased (Mullins, 1962; Findlay and Hope, 1964b). As a result of this change the efflux of chloride ions increases and, according to the Goldmann equation, the potential difference approximates the equilibrium potential for the chloride anion. According to Williams and Bradley (1968b) the increase in the permeability to chloride is followed by a similar increase in the potassium permeability in *N. translucens*, the main permeability changes occurring in the plasmalemma. The presence of the bivalent calcium ions appears to be of importance in the initiation of the action potential. Light was shown to decrease the amplitude of the spike in *N. flexilis* (Andrianov *et al.*, 1969).

18.10. TRANSPORT OF WATER

Apart from the transport of inorganic ions the movement of water is of great importance in plants. Some characteristics of water permeability, of the existence of pores, and of the possible participation of electro-osmosis in the water flow will be briefly mentioned.

The role of the cell wall in the regulation of cell volume or in the turgor pressure has been often discussed. It was shown by Kelly and co-workers (1962) that in *N. translucens* the volume change between zero and full turgor pressure is within the limits of 2.5–5%. This was confirmed for *N. flexilis* (Tazawa and Kamiya, 1965), where the elastic modulus of the cell wall was found to be three times as large along the transverse axis as along the longitudinal axis.

Using the method of transcellular osmosis, water permeability can be expressed in the form of hydraulic conductivity (osmotic permeability) L_p, as described in chapter 13. It was found that L_p is a function of concentration in the outer saline, decreasing with increasing concentration of the solute. For 0.1 M sucrose and 25°C it reaches in *C. australis* and *N. translucens* values of about 10^{-5} cm·sec^{-1}·atm^{-1} (Dainty and Ginzburg, 1964a). Kamiya and Tazawa (1956) pointed to the fact that the L_p for endosmosis differs from that for exosmosis, the first being considerably greater as was confirmed by Dainty and Hope (1959b). According to Ginzburg and Katchalsky (see Dainty and Ginzburg 1964a) this should be expected in a case where two membranes in series (plasmalemma and tonoplast) have different permeabilities.

By the method of transcellular osmosis it was possible to measure the reflection coefficient σ (see sections 3.1 and 4.2) for various solutes. In *Chara* and *Nitella*, positive values of less than one were obtained for lower alcohols, acetate esters and values practically equal to one were found for urea, ethylene glycol, and formamide (Dainty and Ginzburg, 1964b). Hence it was concluded that transport across the membrane could take place via aqueous pores. On the other hand, Gutknecht (1968) presented evidence that in *Valonia* aqueous membrane pores appear to be absent. Permeability to water is apparently very low in this alga and a highly negative reflection coefficient was found for methanol. A comparative study of the permeability properties of various algal cells to nonelectrolytes was performed by Collander (1957).

Another approach serving to provide further evidence for the existence of pores in *Nitella* was due to electro-osmosis. On applying small currents, an electro-osmotic flow of water is expected to appear. The electro-osmotic efficiency, i.e., the amount of water carried by ions, was found to be usually about 100 water molecules per positive charge (100 moles/F) by Fensom and Dainty (1963). Tyree (1968) points to the possibility that perhaps 20% of the value can be contributed by the cell wall. A somewhat lower electro-osmotic efficiency than in *Nitella* was determined in plant roots, but still it was concluded that in view of the existence of biopotentials electro-osmosis could be of some significance for water transport in the root tissue (Fensom *et al.*, 1965). The theory of electro-osmotic phenomena in plants is given in the paper of Dainty *et al.* (1963) who showed that the frictional and the Schmid models are better suited for explaining electro-osmotic fluid flow than is that of Helmholtz and Smoluchowski. Assuming the applicability of the above models to negatively charged pores of small diameter, Fensom and Wanless (1967) calculated that approximately 10^8–10^9 pore sites per cm^2 transporting Na$^+$ or K$^+$ should be located in the membrane.

18.11. TRANSPORT IN HIGHER PLANTS

It is likely that the knowledge of ion and water transport studied in single plant cells will be in principle valid also for higher plants but several new aspects following from the integration of cell systems and the mutual interplay of tissues will be added.

When entering from the soil to the xylem vessels, ions have to pass through several differentiated layers of root tissue, i.e., from epidermis to cortex, endodermis, pericycle, and the central cylinder. Cells appear to be mostly interconnected by protoplasmic bridges, protruding through the cell walls. The resulting protoplasmic continuum, the so-called symplasm, facilitates the passage of electrolytes once they entered the cell interior (Arisz, 1956). The suberized radial cell walls of the endodermis (Casparian strips) extend in older plants, eventually severing the symplast bridges (see Fig. 18.5). Thus in the basal parts a hindrance to transport of water and solutes will appear.

Kramer (1956) points to the fact that the most actively absorbing region in young roots is several centimeters from the apex. Transport of water was also assumed to occur in parts where the suberization has not advanced and yet the xylem system has developed. Many experiments were undertaken to localize the diffusional barriers to the transverse ionic transport into the xylem. Ginsburg and Ginzburg (1968), using stele-less roots of corn seedlings, have come to the conclusion that two principal barriers, one at the epidermal and one at the innermost cortex layer, are involved. Lüttge and Laties (1967) investigated the root absorption of K$^+$ and Cl$^-$

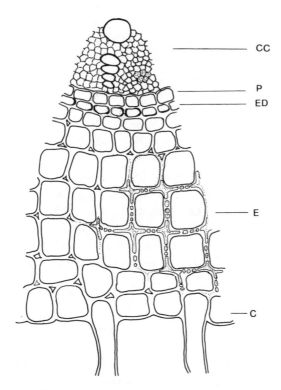

FIG. 18.5. Transverse section of a root. C Cortex, E epidermis, ED endodermis, P pericycle, CC central cylinder with xylem and phloem vessels. Protoplasmic interconnections are shown in a part of the figure.

in maize seedlings over a wide range of external concentrations and assumed that two transport systems can be distinguished. At low concentrations, a high-affinity system is operative in transporting the ions across the plasma membrane into the symplasm. Absorption, as well as the long-distance transport, is under these conditions highly sensitive to inhibitors and un-couplers. In the range of high concentrations the uptake into the symplasm is supposed to be mainly diffusive, and, hence, both the root accumulation and the long-distance transport are less sensitive to inhibitors. On the other hand, Weigl (1969) stresses that the chloride transport is inhibited even at high external ion concentrations and argues that a linear absorption iso-therm need not support the concept of diffusion permeation if the Mi-chaelis–Menten constant is high as compared with ion concentration.

A two-step system of salt accumulation has been proposed for plants since Arisz (1953) showed that the cytoplasmic and vacuolar uptake of chlorides in the water plant *Vallisneria spiralis* can be separated by means of inhibitors. Vesicles filled with the salt solution and discharging into the vacuole are supposed to be involved in the transport across the tonoplast as was already mentioned before. The involvement of mobile carriers in the anion transport has been recently stressed by Weigl (1967).

In a detailed study Higinbotham and co-workers (1967) compared the Nernst equations for the main nutrient ions with the measured vacuolar potential in seedlings of *Avena sativa* and *Pisum sativum*. The following potential differences were found (Etherton, 1963): -110 mV for pea roots, -84 mV for oat roots, -119 mV for pea epicotyls, and -105 mV for oat coleoptiles. Of the ions studied, only potassium seems to approach an electrochemical equilibrium; Na^+, Ca^{2+}, and Mg^{2+} appear to be excluded by an active mechanism. None of the anions (Cl^-, NO_3^-, $H_2PO_4^-$, SO_4^{2-}) is in passive equilibrium, the electrochemical gradients being directed from the tissue to the medium. An electrogenic inwardly directed anion transport is assumed.

Pallaghy (1968) measured the potential difference in guard cells of tobacco and observed oscillations with an amplitude of 12 mV and a period of 6 min. It was found that the permeabilities to Na^+ and K^+ in these cells are identical.

Action potentials have been often investigated in higher plants; a detailed description of the phenomenon in *Mimosa* and *Dionaea* was published by Sibaoka (1966).

The technique of short-circuit current developed by Ussing (see chapter 12) was applied in the study of ion transport in *Nepenthes henryana* (Nemček *et al.*, 1966). The results indicated that active and independent fluxes of Na^+ and Cl^- from the plant pitcher into the tissue take place. The transport of chloride and sulfate in the pitchers of this carnivorous plant was studied by microautoradiography by Lüttge (1965, 1966).

Finally, the effect of growth substances as connected with the transport of nutrients should be mentioned. Brauner and Diemer (1967) tried to show that in the developing tissues of *Helianthus annuus*, *Vicia faba*, and *Zea mays* the auxins act by changing the surface charge load of the membranes (probably due to absorption of the auxin molecule in the membrane material with the acid groups oriented to the outside), thus lowering the permeability to anions and altering selectivity.

It may be stated that, although many of the mechanisms and metabolic interconnections of mineral transport in plants are at least partly comprehended, practically in no respect has a definitive conclusion been reached.

19. ERYTHROCYTES

19.1. INTRODUCTION

The red blood cells represent the simplest animal cells which, particularly for membrane transport studies, possess some highly attractive features. This is especially true of the mammalian erythrocytes which have no nucleus and no defined organelles so that their interior can be considered as a single compartment. Their plasma membrane can be easily prepared practically pure of any interfering protein and, moreover, can be resealed by incubation in an isotonic or slightly hypertonic medium (Gárdos, 1954; Dodge *et al.*, 1963) so that the reconstituted ghosts serve as fine models for transport of solutes both inward and outward.

The greatest amount of work was done on human erythrocytes but important evidence was obtained from beef, horse, sheep, cat, and rabbit blood cells. Among nonmammalian erythrocytes, pigeons and chicks were used as experimental objects.

19.2. SUGARS

19.2.1. Monosaccharides

The transport systems for monosaccharides in human and rabbit erythrocytes represent the kinetically best-analyzed transports by mediated diffusion. The rate of utilization of monosaccharides by erythrocytes is so low that it can be neglected in the transport experiments, there being, moreover, a number of nonmetabolized monosaccharides available which share the glucose uptake system. The uptake of monosaccharides by erythrocytes shows D-kinetics at low and E-kinetics at high (saturation) concentrations (Widdas, 1954; Wilbrandt *et al.*, 1956) and proceeds only up to a diffusion equilibrium. It was in the erythrocytes that some peculiarities predicted by the mobile carrier hypothesis, such as countertransport and competitive acceleration, were first demonstrated (Rosenberg and Wilbrandt, 1957*b*; Wilbrandt, 1961*a*). Table 19.1 shows some of the kinetic parameters of sugar transport in human and rabbit erythrocytes.

The specificity of the carrier for monosaccharides is rather broad, including hexoses as well as pentoses, and aldoses as well as ketoses, the grading of affinities being best accounted for by the stability of the sugar in the *Cl* chair conformation of the pyranose ring (*cf.* Reeves, 1951; LeFevre and Marshall, 1958). Thus, as D-glucose has only one "instability factor" in the α-anomer it has a high affinity for the carrier; L-rhamnose, on the other end of the scale, possesses 3 in the α- and 4 in the β-anomeric form. Ketoses apparently possess structural features which lower their affinity for the monosaccharide carrier but these have not been identified. This type of conformational specificity is likely to be of importance also in other cell types.

The ratio of mobilities of the loaded and of the free carrier (p. 77)

Table 19.1. Transport Parameters of Monosaccharides in Erythrocytes

Species	Sugar	K_M, mM	V, (μmol/ml)/min	Reference
Man	D-Glucose	7.2	500	Wilbrandt (1961*a*)
		4.0	328	Sen and Widdas (1962)
		5.6	590	Miller (1965*a*)
	D-Mannose	12	680	Miller (1965*a*)
		14	710	LeFevre (1962)
	D-Galactose	20	650	Miller (1965*a*)
		12.8	1000	Wilbrandt (1961*a*)
	D-Xylose	50, 71	650	LeFevre (1962)
		28	800	Wilbrandt (1961*a*)
	D-Ribose	2500	770	LeFevre (1963)
	D-Arabinose	5500	620	LeFevre (1962)
	L-Arabinose	220, 250	710	LeFevre (1962)
		90	300	Wilbrandt (1961*a*)
	D-Fructose	9300	124	Miller (1966)
	L-Sorbose	3100	124	Miller (1966)
	L-Rhamnose	3200	—	LeFevre (1961*a*)
Rabbit	D-Glucose	6	0.23	Regen and Morgan (1964)
		4.3	0.12	Park *et al.* (1956)

was found to be in the vicinity of 3–4 in human erythrocytes (Mawe and Hempling, 1965; Levine and Stein, 1966) but equal to 1 in rabbit erythrocytes (Regen and Morgan, 1964). The difference in these mobilities may account for the observation of an "exchange transport" of these sugars, which persists even at low temperatures, as described by Lacko and Burger (1963).

The temperature dependences of the K_M and V values for glucose transport across the red cell membrane are similar ($Q_{10} = 1.8$ for V and 1.6 for K_M) (Sen and Widdas, 1962).

The monosaccharide transport in erythrocytes is blocked by a number of inhibitors, among them $HgCl_2$, p-chloromercuribenzoate, gold chloride, dinitrofluorobenzene, N-ethylmaleimide, phloretin, polyphloretin phosphate, phlorizin, some lachrymators, and some corticosteroids.

The influence of the phloretin group was analyzed in greatest detail, apparently because of its relative specificity, and it was found to exhibit asymmetric effects on sugar transport (Fig. 19.1). This unusual inhibition (observed subsequently in a number of other cell types) indicates that even in the relatively simple transport of sugars in the erythrocyte there may be components other than the carrier involved, such as enzymes catalyzing the attachment of substrate to the carrier (cf. Wilbrandt, 1961b). Studies on the specificity of phloretin inhibition of glucose transport (Rosenberg and Wilbrandt, 1957a; Kotyk et al., 1965) showed that the essential features of the molecule are those shown in heavy type in the accompanying formula

The essential inhibitory features of the phlorizin molecule differ in the

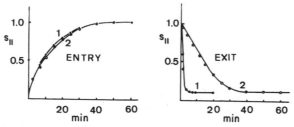

FIG. 19.1. Entry and exit of glucose into and out of human erythrocytes in the absence (1) and in the presence (2) of 0.04% polyphloretin phosphate. s_{II} is expressed in isotones. (According to Rosenberg and Wilbrandt, 1962.)

Table 19.2. Inhibition of Sugar Uptake in Human Erythrocytes *

Inhibitor	Inhibition (mM^{-1} required for 50% inhibition)
Phloretin	160
Phlorizin	1.9
5,5-Di(4'-hydroxyphenyl)nonane	170
5,5-Di(3'-methylphenyl-4'-hydroxy)nonane	500
Stilbestrol	160
3,3'-Di(2-chloroalkyl)-stilbestrol	1600

* From LeFevre (1959).

importance of the hydroxyl group of the B ring. The inhibitory potency of several compounds toward sugar transport in human erythrocytes is shown in Table 19.2. The inhibitors shown in the table may act by sterically hindering sugars from the attachment to their carrier binding site.

There is a considerable interindividual variability in the rates of sugar uptake by human erythrocytes. The same is true for the existence or even prevalence of transport of sugar dicomplexes in these cells (*cf.* Wilbrandt and Kotyk, 1964). The evidence for this type of transport is not altogether conclusive because of this very lack of universality. The experimental evidence for transport of sugar dimers by erythrocytes originally propounded by Stein (1962*b*) was subjected to some critical examination (Graepel, 1966; LeFevre, 1966; Miller, 1966).

19.2.2. Disaccharides

Some disaccharides are also transported into human erythrocytes (isomaltose, α- and β-methylglucoside, and perhaps cellobiose) while others are not (maltose, sucrose, trehalose, lactose) (Yang, 1950). It is of interest that some disaccharides (maltose, cellobiose, isomaltose) inhibit the exchange transfer of glucose for galactose while others (trehalose, lactose, sucrose) do not (Lacko and Burger, 1962).

19.3. AMINO ACIDS

Relatively little is known about the kinetics of amino acid uptake by mammalian erythrocytes, except that they, too, use a facilitated diffusion path to enter the cells (Table 19.3). Apparently, there exist even in the

Table 19.3. Kinetic Parameters of Amino Acid Uptake by Mammalian Erythrocytes *

Species	Amino acid	K_M, mM	V, (μmol/ml)/min
Man, 37°C	Leucine	1.8	0.52
	Phenylalanine	4.3	1.5
	Methionine	5.2	0.56
	Valine	7.0	1.0
	Alanine	0.34	0.0068
	Glycine	0.30	0.0012
Rabbit, 20°C	Leucine	1.1	1.0
	Valine	2.8	1.3

* From Winter and Christensen (1964, 1965).

human erythrocytes multiple carriers for amino acids, as follows from the striking differences in the values of V, although no systematic work seems to have been done on the subject.

The situation is different with pigeon erythrocytes where the transport of neutral amino acids was analyzed in detail and was shown to proceed by four different routes: (1) Na^+-independent, with no apparent exchange diffusion (valine, leucine, methionine, phenylalanine, and isoleucine); (2) Na^+-dependent, with acceleration by exchange diffusion (alanine, serine, cysteine, proline, and possibly threonine); (3) a system for glycine and sarcosine, (4) a system for β-alanine and taurine. The kinetic constants of transport of some of the amino acids are shown in Table 19.4. The variety of transport systems and of the transmembrane rate constants of the carrier is reflected in the rather varied maximum rates of uptake. The uptake of glycine, sarcosine, alanine, β-alanine, serine, valine, and proline is associated with an increase of Na^+ influx but none of the fluxes is inhibited by ouabain.

The stoichiometry of Na^+ activation is different for each amino acid, second-order kinetics apparently being involved (Wheeler and Christensen, 1967). The Na^+-dependent transport system for glycine was studied most thoroughly (cf. Vidaver and Shepherd, 1968) using also restored pigeon erythrocyte ghosts so that concentrations of both glycine and Na^+ could be varied at will at both sides of the membrane. The transport was found to be intrinsically asymmetric and, moreover, the energy for glycine transport was derived from the Na^+ gradient across the membrane while the transport of glycine itself is primarily passive. The best-fitting model for

Table 19.4. Kinetic Parameters of Amino Acid Transport in Pigeon Erythrocytes in the
Presence of Sodium Ions *

Amino acid	K_M, mM	V, (μmol/ml cell water)/min
Glycine	0.18	0.10
Sarcosine	0.09	0.05
β-Alanine	0.16	0.04
Taurine	0.15	0.05
Alanine	0.19	0.20
Serine	0.16	0.19
Cysteine	0.13	0.18
Proline	2.1	0.07
Phenylalanine	0.04	0.018
Methionine	0.08	0.031
Valine	0.23	0.010
Leucine	0.07	0.015
Isoleucine	0.05	0.016

* Average values are given; from Eavenson and Christensen (1967).

this rather complex transport is one with a compulsory path of formation
of the complexes (*cf.* p. 147), thus

$$\text{Carrier} + \text{Na} + \text{Na} + \text{Glycine} = C + A + A + S$$

where the rate constants of movement of the various complexes differ, the
CAA being practically immobile.

19.4. OTHER ORGANIC COMPOUNDS

Few data are available on the mediated diffusion system for glycerol
($K_M = 0.5\ M$ in human erythrocytes), there being apparently a competition
between glycerol and glycols for uptake. It was suggested by Stein (1962a)
that glycerol and glycols are transported in the form of dimers.

The entrance into horse or beef erythrocytes of monocarboxylic acids
(acetate to caproate) seems to be effected by diffusion of the undissociated
molecules through the lipid of the membranes, that of dicarboxylic acids
(oxalate, malonate, tartronate, maleate, fumarate, succinate, malate, tar-
trate, glutarate) by diffusion through aqueous pores about 7–9 Å in diamater
(Giebel and Passow, 1960). One may wonder whether the acids do not

use carrier systems of very low affinity (showing D-kinetics over a wide range of concentrations).

The uptake of purines (uric acid, hypoxanthine, adenine, and some artificial analogues) by human red blood cells (see Lassen, 1962; Lassen and Overgaard-Hansen, 1962*a,b*) seems to take place by two mechanisms of different specificity, at least one of them transporting against a concentration gradient.

19.5. CATIONS

The transport of sodium and potassium ions in red blood cells has attracted much attention of physiologists and its investigation has yielded information that has proved to be applicable to other types of cells.

Unlike the transfer of sugars, the transmembrane movement of cations in erythrocytes is energy-coupled and, under conditions of homeostasis, the distribution of Na^+ and K^+ between erythrocytes and plasma is unequal, as shown by Table 19.5.

Apparently, then, the distribution of the alkali metal ions is asymmetric to different extent, depending on the species. In human erythrocytes, where most of the studies were carried out, it was concluded that metabolic energy for the formation of concentration gradients was provided by metabolism of glucose (Straub, 1953) which proceeds both via fructose-6-phosphate and

Table 19.5. Steady-State Concentrations of Sodium and Potassium in Erythrocytes and Plasma (m-equiv/liter)

Species	Erythrocyte		Plasma		Reference
	Na^+	K^+	Na^+	K^+	
Man	19	136	155	5.0	Bernstein (1954)
Dog	135	10	153	4.8	Bernstein (1954)
Cat	142	8	158	4.6	Bernstein (1954)
Sheep (HK)	37	121	139	4.9	Tosteson and Hoffman (1960)
Sheep (LK)	137	17	139	5.0	Tosteson and Hoffman (1960)
Rabbit	22	142	146	5.8	Bernstein (1954)
Rat	28	135	152	5.9	Bernstein (1954)
Fowl	27	96	137	6.0	Quinn and White (1967)

via ribulose-5-phosphate to lactate as the end product. Moreover, the flow of potassium is specifically regulated by the rate of metabolism of 2,3-diphosphoglycerate in the presence of calcium (Gárdos, 1966).

The interdependence of sodium and potassium fluxes in the erythrocyte has given rise to a host of studies on the subject, by Gárdos, Glynn, Maizels, Post, Straub, Whittam, and Tosteson (in sheep erythrocytes) as are aptly summarized by Passow (1964). It is now generally recognized that the transport of Na^+ and K^+ contains two distinct components. The first of these is a passive leak, the magnitude of which is proportional to the electrochemical gradient of the cation across the membrane. The leak is symmetrical, without preference for either sodium or potassium ions and is insensitive to cardiac glycosides (Post et al., 1967). However, in spite of its passive character, the leak component of transport is affected by vasopressin, acetylcholine, fluoride, iodoacetate, and others. Moreover, both Ca^{2+} and Mg^{2+} can regulate its magnitude. Apparently, then, the effect of all these agents is on the pores of the plasma membrane of the erythrocyte.

The other component of alkali metal transport in human erythrocytes (and most likely in all the species where a similar distribution of sodium and potassium has been found) is an active one. It is rather specifically inhibited by 10^{-5} M ouabain (Schatzman, 1953) and related glycosides and it translocates Na^+ and K^+ in a coupled manner against their respective electrochemical differences, the stoichiometry of the process having been established as 3Na:2K (Whittam and Ager, 1965; Sen and Post, 1964). This ratio is maintained over a wide range of concentrations so that one may speak of a pump operating in fixed gear.

The active coupled system for the transport of Na^+ and K^+ in erythrocytes was found to possess a number of properties similar to the membrane-located ATPase (Post and Albright, 1961):

(1) Both utilize ATP in contrast to inosine triphosphate.

(2) Both require the simultaneous presence of Na^+ and K^+.

(3) Activation by K^+ is completely inhibited by high concentrations of Na^+ in both systems.

(4) Ouabain is inhibitory in both systems.

(5) NH_4^+ can substitute for K^+ but not for Na^+ in both systems.

(6) The half-saturation constants of activation and inhibition by Na^+, K^+, NH_4^+, and ouabain are very similar in both systems, viz., 20–24 mM, 2.1–3.0 mM, 7–16 mM, and 0.3–1.0 mM, respectively.

These findings (aided by analogies from other cell types) laid the foundation for the concept of ATPase as an intrinsic membrane factor involved in the transport of alkali metal ions.

The transport ATPase activity was found to be enhanced by Mg^{2+} but not by Ca^{2+}, 1 ATP molecule being split per transport cycle. The probable mechanism of ($Na^+ + K^+$)-dependent ATPase in transport of alkali-metal ions can be visualized as follows from Fig. 19.2. A calculation proceeding from the Gibbs free energy of ATP in erythrocytes ($\Delta G = -13.9$ kcal/mole) and from the assumption that Na^+ and K^+ are osmotically free intracellularly, one arrives at an efficiency of ATP splitting for running the active transport of Na^+ and K^+ of slightly above 50%.

Work is in progress on the molecular identification of the alkali-metal ATPase from human erythrocytes and it appears that the salient features of the ATPase are similar to those from other sources (e.g., the existence of a phosphorylated intermediate; Blostein, 1966).

Alkaline earth cations are also distributed unevenly between the red cell interior and plasma (Table 19.6) but the mechanism underlying this distribution has not been investigated in detail. Olson and Cazort (1969) described a clear case of active transport of both calcium and strontium out of human erythrocyte ghosts. There is apparently a Ca-activated ATPase involved in the transport (Schatzmann, 1966; Vicenzi and Schatzmann, 1967).

The paths for entry of K^+, Mg^{2+}, and Ca^{2+} into erythrocytes are different. The mutual effects observed are to be ascribed to activation (or inhibition) of membrane enzymes regulating the membrane permeability toward the bivalent cations (Passow, 1961).

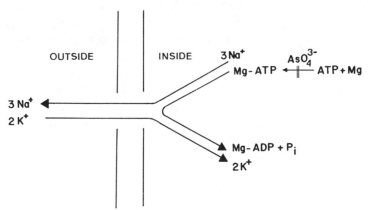

FIG. 19.2. Schematic representation of ATPase transport of Na^+ and K^+ in human erythrocytes.

Table 19.6. Distribution of Ca^{2+} and Mg^{2+} between Plasma and Erythrocytes (m-equiv/liter) *

Species	Erythrocytes		Plasma	
	Ca	Mg	Ca	Mg
Man	2.2	4.6	4.6	0.8
Sheep	1.3	2.8	4.8	1.8
Dog	0.8	2.7	2.9	0.8
Rabbit	2.0	8.0	6.8	2.0
Fowl	4.0	6.8	5.8	1.8

* From Quinn and White (1967).

19.6. ANIONS

The distribution of chloride, hydroxyl, bicarbonate, and sulfate anions into erythrocytes seems to obey the Gibbs–Donnan ratio and hence the movement of these anions is assumed to be passive. Considerations of penetration of chloride and sulfate anions through beef and horse erythrocyte membranes led Passow to conclusions as to the amount of fixed charges in the water-filled pores of the membranes, the probable value reached being 3 M with respect to amino groups active in sulfate transport (Passow, 1961).

The Gibbs–Donnan distribution of the anions, as well as the intracellular concentrations of cations, can be shifted by introducing CO_2 into the erythrocyte suspension (the so-called Hamburger shift). As carbon dioxide enters the cells very rapidly it reduces the intracellular pH so that some potassium-hemoglobin is converted to undissociated hemoglobin and the K ions released are associated with HCO_3^-. This, in its turn, reduces the concentration of H_2CO_3, more CO_2 enters the cells, is ionized to HCO_3^-, some of which moves out of the cell. The consequent disturbance of electroneutrality is removed by entry of chloride ions so that the result is an increase in the cell content of both bicarbonate and chloride ions.

20. MUSCLE

20.1. INTRODUCTION

Participation of ions in the specific function of muscles is of dual character. First, it is the involvement of ions in the mediation of signals for muscle work, and, second, it is the role of ions in the actual process of muscle contraction. Furthermore, various cations take part in the metabolic regulations also in the muscle and, as will be mentioned in the last part of this chapter, the unequal distribution between the outside and the inside of the muscle appears to be a decisive factor for the transport of some organic metabolites. Such phenomena are, nevertheless, observed with numerous other cells and are not concerned with the specific task of muscles as organs of body movement.

Let us first turn our attention to the role of the membrane surrounding the muscle fiber, the sarcolemma, in the distribution and transport of ions, to be followed by a discussion of the significance of the action potential and by a review of the present ideas on the transport of saccharides and amino acids.

Most of the studies were performed with striated muscles, particularly with frog sartorius or gastrocnemius and with mammalian diaphragm. Of the smooth muscles, those used most are the muscles of the alimentary tract and the uterine muscle. Experiments with the heart muscle related to transport problems contributed to the elucidation of some mechanisms underlying the function of this muscle. The heart muscle is not fully identical with either the striated or the smooth muscle but rather, due to its morphological anomalies which are reflected in the anomalous membrane characteristics, forms a class of its own. In some experiments, use of isolated muscle fibers was found to be expedient.

20.2. MORPHOLOGY

Before the characteristics of membrane transport in the muscle fiber are examined, a morphological description of the sarcolemma (the cell membrane of the muscle fiber) will be helpful. A detailed electronmicroscopic study of *Rana pipiens* m. sartorius was carried out by Peachey (1965). The muscle fiber, surrounded by a plasma membrane, consists of a number of myofibrils which display the well-known transverse striation. This arises by alternation of light isotropic bands (*I*), in the middle of which a dense line (*Z*) may be observed with optically denser anisotropic (*A*) bands. The central part of the *A* bands is formed by the so-called *H* zone, optically lighter, where the chains of myosin and actin molecules do not overlap. The outer plasma membrane (the sarcolemma) appears to be interconnected across the muscle fibers by connections situated along the *Z*-lines. The volume of these transverse tubules forms about 0.3% of the fiber volume, their surface, on the other hand, being about 7 times larger than that of the cylindrical fiber with a diameter of 100 μ. The continuity of the sarcolemma with the tubules aligned perpendicularly to the longitudinal axis of the fiber is suggested also by experiments of other authors, e.g., by Franzini-Armstrong and Porter (1964), Huxley (1964), and Zadunaisky (1966). Their results indicate that the intratubular and the extracellular fluids are of identical composition. On the other hand, a direct communication between the sarcolemma and the sarcoplasmic reticulum does not seem probable on the basis of electron microscopy. This reticulum is formed by a longitudinal network, originating in the center of one *I*-zone, proceeding across the *A*-zone to the center of the neighboring *I*-zone, and terminating in the terminal vesicles. The vesicles are joined to form continuous channels located in the close vicinity of transverse tubules at the *Z*-line. Even if the sarcoplasmic reticulum and the so-called *T*-system of the transverse tubules are not connected, their electrical coupling in, for instance, the course of a depolarization pulse may be envisaged. It may be seen in a longitudinal section of a striated muscle that due to a flattening of the transverse tubules (the longer axis being situated in the plane of the *Z*-line) the substantial part of their surface is reasonably close to the terminal cisternae which surround the tubules from both sides. This so-called triad system, described first by Porter and Palade (1957) seems to be a connective link between the electrical pulse spreading over the membrane and muscle contraction, as will be mentioned later. In Fig. 20.1 the muscle fiber structure is shown.

Morphological studies can assist in assessing the size and the localiza-

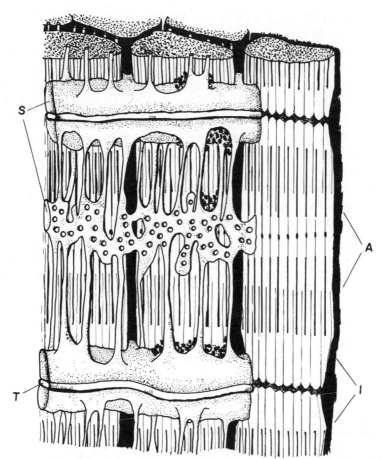

FIG. 20.1. Structure of the muscle fiber (Peachey, 1965). **A** Anisotropic band with *H* zone, **I** isotropic band with *Z* line, **T** transverse tubules, **S** sarcoplasmic reticulum.

tion of spaces in the tissue which may differ in its ionic composition and appear to be enveloped in a diffusion barrier. The interpretation of electron micrographs in terms of actual volumes of these spaces should be done with caution in view of possible fixation artifacts.

20.3. DISTRIBUTION OF IONS AND MEMBRANE POTENTIALS

Spaces of the greatest importance are those of intra- and extracellular tissue. The exact determination of the latter is an essential step in the estimation of the distribution of solutes between the outer and inner medium.

A great variety of compounds was used for the determination of the extracellular space in muscle, such as chloride, thiosulfate, sucrose, mannitol,
raffinose, inulin, etc. The principle of applying chloride is based on the
determination of the total tissue content of chloride anion and on the
calculation of its distribution according to the Donnan equilibrium (Conway, 1957). The various sugar spaces vary considerably, as shown in Table
20.1. This may be due either to imperfect impermeability or to adsorption
of the solute. It is generally agreed that inulin is most reliable for determining the extracellular tissue space. An accurate determination of the intra-
and extracellular spaces is even more difficult with the smooth muscle.
Here the problem of tissue separation may arise as the smooth muscle is
often accompanied by other material (e.g., secretory cell layers) and is
often attached to large amounts of connective tissue or nerve cells.

Table 20.2 shows several typical examples of the internal ionic composition of striated muscle together with the measured potential differences.
The muscle cell was first considered as a classical example of the Donnan
system (Boyle and Conway, 1941). The impermeability of the membrane
toward the sodium cation outside and the existence of nondiffusible anions,
such as proteins, phosphate esters and various coenzymes, inside the cell
were held to be the reason of intracellular potassium accumulation. Highly
permeant potassium ions compensate the negative charge of the intracellular
anions, the permeant chloride anions being present in small amounts.

Table 20.1. Extracellular Space of Different Muscles

Muscle	Estimated with	Extracellular space, % wet weight	Reference
Rat	sucrose	21.5	Kipnis and Parrish (1965)
diaphragm	mannitol	21.2	
	raffinose	20.1	
	inulin	15.5	
	thiosulphate	22.2	
Frog sartorius			
muscle	inulin	12.5	Desmedt (1953)
Frog stomach	inulin	28	Bozler and Lavine (1958)
Guinea-pig	inulin	33	Goodford and
taenia coli	polyglucose	22	Hermansen (1961)

Table 20.2. Internal Concentrations of Ions in Different Muscles *

Muscle	Na, mM	K, mM	Cl, mM	Reference
Frog sartorius muscle				
(freshly excised)	18.6	131.2	2.3	Sorokina (1964)
(incubated)	24.7	127.4	4.2	
	(116.2)	(2.5)	(117.1)	
Frog stomach	66	129	14	Bozler *et al.* (1958)
	(120)	(2)	(124)	
Guinea-pig	56–85	98–119	—	Goodford and
taenia coli	(137)	(5.9)	(134)	Hermansen (1961)

* The values in parentheses refer to the external media.

According to the Donnan equilibrium the products of diffusible ion concentrations at the two sides of the membrane are identical

$$[K^+]_i[Cl^-]_i = [K^+]_o[Cl^-]_o \qquad (1)$$

the membrane potential being equal to the Nernst equilibrium potentials of the diffusible ions

$$E_m = \frac{RT}{F} \ln \frac{[K^+]_i}{[K^+]_o} = \frac{RT}{F} \ln \frac{[Cl^-]_o}{[Cl^-]_i} \qquad (2)$$

Equation (1) was verified by Boyle and Conway (1941) and later confirmed by Adrian (1960) for varying concentrations of KCl in the outer saline. Changes in the value of $[K^+]_o[Cl^-]_o$ brought about corresponding changes in $[K^+]_i[Cl^-]_i$ within the range of experimental error. If only one factor of the total product is increased with a concomitant decrease of the other, the value of $[K^+]_o[Cl^-]_o$ remains constant and no KCl uptake takes place.

It was not necessary to reject eq. (2) even when it was demonstrated that the original assumption on sodium impermeability was wrong and that there doubtless exists an active sodium pump as was first postulated for muscle tissue by Dean (1941). Conway (1960a) emphasized that the operation of a pump does not invalidate the initial concept as it will do to assume virtual instead of absolute impermeability to sodium ions.

Nevertheless, when the dependence of the membrane potential on external concentration of potassium ions was investigated by the microelectrode technique (Adrian, 1956; Hodgkin and Horowicz, 1959), it was clearly demonstrated that the potential measured corresponded to the Nernst potential of potassium only at concentrations higher than 10 mM. With lower concentrations of potassium the E_m–log K_o^+ relationship deviates from a straight line, the potential measured at 2.5 mM K$^+$ being about 10 mV lower than the calculated one. Two suggestions were advanced to explain this discrepancy:

(1) In analogy with the nerve tissue, the Goldman equation, expressing the transmembrane potential as a function of concentrations of the principal permeant ions, was applied. On the assumption that chloride ions are in a passive flux equilibrium, the following relationship may be obtained (cf. section 4.1):

$$E = \frac{RT}{F} \ln \frac{[K^+]_o + \alpha[Na^+]_o}{[K^+]_i + \alpha[Na^+]_i} \tag{3}$$

At low [K$^+$]$_o$ the α[Na$^+$]$_o$ term may account for the deviations from linearity, as suggested by Adrian (1960).

(2) Conway (1957), on the other hand, proposed that under the experimental conditions used a true steady state does not obtain in the muscle at low [K$^+$]$_o$ and points out that in isolated muscle a net entry of sodium and a loss of potassium take place. It was demonstrated later by Kernan (1960, 1963) that on incubating the excised muscle in blood plasma a linear relationship between the potential and the logarithm of the outer potassium concentration may be obtained even at low [K$^+$]$_o$. Results agreeing within 1 mV of the Nernst formula were obtained for frog sartorius muscle and for the extensor digitorum muscle of rat.

It appears, however, that, especially for the smooth muscle, an equation of the Goldman type will be more applicable than the Nernst formula. The smooth muscle differs in several respects from striated muscle. The dimensions of the smooth muscle cells do not usually exceed 5 μ. According to Bozler (1948) two groups may be differentiated: (a) multiunit smooth muscles, excitable by extrinsic motor nerves (e.g., ciliary muscles, iris, nictitating membrane), (b) single-unit smooth muscles, which frequently exhibit continuous rhythmic activity and are not obligatorily dependent on their extrinsic nerves. As is shown in Table 20.2, the striated and smooth muscles differ also in the relative amounts of intracellular cations. A high intracellular content of sodium is found particularly with that group of smooth muscles which display spontaneous rhythmic activity (stomach and other muscles of the gastrointestinal tract, uterus, and urinary bladder muscles). A rapid sodium exchange across the membrane is linked with the high sodium concentration. It was found by Goodford and Hermansen (1961) that the value of 2×10^{-10} mol·cm^{-2}·sec^{-1} for sodium flux exceeds hundred times the potassium flux in taenia coli. Also the chloride content in smooth muscle is higher than in the skeletal muscle. It is worthy of notice that the cation content of the uterus may be remarkably influenced by hormones like estrogen, progesterone, or oxytocin. Also the values of the resting potential are influenced by sex and neurohypophyseal hormones.

The resting potential of the smooth muscles is generally lower and less stable than that of the skeletal muscle. For spontaneously active muscles the maximal transmembrane potential difference is defined by most authors as the resting potential value. Marshall (1959) found for rabbit uterus a value of about 40 mV but after estrogen alone or in combination with progesterone the figure increased by approximately 10 mV. The measured potential differences of smooth muscles vary between 40 and 60 mV, reaching exceptionally 70 mV. Membrane potentials of this muscle group are considerably affected by muscle tension, this being again in contrast to striated muscles.

It was most important to establish the actual intracellular activities of the principal ions, not only for calculating the relationship between the membrane potential and ion activities (instead of concentrations) if some of the ions were bound, but as one of the principal arguments in favor of either the membrane or the sorption theory. As reviewed by Ernst (1963), it has been suggested by a number of authors that at least a part of the intracellular potassium is bound; Ernst himself is of the opinion that all potassium is present in a bound form. Even if it cannot be denied that transient associations of ions with macromolecular structures play an important role in the membrane phenomena, it appears that potassium occurs free in the protoplasm. This view is supported by measurements of potassium mobility under the influence of an electrical field (Harris, 1954) as well as by the determinations of intracellular ionic activities by means of selective microelectrodes (Lev, 1964a). The activity of potassium ions was found to be 0.094–0.099 M, the corresponding activity coefficient being 0.75–0.80. Such an activity coefficient is just to be expected for a solution of K ions found within the muscle cells. On the other hand, the very low activity of 0.006 M Na$^+$ was found. The author notes that not only the binding but also the accumulation in some compartments with a resulting decrease of sodium activity in the rest of the sarcoplasm could be considered. Thus, some organelles such as mitochondria (Ulrich, 1959) or nuclei (Naora et al., 1962) contain more sodium than the bulk protoplasm. However, an analysis of the intracellular sodium state by NMR spectra performed by

Cope (1965) in the hind leg muscle of *Rana catesbiana*, indicates that 70% of the total sodium is found in a complexed form with macromolecules. Recently, the NMR analysis of sodium-loaded muscles was described (Ling and Cope, 1969), and since the newly entered sodium gave spectra characteristic for complexation, it was argued that the potassium previously present should also have been bound. This appears to be in contradiction with the measurements of activities and mobilities.

Another important problem tackled first in muscle tissue was that of electroneutrality or electrogenicity of the sodium pump. With an electroneutral pump, the tight coupling of the inward movement of potassium to the outward movement of sodium, representing an active transport of both these ions, would cause no separation of charge across the membrane. On the other hand, one can conceive that only sodium is actively transported and due to this transfer a potential difference is created. The true electrogenicity of the sodium pump could be under most circumstances masked by the high permeability of the K ion. Keynes (1954) suggested a model of a chemically coupled sodium–potassium pump, which would exchange the cations in a 1:1 ratio. The author made use of the observation that sodium efflux appears to be regulated by the external potassium, increasing with higher and decreasing with lower $[K^+]_o$. Nevertheless, there were indications that a 1:1 exchange need not hold under all conditions. Ussing (1960), referring to the finding that more sodium is taken up than potassium lost in an ischaemic muscle and hence during recovery more sodium pumped out than potassium gained, wrote "... it thus would seem that a sodium pump would be better suited than a coupled Na–K pump to bring about normal ionic distribution." To clarify the situation the following test was used by Kernan (1962) and Keynes and Rybová (1963): Sodium-loaded muscles were transferred from a potassium-free to a recovery solution containing 10 mM K^+. In the latter solution, an active extrusion of sodium with a simultaneous uptake of potassium takes place. The transmembrane potential E_m, measured with glass microcapillary electrodes until a new steady state was reached was compared with the Nernst diffusion potential for potassium ions E_K. Now, if at any time the potential E_m was smaller than E_K or equal to it, an electroneutral pump might be in operation. If, however, E_m was greater than E_K, indicating that passively moving potassium is driven inside the cell by electrical forces, an electrogenic pump should be considered. The results were clearly in favor of an electrogenic pump (Fig. 20.3), the E_m being substantially above E_K during the initial period of recovery. Cross and co-workers (1965) observed that, under otherwise identical experimental conditions with Ca^{2+} and Mg^{2+} ions

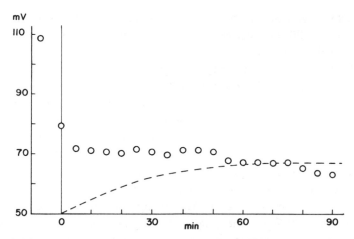

Fig. 20.2. Time change of the membrane and of the Nernst potassium potentials of Na-loaded frog sartorius muscle in a 10 mM K$^+$ solution. The points represent the measured membrane potential, the dashed line expresses the calculated E_K. At zero time, the muscles were transferred to the recovery solution. (According to Cross *et al.*, 1965.)

omitted from the incubation saline, the muscles gained potassium at E_K higher than E_m. A partly electrogenic pump is preferred by Adrian and Slayman (1966) who compared the Na–Rb and Na–K exchange in Na-loaded muscles. A sodium pump capable of transferring electrical charge in muscle tissue was also supported by the results of Mullins and Awad (1965) and Hashimoto (1965).

Let us now analyze some specific features of the fluxes of individual ions across the muscle fiber membranes.

The sodium flux in the resting state of muscle reaches usually values of about 10^{-11} mol·cm^{-2}·sec^{-1}. The sodium efflux from the frog sartorius muscle appears to be (for a certain range of concentrations) proportional to the third power of the internal sodium concentration, as shown by Keynes and Swan (1959). The finding was confirmed by Harris (1965a) who, studying the effects of external ion concentrations on the sodium efflux, suggested that an active sodium–potassium exchange, as well as a passive sodium–sodium exchange, is of importance in this process. Likewise, the intracellular pH of the fiber affects sodium efflux (Keynes, 1962). Decrease of pH inside the fiber by increasing p_{CO_2} in the external saline reduced the sodium efflux and it was concluded that a competition between sodium and hydrogen ions for the carrier sites is the likely explanation. Such an explanation is in agreement with the findings of Caldwell (1954) and Kostyuk and Sorokina (1961) who demonstrated that hydrogen ions are not in equilibrium across the muscle fiber membrane but rather that they are actively extruded.

The potassium permeability of single muscle fibers was studied in more detail by Hodgkin and Horowicz (1959) who showed that the changes in potassium conductivity are incompatible with the equations according to Hodgkin and Katz (1949), if the permeability coefficient for K^+, P_K, is considered to be constant, and concluded that potassium permeability varies, depending on the magnitude and the direction of the driving forces. Movements of potassium chloride under conditions where a net efflux of potassium and chloride ions takes place were investigated by Adrian (1960) who concluded that the P_K decreases during outward potassium movement, being inversely related to the magnitude of the outward driving force. Later the phenomenon of anomalous rectification and its correlation with potassium conductivity was studied by Adrian and Freygang (1962a,b), using the cable theory and the voltage-clamp method (see chapter 21). The authors explained their results by assuming the existence of two membranes with differing properties, across which potassium is exchanged. One has a low potassium conductance when at rest, giving a linear relationship of current–voltage curves and was localized at the fiber surface; the other is responsible for anomalous rectification, has a greater permeability to potassium, and was identified as a probable membrane on the border of endoplasmic reticulum.

Harris (1965b) gives data on the temperature dependence of the permeability: $P_K = 0.4 \times 10^{-6}$ cm·sec^{-1} at 0°C and 0.7×10^{-6} cm·sec^{-1} at 20°C. It is of interest that for chloride ions the corresponding values were 0.4×10^{-6} cm·sec^{-1} and 2.3×10^{-6} cm·sec^{-1}, respectively, indicating a greater sensitivity of chloride permeability to temperature.

Under resting conditions, two-thirds of the membrane conductance are due to chloride ions, the remaining third being due to potassium ions (e.g., Hutter and Noble, 1960). On the basis of flux experiments Adrian (1961) expressed the view that the outward and inward fluxes of chlorides are interdependent. Moreover, the efflux of chloride is impaired by replacement of external chloride with a foreign anion, such as Br^-, NO_3^-, I^-, ClO_4^-, or CNS^- (cf. Harris, 1958; Adrian, 1961), which might emphasize the fact that the efflux of chloride is not independent of the external anion. Harris (1963), investigating distribution of muscle chloride, came to the conclusion that two compartments exist in the muscle, differing with respect to the composition and exchangeability of chlorides. One might speculate that in one compartment all the principal ions are dissolved and the efflux of chlorides from this region is independent of the external anions. The second compartment does not admit sodium and the more slowly exchanging chloride from this region depends on foreign anions and on the counter-flow of chlorides.

Considering the dependence of chloride conductance on pH changes, the results of Hagiwara and co-workers (1968) are relevant. Whereas at external pH of 7.7 the ratio of Cl$^-$ to K$^+$ conductance is about 1:6, it is reversed under acid pH values. This decrease in cationic and increase in anionic conductance with decreasing pH is consistent with the concept of amphoteric fixed-charge groups in the membrane.

It should be stressed that the electrical parameters of skeletal muscle (e.g., membrane resistance, ionic permeabilities) do not substantially differ from those found for other excitable animal cells with the exception of membrane capacity. The striated muscle capacitance is higher in comparison with the value found in the nerve (about 1 μF·cm^{-2}). This was explained by Falk and Fatt (1964) as follows: In addition to the capacitance in parallel with resistance there should be a capacitance component in series with resistance which might be due to the current flow in the transverse tubules.

In concluding the section on ion fluxes, let us consider the amount of energy consumed for maintaining the unequal distribution of ions and sources of this energy.

The energy required by the active transport of one mole of actively excreted Na ions is given by

$$\frac{dG}{dn} = \bar{\mu}_{Na_o} - \bar{\mu}_{Na_i} \tag{4}$$

which may be rewritten in the form:

$$\frac{dG}{dn} = RT \ln \frac{[Na^+]_o}{[Na^+]_i} + FE_m \tag{5}$$

where dG represents the input of free energy necessary for the reaction.

Levi and Ussing calculated first (1948) the energy expenditure for active transport in muscle. Multiplying the right-hand side of eq. (5) by the total amount of sodium leaving 1 kg of muscle fibers per hour, the authors obtained for the work performed 50 cal·hr^{-1}·kg^{-1}. Comparing this value with the energy derived from oxygen consumption, Levi and Ussing concluded that about 30% of metabolic energy in the resting muscle can be used for sodium transport. In an analogous treatment, a somewhat lower yield of energy (about 10%) was found to take part in the active process by Keynes and Maisel (1954). Objections may be raised against such calculations as to certain inaccuracies which seem to be included (*cf.* Ussing, 1960). One complication is connected with the employment of the actual activities of intracellular sodium, because in eq. (5) activities rather than concentrations should be used. Further, it should be taken into account that a part of the Na efflux performed by exchange diffusion (Levi and Ussing, 1948) ought to be omitted from the calculations. In spite of these imperfections such computations are worthy of note, giving a rough idea on the distribution of metabolic energy.

The calculations of the critical energy barriers in the active transport of ions as put forward by Conway (1960*b*) are instructive. Sodium-loaded muscles will not pump sodium ions outward in a recovery solution with 10 mM K$^+$ unless either (i) the external sodium concentration is decreased below a certain limit or (ii) the membrane potential E_m is decreased, for instance by raising the external potassium concentration (*cf.* eq. (5)). The critical energy barrier in *Rana pipiens* was found to be 2.4 cal/m-equiv Na$^+$ (Dee and Kernan, 1963). The concept of the critical energy barrier due to Conway suggests that the pumping rate of sodium depends not only on the concentration gradient for sodium but also on the electrical forces, i.e., on the membrane potential. Fozzard and Kipnis (1967) actually demonstrated that on changing the concentration of potassium ions in the medium both the membrane potential and the intracellular concentration of sodium change in a way corresponding to a sodium pump transporting up to a certain constant electrochemical potential gradient of the sodium ions.

The reaction mechanisms delivering energy for the maintenance of the internal ionic milieu in muscle appear to be different from those in other tissues (e.g., nerve; *cf.* chapter 21) and, moreover, a variability is found in muscles from various animal species. Thus, whereas mammalian muscles appear to be sensitive to the lack of oxygen, the frog muscles are capable of transporting actively even under anaerobic conditions, losing this property only if iodoacetic acid combined with anaerobiosis is applied. Dydynska and Harris (1966) investigated the level of high-energy phosphates under conditions when the semitendinosus and sartorius muscles of *Rana pipiens* extruded sodium and reaccumulated potassium. Under anaerobic conditions, the movements of ions were accompanied by phosphocreatine decomposition and lactate synthesis. An ATP breakdown, together with lactate formation can be observed when creatine phosphotransferase is inhibited by dinitrofluorobenzene; ATP appears therefore to be more directly involved in the active pumping processes. The authors found that about 2.5 sodium ions are transported per ATP molecule.

Skou (1965) proposed that the membrane ATPase could be the system that might account for a direct connection with ion transport. In 1965, Skou stressed that this ATPase fulfils numerous requirements of such a hypothesis. The enzyme system is localized in the cell membrane, it has a higher affinity for sodium than for potassium inside, the opposite being true outside. The system is ATP-driven, containing enzymes for ATP hydrolysis and being potentially capable of energy conversion into cation movement. The rate of hydrolysis should depend on $[Na^+]_i$ and $[K^+]_o$. Finally, the system is present wherever a transport-linked system was demonstrated. A membrane ATPase, selectively binding cations, was actually demonstrated in various tissues and in organisms of various phylogenetic stages. In muscle, the existence of the enzyme system was demonstrated by Bonting and co-workers (1961). Kinetic analyses of the ATPase system showed that there are two different sites with affinity for sodium and potassium ions, it being not decisive for the merit of the hypothesis whether the sites are actually two or only one with a changeable affinity according to its position in the membrane. The finding that ouabain is without effect if added from the inside (Caldwell and Keynes, 1959) supplies evidence in favor of the asymmetry. The transport ATPase system of Skou is in best accordance with contemporary ideas of the Na–K-linked active movement across the membrane. A powerful argument in favor of the ATPase hypothesis was brought forth by Garrahan and Glynn (1967) who demonstrated that when reversing the sodium and potassium concentration gradients, ATP is being synthesized in resealed red cell ghosts.

Another hypothesis pertinent to the coupling of ion transport with enzyme activities is the redox-pump theory of Conway, originally applied to the secretion of H^+ by the oxyntic cells of the gastric mucosa (Conway, 1951) and later also to muscle tissue (e.g., Conway, 1964). In muscle, Na^+ is supposed to be bound to a negatively charged reduced enzyme cofactor which acts as a carrier. The transferred cation is released when the enzyme is converted into the oxidized form (*cf.* p. 316).

The phenomenon of the action potential and its relation to movements of ions is described in some detail in chapter 21. In principle, the mechanism is the same in muscle tissue and only a few deviations and the role of calcium as a joining link between the action potential and muscle contraction will be dealt with here.

The excitation pulse proceeds from the nerve endings into the muscle fibers via the neuromuscular junction. Acetylcholine, released from miniature nerve vesicles, causes a depolarization of the muscle fiber in a limited region and once the action potential is formed it propagates quickly along the fiber, so that the delay between individual fibers at distant places is negligible and a synchronization of tension in the whole muscle is made possible. The form of the action potential spike in the striated muscle does not differ from the one in nerve, but another type of waves is typical of cardiac and some smooth muscles (e.g., in ureter: Burnstock and Prosser, 1960). The action potential in the heart muscle is represented by a wave of comparably longer duration (some 500 msec), this prolongation being presumably due to a plateau of constant potential level, which appears after the top spike potential has decreased by several millivolts. The appearance of the plateau in the action potentials of this type was in most cases explained in accordance with the Hodgkin–Huxley theory (see chapter 21). A delay in the increase of potassium permeability, while simultaneously a certain fraction of raised sodium permeability is maintained, was proposed. More recently, Dudel and co-workers (1967) stressed the role of chloride ions in the early outward-directed current which brings about the termination of the plateau. Other results show, however, that this need not necessarily be decisive for the duration of the action potential (Noble and Tsien, 1969).

Experimental results of the last ten or fifteen years concerning the calcium translocations across the muscle membrane and in the sarcoplasmic reticulum and the various effects of this ion in muscle tissue led to a more unified hypothesis on the role of calcium ions in the stimulation-contraction coupling, as was summarized by Podolsky (1965) and Caldwell (1968), among others.

When the electrical pulse spreads over the muscle fiber and, very probably, also into the transverse tubules, Ca ions begin to be released from the sarcoplasmic reticulum. The calcium concentration increases from the resting value of 10^{-8}–10^{-7} M to a value of 10^{-6} M. This concentration is capable of activating the actomyosin system and ATP-breakdown together with contraction occur. The contractile proteins require Ca^{2+} in addition to ATP to contract. For the muscle to return to the resting state, the Ca^{2+} concentration must be decreased below 10^{-6} M; thus, during relaxation, calcium is reaccumulated in the sarcoplasmic reticulum by an active process requiring ATP and Mg^{2+}.

20.4. TRANSPORT OF SUGARS AND AMINO ACIDS

Transport of sugars and amino acids as of principal natural metabolites is mostly carrier-mediated, but it appears that there exist substantial differences between carrier systems for amino acids and those for sugars.

Akedo and Christensen (1962) showed that most amino acids are accumulated actively in rat diaphragm, the s_{II}/s_I being greater than one. The amino acids L-valine and L-serine appear to compete with the non-metabolizable α-aminoisobutyric acid for the entry into this muscle. The penetration of amino acids into the sartorius muscle of *Rana catesbiana* was investigated by Padieu (1965). Serine and methionine enter the muscle readily, whereas the entry of glycine, valine, phenylalanine, and tyrosine is considerably slower. Glutamic and aspartic acid penetrated only slightly.

Kipnis and Parrish (1965), using nonmetabolizable α-aminoisobutyric acid (AIB) which does not exhibit any exchange diffusion, succeeded in demonstrating the Na^+ dependence of active amino acid transport. When sodium was replaced with choline in the external solution the s_{II}/s_I ratio was always below one. Nevertheless, competition with glycine was preserved which pointed to the fact that the penetration was still mediated by a carrier. If 1/AIB is plotted against $1/Na^+$ a linear relationship is obtained, this indicating that the acid and sodium react with the carrier molecule in a 1:1 proportion. The participation of an active sodium pump in the amino acid accumulation was excluded since no sensitivity to strophantin was found. The increased potassium concentration in the outer saline caused changes in the AIB fluxes. The authors attributed the most important role in amino acid accumulation to the unequal distribution of cations, this being reflected in an asymmetry of carrier activity at the outer and the inner surface of the membrane, as Na ions activate and K ions inactivate the carrier system.

In contrast to the transfer of amino acids, the carrier-mediated penetration of sugars does not appear to be an active process as it does not occur against a concentration gradient. It should be, however, realized that the apparent and the real concentration of a substance inside the fibers are not necessarily identical, a compartmentation of the substance inside the fiber being generally assumed. This situation may modify the apparent concentration gradients considerably.

Kipnis and Cori (1959) on the basis of competition between glucose and mannose, on the one hand, and no interference between glucose and xylose or 2-deoxyglucose and galactose, on the other, concluded that there might exist different pathways for metabolizable and nonmetabolizable sugars. A competitive inhibition between the group of sugars comprising D-arabinose, D-glucose, D-mannose, D-lyxose, D-xylose, and 3-O-methyl-D-glucose was later found by Battaglia and Randle (1960) who also described the inactivity of the carrier system to D-galactose, D-fructose, and D-sorbitol and its sensitivity to —SH inhibitors. The above-mentioned findings about sugar competition and stereospecificity, together with the observation by

Narahara and co-workers (1960) or Morgan and co-workers (1961b) on the saturation kinetics of substrate penetration, indicated a binding of the transported sugars to a membrane component. The ocurrence of sugar counterflow in the heart muscle (e.g., Morgan et al., 1961a) shows that this carrier is mobile. Countertransport of nonmetabolizable 3-O-methylglucose against glucose was also demonstrated in skeletal muscles (m. gastrocnemius, diaphragm) by Morgan and co-workers (1964). The Michaelis–Menten constant for glucose transport (11 mM) calculated by these authors from findings with two competing sugars was in good agreement with the one (9 mM) found by Post and co-workers (1961).

It is of interest to ponder the problem of regulation of the intracellular sugar level in the muscle tissue. Randle and Smith (1958) demonstrated that sugar transport is stimulated by anoxia and in the presence of inhibitors of oxidative phosphorylation. Recently, evidence was provided by Bihler (1968) that the 3-O-methylglucose entry is enhanced under conditions of Na-pump inhibition, i.e., after application of cardiac glycosides at a concentration known to impair the active sodium transport, or in potassium-free medium. This finding fits in with the observed decrease in the intracellular sugar space in high potassium saline (Bhattacharya, 1961; Rybová, 1965). It should be noted further that increased external potassium concentrations stimulate the muscle sodium pump (e.g., Horowicz and Gerber, 1965). Bihler (1968) postulated that the sodium pump might regulate the sugar transport in muscle by a negative feedback, the link between pump and influence on the transmembrane sugar movement being yet unclear.

Much has been done on the action of hormones, especially of insulin, on the transport of sugars and amino acids. Although there are indications that insulin might stimulate protein synthesis (Wool, 1965) it is now without doubt (Levine et al., 1949) that insulin facilitates the uptake of sugars (Fisher and Lindsay, 1956; Morgan et al., 1959; Kipnis, 1959) and amino acids (Battaglia et al., 1960; Manchester and Young, 1961; Wool, 1965) across the cell membrane. It appears that insulin acts by accelerating somehow the operation of the carrier.

When touching on the problem of hormone action on the transport of nonelectrolytes, the possible participation of estrogen in increasing the rate of sugar and amino acid transport across the membrane should be mentioned (Roskoski and Steiner, 1967a,b).

For more detailed information about the phenomena dealt with in this chapter, the following literature may be consulted: Harris (1960), O'Connor and Wolstenholme (1960), Ussing (1960), Burnstock et al. (1963), Kernan (1965), Katz (1966), and Noble (1966).

21. NERVE

21.1. INTRODUCTION

Nerves represent a favorite object of studies of ion transport across plasma membranes in the animal tissue. Transport of sodium and potassium ions is directly related to the specific function of the nerve fibers, *viz.*, excitability and rapid information transfer within the animal body, the velocity of which may reach some 100 m·sec^{-1}.

The study of transport processes in nerves has contributed to our understanding of various facets of membrane phenomena. The volume and the surface of the studied material can be readily estimated, just as the internal ionic composition. The giant axons of marine cephalopods, such as sepia or squid, appear to be especially suited for such purposes. These axons are about 1 mm in diameter and several cm in length and their gel axoplasm can be extruded by gently squeezing the integument so as to obtain intact material for analysis. The giant axons do not have the myelin sheath and represent therefore a suitable object for kinetic studies.

The important finding of Hodgkin and Keynes (1953) that the mobility of labeled potassium ions inside the nerve is about the same as in a pure salt solution is relevant to the controversy between the sorption and the membrane theories; the free mobility inside the nerve and the resistance encountered between the axoplasm and the external medium support the concepts of the membrane approach. An even more powerful argument in the same direction derives from experiments with perfused axons to be discussed later (Baker *et al.*, 1961).

The development of the microelectrode technique made it possible to consider the mutual relationship between the electrical potential difference across the cell membrane and the concentrations, or rather activities of ions at both sides of the cell membrane. Originally, the potential difference across the axon membrane was measured as the so-called injury potential: one electrode approached an injured part of the axon or a part depolarized by isotonic KCl and the second one an intact part. Using special techniques (e.g., isolating gaps; Huxley and Stämpfli, 1951) satisfactory results may be obtained in this way. Glass microcapillary electrodes with a tip diameter of about 1 μ or even less, functioning as salt bridges, allowed to measure the potential difference practically in an intact state (Ling and Gerard, 1949). A number of experiments has been carried out with the microelectrodes of the Ling–Gerard type, in which the transmembrane differences, the dependence of cell potential on the ionic composition, etc. were studied.

21.2. DISTRIBUTION AND FLUXES OF IONS

Let us turn our attention first to the characteristic features of the steady-state distribution of ions between the external physiological medium which is rich in sodium and the intracellular milieu of the nerve cell (neuron) rich in potassium. The concentrations of the main ions inside and outside the nerve cell pictured in Fig. 21.1 are shown schematically in Fig. 21.2. The relation between the electrical potential of the cell and the unequal distribution of ions has been discussed for a number of years. It appears that the simplest correlation of these parameters for the nerve resting state is given by the formula

$$E_m = -\frac{RT}{F} \ln \frac{[K^+]_i}{[K^+]_o} \tag{1}$$

Following the suggestion of Bernstein (1902) this expresses the fact that the resting membrane potential is determined by the ratio of potassium concentrations and that thus the cell membrane behaves analogously to an ideal potassium electrode. It will be seen later that this equation does not hold under all conditions and also that other factors must be taken into consideration as playing a role in the formation of the membrane potential (cf. eq. (21)). Nevertheless, formula (1) is suitable as a first approximation for the nerve in an inactive state and under physiological conditions (Moore and Cole, 1960).

The validity of the formula is suggested also by analogy with muscle tissue, where the Nernst potential of potassium is in good agreement with experimental results for a wide range of external potassium concentrations. An essential impermeability of the cell membrane to sodium was originally assumed by Boyle and Conway (1941) as a prerequisite for formulating eq. (1). However, it was demonstrated later by means of measurements of

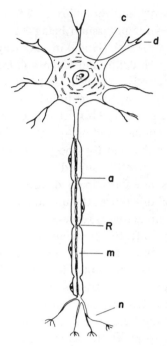

FIG. 21.1. Structure of the nerve cell. **c** Nerve cell body with nucleus, **n** nerve ending, **a** axon, **d** dendrite, **m** myelin sheath, **R** node of Ranvier.

ionic fluxes that in fact a rapid exchange between the intracellular and the extracellular sodium takes place (Rothenberg, 1950; Keynes, 1951). Moreover, it was obvious that the influx and efflux of sodium in the resting state is comparable in size to the movements of potassium. Thus, for example, in the giant axon of *Sepia officinalis* the influx of potassium amounts to

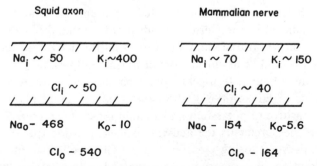

FIG. 21.2. Illustration of the internal and the external ionic concentrations of the nerve cell.

3.1×10^{-11} mol·cm^{-2}·sec^{-1} and the efflux to 2.8×10^{-11} mol·cm^{-2}·sec^{-1}, the corresponding values for potassium being 3.2×10^{-11} mol·cm^{-2}·sec^{-1} and 3.9×10^{-11} mol·cm^{-2}·sec^{-1} (Hodgkin and Keynes, 1955a). The contradiction was solved by Hodgkin and Keynes (1955a), who demonstrated that sodium ions are actively extruded from the axon at the expense of metabolic energy. Moreover, potassium influx was affected by the inhibitors in the same way as the sodium efflux. It was concluded that the sodium pump is operative as a coupled sodium–potassium pump, one sodium ion extruded being always exchanged for one potassium ion. Such an electroneutral pump presumes a tightly coupled active transport of both sodium and potassium. It does not participate directly in the formation of the transmembrane potential difference; the unequal distribution of ions (which is mostly due to the pump activity) represents the source of the membrane potentials, generated by different permeabilities of the membrane toward sodium and potassium ions. The pump, however, does not appear to be always electroneutral. It was mentioned in chapter 20 that, at least for the muscle and possibly some other tissues, the concept of an electrogenic sodium pump is justified. Recently, evidence has been presented by Rang and Ritchie (1968) that the hyperpolarization, which follows the period of activity in nerve, is due to an electrogenic sodium pump.

Curtis and Cole (1942) measured the response of the membrane potential to the changes in the concentration of external potassium. The agreement of the Nernst potential for potassium with the membrane potential was obtained with higher concentrations of the external potassium. However, in the range of physiological concentrations, the potential was less dependent on the outer potassium concentration, the slope of the line being substantially lower than should correspond to an ideal relationship. The permeability of the membrane to Na and Cl ions is often used to explain this behavior which can be described by the Goldman equation (21).

As in the nerve the potential difference found is usually lower than would be predicted by the Nernst equilibrium potential for potassium ions, the explanation was occasionally put forth that the activity inside the axon is lower than would be assumed from chemical analyses, i.e., that a part of the potassium is bound. It appears that such an assumption is contradicted by the results of Hinke (1961) who penetrated the squid giant axon with electrodes sensitive to potassium and sodium ions (cf. chapter 12). According to these measurements the activity coefficient inside the axon is equal to 0.605, this value being close to the activity coefficient for potassium chloride solution at a concentration analogous to that of the internal medium. This finding is also in accordance with the earlier observation of

Hodgkin and Keynes (1953) about potassium mobility in the axoplasm where practically all the intracellular potassium appeared to participate in the diffusion through the axon. A lower activity was, however, found for sodium ions (Hinke, 1961), either binding or compartmentation being thus suggested.

Further progress in the examination of the role of ions in the generation of the membrane potential was achieved by the technique of axon perfusion, where not only external but also internal ionic composition may be arbitrarily changed. Baker and co-workers (1961) used the method of complete removal of the axoplasm by gentle squeezing, replacing it with a solution of the required composition. An analogous method was applied by Oikawa and co-workers (1961) to the giant axon of *Loligo pealii*, who perfused a part of the axon by means of inserted cannulae. With the above techniques the possibility was opened for examining whether eq. (*1*) predicts correctly the changes in the membrane potential brought about by varying the intracellular ion concentrations. On replacing partly the internal solution of potassium sulfate with a saccharide solution (Tasaki *et al.*, 1962; Baker *et al.*, 1962) the potential was not reduced to the expected degree. The membrane potential of the axon appeared to be insensitive to changes in the internal potassium concentration over a wide range. On the other hand, the transmembrane potential decreased practically to zero (as would correspond to the Nernst potassium potential) when potassium was completely removed and replaced with sodium. The authors of the above studies tried in fact to test the validity of the Goldman equation (eq. (*21*)) which will be discussed later and which (see p. 385) reduces to eq. (*1*) if (i) the nerve permeability to sodium in the resting state is negligible as compared to that to potassium and (ii) chlorides are distributed passively (or, alternatively, their permeability is also negligibly small in comparison with potassium permeability). Baker and co-workers (1962) offer an explanation for the disagreement between the changes in $[K^+]_i$ and the experimental values of E_m by assuming that alterations in internal potassium concentration cause changes in the permeability in such a way that the product of the two $(P_{K^+}[K^+]_i)$ remains constant. A similar conclusion was made by Stämpfli (1959).

Summarizing, the resting nerve fiber under optimal conditions displays an efflux of sodium accounting for the energy-driven flux and a small fraction of passive leak, and a corresponding influx, and a considerably larger flux of potassium (due to higher permeability) including, in the influx, probably a smaller fraction of active transfer. Hodgkin and Keynes (1955*b*) found in a cyanide-poisoned fiber an anomalous dependence for the ratio of passive K^+ fluxes, where n was approximately 3:

$$RT \ln \frac{\Phi_{in}}{\Phi_{ex}} = n(\mu_i - \mu_o) \qquad (2)$$

They explained this discrepancy with the Ussing–Teorell flux equation by suggesting a single-file diffusion in a pore (see chapter 5).

Another cation which appears to be actively transported in the nerve is the hydrogen ion. Caldwell (1958) succeeded in measuring the internal pH of the nerve fiber with a glass electrode and found it to be equal to 7.2. This value shows that the distribution of hydrogen ions across the membrane does not correspond to the Donnan equilibrium according to which E_m, the membrane potential, would have to be equal to 58 (pH$_i$ – pH$_o$). When increasing the partial pressure of CO_2 in the external medium, the internal pH decreases to a new value suggesting (a) the presence of a bicarbonate buffer system with the membrane being practically impermeable to bicarbonate, (b) an active transport of hydrogen ions outward.

Let us now consider the most readily permeating anions taking part in the ionic equilibria between cells and their surroundings, the chloride ions. The assumption of a passive equilibrium distribution of the ions across the cell membrane was already encountered when dealing with the transport of ions in the muscle fiber. According to Boyle and Conway (1941) the distribution of chloride, as well as of potassium, corresponds there to the Donnan equilibrium. The same was presumably true in the nerve and it was assumed that there, too, the resting potential of the membrane could be described by

$$E_m = \frac{RT}{F} \ln \frac{[Cl^-]_i}{[Cl^-]_o} \qquad (3)$$

Indeed, when using the value of Steinbach (1941) for the intracellular concentration of chlorides the value of -66 mV was derived, in agreement with the *in vivo* resting potential in the nerve (Moore and Cole, 1960). However, when analyzing the content of chloride anions in the squid axon (*Loligo forbesi*) Keynes (1963) found values exceeding 100 m-equiv/kg fresh tissue which is in contradiction with the condition of equilibrium distribution of chloride ions since a membrane potential of only -39 mV would correspond to such an equilibrium distribution. With Ag/AgCl electrodes the activity coefficients of about 0.7 were found and the possibility of an intracellular binding of chloride ions in the nerve was thus excluded. Moreover, 2,4-dinitrophenol was found to inhibit the entry of chlorides by about 50%, whereas ouabain was shown to be without effect. Consequently, an inward directed active chloride pump independent of the active sodium–potassium pump was suggested to operate in the nerve membrane. The independence of the two pumping mechanisms was further demonstrated by the finding that in potassium-free media the transport of chloride ions is not affected. The very important demonstration of an active chloride pump in animal tissue due to Keynes was recently followed by similar demonstrations in other animal tissues (e.g., in the frog skin, see chapter 22), whereas previously anionic pumps were considered as an exception in the animal kingdom, limited to specialized tissues like the gastric mucosa.

The actively transported anions most probably include also phosphates. Caldwell and Lowe (1966) showed that the phosphate influx in the giant sepia axon which is in the resting state amounts to 0.0196 pmoles$\cdot$cm$^{-2}\cdot$sec^{-1} in media containing 0.1 M orthophosphate, is markedly inhibited by cyanide and dinitrophenol. It was interesting to observe the inhibition by 10^{-4} M ouabain and the authors do not exclude the possible connection between phosphate transport and the Na–K pump mechanism.

A rather detailed analysis of metabolic sources which are required by the nerve cell for active sodium extrusion, was carried out by Caldwell and co-workers (Caldwell and Keynes, 1957, 1959; Caldwell et al., 1964). The difference between the action of metabolic inhibitors and the specific action of cardiac glycosides on the sodium pump was demonstrated in experiments where, in cyanide-poisoned axons, sodium extrusion was restored by injecting energy-rich compounds, whereas in ouabain-treated axons no restitution of the pump activity was achieved. In the course of studies also the interesting observation that ouabain is not effective when applied to the inner membrane surface was made. At first it seemed puzzling that the activity of the pump in poisoned axons could be restored by arginine phosphate or phospho-enol-pyruvate, ATP being practically ineffective. An analysis of phosphorus compounds showed that after addition of arginine phosphate ATP is quickly synthesized and, after phospho-enol-pyruvate injection into cyanide-poisoned nerve trunks, ATP as well as phosphocreatine are formed. It remains, however, not clear, whether the active pump requires the presence of both ATP and phosphocreatine or, in case that phosphocreatine is not necessary, why only ATP formed de novo is active.

21.3. PASSIVE ELECTRICAL PROPERTIES OF NERVE FIBERS

By its shape (an elongate cylinder), its conductivity due to axoplasm, the presence of an external insulating layer, and its being surrounded by conductive environment, the nerve axon reminds one of a submarine cable.

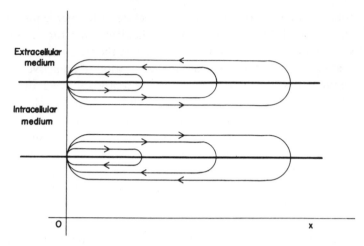

FIG. 21.3. Spread of subthreshold potential changes along a nerve fiber (or another cable-like structure) by local currents.

The particular electrical parameters also comply with such a concept even if the conductivity of the internal protoplasm is not comparable with the conductivity of a metal conductor and the resistance of the axon sheath is insignificant as compared with the resistance of a cable insulator. Moreover, the current is not effected by electrons but rather by ions. Nevertheless, the so-called cable theory provides a fine interpretation of the electrical phenomena and definitions of electrical parameters not only for nerves, but also for muscles or some single plant cells of elongate shape (e.g., algae belonging to the *Characeae*).

Let us assume that a change of the membrane potential, of an arbitrary time course too small to excite the membrane, occurs in the region of the point $x = 0$ of the nerve fiber in Fig. 21.3. Local currents are caused by this potential change and, through them, the potential change spreads along the fiber. The dependence of the potential change on the distance as well as on time is often satisfactorily expressed by a partial differential equation called the cable equation. Solutions of the cable equation for various special cases make it possible to obtain good approximations to the values of membrane resistance and capacity. The membrane resistances and capacities thus obtained may be compared with the values measured on other cell membranes or artificial membranes, and various conclusions concerning the membrane permeability to ions may be drawn from them.

There is a number of papers dealing with the cable theory and its applications in biology (see Hodgkin and Rushton, 1946; Taylor, 1963;

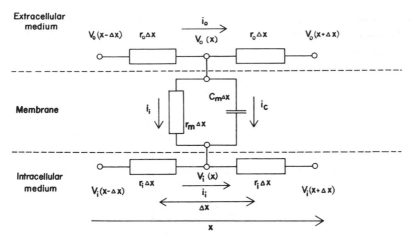

FIG. 21.4. An element of an equivalent circuit of the cable membrane.

Noble, 1966). Here only a simplified derivation of the cable equation will be shown, and one of its simple solutions valid under steady-state conditions will be briefly discussed.

An equivalent circuit of the cable membrane is shown in Fig. 21.4. Here

i_o is the current flowing through the extracellular medium (in μA),

i_i is the current flowing through the core of the cablelike structure (in μA),

i_m is the current flowing through the membrane per unit of the cable length (in $\mu A \cdot cm^{-1}$); i_m has two components, $i_m = i_i + i_c$, i_i corresponding to the flow of ions through the membrane resistance, i_c flows into the membrane capacity,

r_m is the resistance of the membrane, corresponding to the unit length of the structure (in $\Omega \cdot cm$),

r_o is the resistance of the unit length of the extracellular medium,

r_i is the resistance of the unit length of the intracellular fluid,

c_m is the capacity of the membrane per unit length of the structure (in $\mu F \cdot cm^{-1}$),

V_m is the change in the membrane potential brought about by the flow of current across the membrane, $V_m = E_{m,i=0} - E_m$, where $E_{m,i=0}$ is the open-circuit membrane potential, assumed to be a constant and E_m is the actual membrane potential at distance x and time t. V_m may be written as a difference of two potentials, $V = V_o - V_i$.

By applying Ohm's and Kirchhoff's laws to the circuit in Fig. 21.4, the cable equation may be derived as follows. For currents i_o and i_i Ohm's law may be written as

$$i_o = - \frac{1}{r_o} \frac{\partial V_o}{\partial x} \tag{4a}$$

$$i_i = - \frac{1}{r_i} \frac{\partial V_i}{\partial x} \tag{4b}$$

By differentiating the change in the membrane potential $V_m = V_o - V_i$, we obtain

$$\frac{\partial V_m}{\partial x} = \frac{\partial V_o}{\partial x} - \frac{\partial V_i}{\partial x} \tag{5}$$

Introduction of eqs. (4a) and (4b) into eq. (5) yields

$$\frac{\partial V_m}{\partial x} = - r_o i_o + r_i i_i \tag{6}$$

Since the local currents are closed in a loop

$$i_o + i_i = 0 \tag{7}$$

combination of eqs. (7) and (6) results in

$$\frac{\partial V_m}{\partial x} = - (r_o + r_i) i_o \tag{8}$$

The change of current i_o with distance is current i_m:

$$\frac{\partial i_o}{\partial x} = - i_m \tag{9}$$

(the greater the i_m, the sharper the decrease of i_o with x). By differentiating eq. (8) we obtain

$$\frac{\partial^2 V_m}{\partial x^2} = - (r_o + r_i) \frac{\partial i_o}{\partial x} \tag{10}$$

and from a combination of eqs. (10) and (9) it follows that

$$\frac{\partial^2 V_m}{\partial x^2} = (r_o + r_i) i_m = (r_o + r_i)(i_i + i_c) \tag{11}$$

i_i may be expressed by Ohm's law,

$$i_i = \frac{V_m}{r_m} \tag{12}$$

and i_c is equal to

$$i_c = c_m \frac{\partial V_m}{dt} \tag{13}$$

By introducing eqs. (12) and (13) into eq. (11) we obtain

$$\frac{\partial^2 V_m}{\partial x^2} = \frac{r_o + r_i}{r_m} V_m + (r_o + r_i)c_m \frac{\partial V_m}{\partial t} \tag{14}$$

which may be written as

$$\frac{r_m}{r_o + r_i} \frac{\partial^2 V_m}{\partial x^2} = V_m + r_m c_m \frac{\partial V_m}{dt} \tag{15}$$

or

★ $$\lambda^2 \frac{\partial^2 V_m}{\partial x^2} = V_m + \tau \frac{\partial V_m}{\partial t} \tag{16}$$

This is the cable equation, where λ (called the space constant) is equal to $(r_m/r_o + r_i)^{1/2}$ (with the dimension of length) and τ $(= r_m c_m)$ is referred to as the time constant. In Table 21.1 values of cable constants for squid and crab nerve, frog muscle and the alga *Nitella* are shown.

Solutions of the partial differential equation (16) are very complicated and may be found for various special cases in the references mentioned above. However, under steady-state conditions the last term on the right-hand side of the eq. (16) disappears and an ordinary differential equation, describing the time-invariant distribution of the membrane potential change, results:

$$\lambda^2 \frac{d^2 V_m}{dt^2} = V_m \tag{17}$$

As may be easily verified, the general solution of this equation is

$$V_m = A e^{-x/\lambda} + B e^{x/\lambda} \tag{18}$$

The values of the constants **A** and **B** must be determined from the boundary conditions. At distant regions, when x is very large, V_m should be zero and hence **B** = 0. If $V_m = V_0$ at $x = 0$, **A** = V_0. Thus

$$V_m = V_0 e^{-x/\lambda} \tag{19}$$

Table 21.1. Cable Constants of Various Cell Types

Object examined	Space constant λ, cm	Time constant τ, msec	Membrane resistance R_m, $\Omega \cdot cm^2$	Membrane capacity C_m, $\mu F/cm^2$	Reference
Squid nerve	0.5	0.7	0.7	1	
Crab nerve	0.25	5.0	5.0	1	Katz (1966)
Frog muscle	0.2	24.0	4.0	6	
Nitella translucens	2.6	—	21.4	1	Williams *et al.* (1964)

Hence, under steady-state conditions, the membrane potential change decreases exponentially with distance: as distance reaches the value of the length constant, V_m drops to $1/e$ of the original value. The value of λ can thus be calculated, and λ, in its turn, yields the resistance of the membrane.

21.4. ACTION POTENTIAL AND ITS PROPAGATION

The occurrence of the action potential is intimately linked to the nerve function proper; however, the phenomenon is not restricted to excitable tissues, such as nerves and muscles, but can also be observed with other cells, particularly in several algae. It thus appears that this phenomenon might be of more general significance and, as it represents a rather interesting membrane property, it will be described in more detail.

If the axon is depolarized by a current pulse of subthreshold intensity and duration, the membrane potential is shifted to the corresponding value after a time period sufficient to charge the membrane capacity. If the impulse is of threshold intensity (usually on depolarization of the membrane potential by 20 mV) the initial stimulus is suddenly spontaneously amplified, a transient change in the permselective properties of the membrane occurs and the so-called action potential appears (Fig. 21.5). Whereas the subthreshold response dies away after passing a certain distance, the action potential is conducted over the entire nerve length. After the passage of the action potential wave a refractory phase sets in, the nerve being for several milliseconds insensitive to further stimulation. The duration of a propagated action potential is approximately 3 msec.

As early as 1902, Bernstein pointed to the role of ionic currents in the mechanism of nerve conduction and predicted also the membrane permeability changes with respect to sodium and potassium. Likewise, Overton (1902) suggested the importance of the presence of sodium ions in the external saline and demonstrated that they can be replaced by lithium ions. But only the further development of experimental techniques has provided support for these ingenious hypotheses. Hodgkin and Huxley (1939) showed that what takes place during the passage of an impulse is not a mere depolarization of the membrane potential difference to zero, but rather the potential step across the membrane is reversed. This corresponded to the suggestion that during the depolarization phase the membrane permeability toward sodium is many times increased and the potential tends toward the situation expressed by

$$E_m = -\frac{RT}{F} \ln \frac{[\text{Na}^+]_i}{[\text{Na}^+]_o} \tag{20}$$

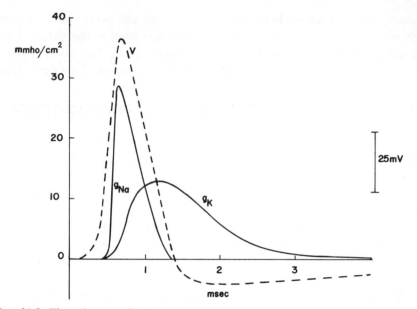

FIG. 21.5. Time changes of sodium and potassium conductances and the reconstructed propagated action potential. (According to Hodgkin and Huxley, 1952c.)

which is just approached at the peak of the action potential. The maximum potential change having been reached, a strong current directed outward brings the potential through a mild overshoot to the original value. Hence, in the repolarization phase the permeability toward potassium is increased (Fig. 21.5).

Using the "voltage-clamp" technique, Hodgkin and Huxley (1952a,b,c) succeeded in measuring the ionic conductances in the course of gradual changes of the membrane potential. The principle of the method is a fixation of the transmembrane potential difference at a selected value, using a feedback attachment with an (e.g., differential) amplifier and measurement of currents necessary for the maintenance of the desired potential level. The membrane potential thus figures here as an independent variable. The technique was introduced by Cole (1949) and applied successfully by Hodgkin, Huxley, and Katz (1949), and Hodgkin and Huxley (1952a,b,c) to explain the basis of ionic currents in the course of the action potential. The measuring circuit applied in the method is shown in Fig. 21.6.

The time changes of sodium and potassium conductances which may be calculated by analyzing the measured currents are shown in Fig. 21.5 together with the constructed action potential. Hodgkin and Katz (1949) proceeding from Goldman's suggestion of the existence of a constant field

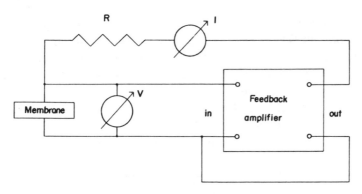

FIG. 21.6. Voltage-clamp circuit.

in the membrane and from his expression for the transmembrane potential difference (see section 4.1) derived the following equation for the nerve cell potential:

$$E_m = - \frac{RT}{F} \ln \frac{P_{K^+}[K^+]_i + P_{Na^+}[Na^+]_i + P_{Cl^-}[Cl^-]_o}{P_{K^+}[K^+]_o + P_{Na^+}[Na^+]_o + P_{Cl^-}[Cl^-]_i} \qquad (21)$$

This formula is applicable under the conditions of zero current flow across the membrane.

The ratios of permeabilities $P_{K^+} : P_{Na^+} : P_{Cl^-}$ in *Loligo* axons were found to be 1.0 : 0.04 : 0.45 in the resting state and 1.0 : 2.0 : 0.45 during activity. The permeability constants may be calculated from eqs. (5.*11*) and (5.*12*).

The theory of Hodgkin, Huxley, and Katz received much support in the literature. One of the most important proofs were the results of Keynes and Lewis (1951*a,b*) who used activation analysis to calculate the amount of sodium entering the giant nerve fiber during the impulse and the amount of potassium leaving the fiber. The agreement between the two values (3.8 μmol·cm^{-2} per impulse for sodium and 3.0 μmol·cm^{-2} per impulse for potassium) was good. Analogous results were obtained when the changes of ionic activities were measured by means of selective glass electrodes (Hinke, 1961).

The role of the external sodium was confirmed in experiments in which the outer sodium concentration was changed. On omitting sodium and replacing it with choline the impulse conduction was blocked (Hodgkin and Huxley, 1952*a,b*); when sodium concentration was raised, the amplitude of the action potential was increased. When, on the other hand, the internal potassium was replaced with sodium, the depolarization was decreased or even no action potential was generated (Baker *et al.*, 1962; Tasaki *et al.*, 1962), this again being in accordance with eq. *(21)*.

It was known from earlier measurements of membrane resistance that the membrane conductance increases in the course of the action potential, the resistance of the membrane of the *Loligo* axon being approximately 1000 $\Omega \cdot cm^2$ in the resting state and 25 $\Omega \cdot cm^2$ during activity (Cole and Curtis, 1939). Very important was the finding that the sodium permeability changes continuously in dependence on the membrane potential (Hodgkin, 1958).

In the voltage-clamp experiments the separation of the sodium and potassium currents was also achieved in the presence of the puffer fish poison (Nakamura *et al.*, 1965). Tetrodotoxin blocks the sodium current activation so that the sodium conductance component disappears and only the component connected with potassium conductivity changes remains. This specific block of spike electrogenesis with the maintenance of the potassium response to a depolarizing impulse appears to demonstrate that in the course of the action potential two different conductance paths exist in the membrane which do not interfere with each other.

An important role in the conduction of nerve stimulation is ascribed to calcium ions. A decrease in the external calcium concentration causes an increase in nerve excitability. This could be perhaps attributed to a rise in sodium conductivity (Weidman, 1955; Frankenhauser, 1957). Calcium seems to fulfil still another task, the presence of bivalent cations being very important for the generation of the action potential. Hodgkin and Keynes (1957) showed that a minute amount of calcium (0.006 $pmol \cdot cm^{-2}$) enters the membrane during nerve activity. It may be visualized that calcium changes the selectivity of membrane polymers, thus taking part in the change of permeability toward sodium and potassium ions.

The physical concepts of the Hodgkin and Huxley theory of the nerve impulse conduction are at variance with the heat changes of the stimulated nerve (Hodgkin, 1964). It may be expected that during the depolarization transfer of ions the nerve should cool down while during the work on restoration of the ionic gradient heat should be released. However, two-stage heat changes were described by Abbott and co-workers (1958), a rapid heat release of about 9×10^{-6} cal/g nerve being connected with the initial active phase of action potential and the subsequent cooling lasting approximately 300 msec, representing a heat consumption of about 7×10^{-6} cal/g nerve. According to its duration, this negative heat cannot be connected with the recovery potential but rather with a process which follows after the permeability changes are terminated.

The nature of events giving rise to the action potential was described; it remains to be mentioned how the stimulus is spread along the nerve. In a zone through which a wave of action potential is just passing a transient change of membrane polarity takes place. The axon surface becomes negative, the inside positive. An electrical circuit is set up between this and the neighboring zone and it may be envisaged that the local current is sufficient for depolarizing a further section of the membrane, and so forth. This mode of conduction is suggested for both nonmyelinated and myelinated nerve fibers; it was with myelinated nerve fibers that good evidence for the impulse conductance by local current was obtained. The myelin sheath forms a perfect insulation which is interrupted only at the Ranvier nodes. The impulse conduction is then mediated always between two neighboring nodes. The advantage of this arrangement is to be seen in speeding up the impulse conduction; thus the nerve system of higher animals has been perfected. Huxley and Stämpfli (1949) demonstrated in a simple experiment that if the center of one fiber internode (bathed in a Ringer solution) is insulated by an air gap, it is necessary to remove the insulation by a conducting bridge in order that the impulse conduction may proceed further.

In the synapses, where the nerve endings are found in a close vicinity of the dendrites of adjacent cells, specific substances (the so-called transmitters) take part in propagating the impulse. Of the widespread natural transmitters let us mention acetylcholine and adrenalin (or rather a mixture of adrenalin and noradrenalin). The transfer of impulses mediated by acetylcholine was investigated most. It was found by using suitable inhibitors that the presynaptic release of acetylcholine in the form of small droplets from the nerve endings causes, within a tenth of a millisecond, a postsynaptic excitation, i.e., an increase of membrane permeability to all ions with a subsequent depolarization, the impulse being then conducted further as usual. Acetylcholine functions also as a transmitter in nerve–muscle endplates. The inhibitory transmitters are of no less importance than the excitatory ones; by increasing the chloride permeability they prevent the resting potential changes. In mammals they increase the permeability to potassium (which results in a higher potential of the postsynaptic membrane), as well as to chloride, thus counteracting the effect of an excitatory postsynaptic potential. The nature of the inhibitory substances is mostly unclear; the properties of such substance were attributed to γ-aminobutyric acid for the inhibition in crustacean stretch receptors. It is of interest that acetylcholine can obviously in some cases act as an inhibitory substance. A detailed description of synaptic phenomena can be found in the book by Eccles (1964) to whom much of our knowledge in this field is due.

There exist various concepts of the actual mechanism of the excitability phenomena on the molecular level. Tasaki and Singer with a group of co-workers (Tasaki et al., 1965, 1969; Singer and Tasaki, 1968) founded their hypothesis on the ion exchange properties of the membrane. The macromolecules within the membrane are supposed to exist in two stable conformational states. One state corresponds to the resting conditions, in which the membrane ampholytes (proteins, phospholipids) are bound together presumably by salt linkages (attraction of negatively and positively charged ampholytic groups) and partly by Ca^{2+} ions. Now, the formation of the second stable conformational state is triggered by penetration of intracellular univalent cations into the membrane. The salt linkages and the bridges are disrupted by the permeating univalent cations, the ionic sites being occupied by them. Once the macromolecules in a certain region are raised to an excited stable state, the activity may spread. In the excited conformation, the increase of the density of the ionized groups in the membrane is connected with an increase of conductance, the external univalent cations being allowed to enter the axon. A local electrical circuit between the excited and the resting region across the membrane is set up, the internal cations are driven into the membrane, the macromolecular conformation is changed, the whole process being thus propagated along the axon. The hypothesis rests upon the study of the so-called bi-ionic potentials. It was demonstrated that an all-or-none response can be obtained with perfused axons in an artificial system where only salts of two cations are present when the external cation is divalent and the internal one univalent. An action potential with increased amplitude and a typical repolarizing phase together with a decrease in membrane resistance are caused by addition of univalent cations to the outer saline. The affinity of the internal cations for the membrane binding sites was measured and found to decrease in the sequence: Rb^+, K^+, NH_4^+, Na^+, Li^+. The ability of the accompanying anions to increase the amplitude of the action potential was in the order $F^- > SO_4^{2-} > Cl^- > NO_3^- > Br^- > I^- > SCN^-$ which is in agreement with the lyotropic series. The termination of the excitation process sets in when Ca^{2+} is bound within the membrane and the decreased conductivity restored. The finding by Hodgkin and Keynes (1957) on calcium uptake during activity gives credence to such ideas.

Conformation changes of macromolecular compounds in the membrane during excitation appear to be suggested by two recent findings: The observation of birefringence changes in the axon surface (Cohen et al., 1968) and appearance of infrared emission (Fraser and Frey, 1968). Wei (1969) employed these results, together with the occurrence of free negative charges at the surface of the axon (Segal, 1968), to support his hypothesis as to the role of dipoles in the conduction of the impulse. The author believes that a rotation of dipoles by 180°, comprising a replacement of free negative charge with positive charge at the cell surface (caused by the energy of the applied electrical stimulus), breaks the energetic barrier for the entry of the external cations and thus represents the initiation of stimulation.

Much useful information on the transport processes in the nerve may be found in the following reviews and monographs: Shanes (1958a,b), Ussing (1960), Eccles (1964), Hodgkin (1964), Mullins (1965), Katz (1966), Noble (1966), Caldwell (1968).

21.5. PERMEABILITY TO NONELECTROLYTES

Whereas much attention has been devoted to various aspects of electrolyte movements in nerves, only a few papers are concerned with the transport of nonelectrolytes. Tasaki and Spyropoulos (1961) compared the permeabilities to cationic compounds, such as choline and guanidine, with the permeability to the neutral molecules of thiourea, urea and sucrose. Fluxes of choline and guanidine were increased by stimulation but no conclusive results were obtained in this respect with neutral compounds. Thiourea and urea crossed the membrane more readily than choline and guanidine, the permeation of the two last-named compounds being comparable with the permeation of sodium and potassium. On the other hand, sucrose efflux was very low. Krolenko and Nikolsky (1967) found the permeability coefficient for sucrose to be 0.35×10^{-7} cm/sec, while for D- and L-arabinose and D-fructose this coefficient varied between 1.16×10^{-7} and 2.36×10^{-7} cm/sec. Villegas and co-workers (1965) observed an increase in the penetration of erythritol, mannitol, and sucrose during stimulation which the authors ascribe either to a change in membrane permeability or to the drag effect due to sodium ions, supporting this by the finding that increase in sodium concentration from 135 to 445 mM caused a rise in erythritol permeability from 3.7 to 5.3×10^{-7} cm/sec.

22. EPITHELIAL LAYERS
OF ANURANS

22.1. ACTIVELY TRANSPORTED IONS

An important part of the present knowledge about the active transport of ions derives from experiments with anuran epithelial membranes, of which the frog skin is a typical representative. Membranes of this kind are easily mounted into split chambers or studied in the form of vesicles (see chapters 12 and 13) and they survive for many hours in various simple solutions containing only inorganic salts at arbitrary temperatures. Bladders of the toad and of the frog, although not always so easy to handle, are morphologically simpler than frog skin and have been extensively studied for more than a decade. The literature concerning the anuran membranes is now very extensive, a review of the work done on the frog skin having been prepared by Ussing (1961) and that of investigations of the toad bladder by Leaf (1965). The main part of this chapter will be therefore devoted to only some of the recent observations on the frog and toad skin and on the frog and toad bladder.

The ability of the frog skin to generate an electrical potential difference between the media at its two surfaces has been known for more than a hundred years. The difference may sometimes well exceed 100 mV. The most direct evidence that this potential difference is brought about in some way by the active transport of ions is due to Ussing and Zerahn (1951), who invented the technique of the short-circuit current (see chapter 12) for this purpose; those of the actively transported ions which carry the current across the short-circuited preparation are also responsible for the spontaneous potential difference found in the open-circuit conditions. In most cases it is the transport of sodium ions from the morphologically outer (epidermal) side to the inner (corium) side, which is almost entirely responsible for the short-circuit current and the potential difference. Unless the concentration of sodium ions at the epidermal side of the skin is very dilute (sometimes considerably less than 1 mM), this side is electrically negative with respect to the other. When short-circuiting the surviving skin of, say, *Rana temporaria*, with the same frog Ringer solutions on both sides, the short-circuit current corresponds within the limits of experimental error to the net flux of the sodium ions (Ussing and Zerahn, 1951). The sodium ions thus appear to be the only ionic species actively transported across the surviving skin of this frog in normal media, the transport proceeding from the external medium inward. The same situation obtains in the isolated toad bladder (Leaf *et al.*, 1958). In artificial media containing lithium ions which are very similar to sodium, the ions were shown to be also transported across the frog skin to a degree actively (Zerahn, 1955). The same is true of the frog bladder (Leontyev and Natochin, 1964). Chloride ions are occasionally actively transported across the anuran skins in an appreciable extent; e.g., *in vivo*, where an active uptake of chloride ions was demonstrated by Jørgensen and co-workers (1954) in the toad *Bufo bufo* as well as in the frogs *Rana esculenta* and *Rana temporaria*. The same process was suggested to be the cause of the inequality of the short-circuit current and the net flux of the actively transported sodium ions across a piece of the skin of anaesthetized frogs (*Rana pipiens*), found in cases where the spontaneous potential difference across the skin was rather low (Watlington *et al.*, 1964). The skin of the South American frog *Leptodactyllus ocellatus* shows a considerable active transport of chloride ions in the inward direction even *in vitro*, as demonstrated by the short-circuit current technique on the isolated skins (Zadunaisky *et al.*, 1963) whereas no such transport was demonstrated under the same conditions across the skin of the toad (*Bufo arenarum* Hensel) from the same surroundings (Zadunaisky and De Fisch, 1964). According to Martin and Curran (1966) the differences between various anuran species with respect to the active transport of chloride ions may be primarily of quantitative, rather than of qualitative, character and the active transport of chloride ions is in most cases simply too small to be detected. Martin and Curran (1966) demonstrated a small active flux of chloride ions across surviving skins of the frogs *Rana pipiens* and *Rana esculenta* in solutions of low chloride concentrations, so that the small active chloride transport was not masked by relatively large passive fluxes of these ions.

Besides this active transport of chloride anions proceeding across some skins in the inward direction, i.e., in the same direction in which the active transport of sodium ions takes place, an active transport of chloride ions across the frog skin in the opposite direction was observed under special conditions: Koefoed–Johnsen and co-workers (1952) observed this transport phenomenon in the adrenalin-stimulated frog skin and explained it as being due to the action of the hormone on the mucus glands. This view was recently supported by House (1969) who found that the volumetric record of the gland secretion rate and the short-circuit current have the same time course in the isolated skin of *Xenopus laevis*.

22.2. ORIGIN OF THE SPONTANEOUS TRANSEPITHELIAL POTENTIAL AND LOCALIZATION OF THE POTENTIAL GRADIENTS

As demonstrated by the short-circuit current technique, the electromotive force in the epithelial membranes of anurans is generated by the inward active transport of sodium ions from which the active transport of chloride ions may sometimes be subtracted in anuran skins. Phenomenologically, this situation may be described by a simple equivalent circuit suggested by Ussing and Zerahn (1951) and shown in Fig. 22.1. Here R_{Na} is the resistance of the epithelial membrane toward the actively transported sodium ions and R_{Σ} is the resistance of the shunt through which the flow of ions, transported across the layer passively, takes place. When the

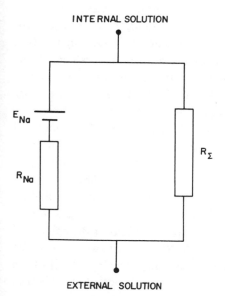

INTERNAL SOLUTION

E_{Na}

R_{Σ}

R_{Na}

EXTERNAL SOLUTION

FIG. 22.1. The equivalent circuit of the frog skin with the same solution on both sides according to Ussing and Zerahn (1951). $\mathbf{E}_{Na}$ electromotive force of sodium pump, $\mathbf{R}_{Na}$ skin resistance toward the actively transported sodium ions, $\mathbf{R}_{\Sigma}$ skin resistance toward the passively transported ions.

chloride ions are transported actively across the layer, another electromotive force with an appropriate resistance must be added to the circuit. A number of interesting phenomena of the epithelial membranes may be conveniently described using the above scheme. Thus the increase in the spontaneous potential across the frog skin obtained when chloride ions in the media bathing its surfaces are replaced with sulfate ions, to which some skins are very poorly permeable, or when the permeability of the skin toward chloride ions is reduced by treating the skin with cupric ions (Ussing and Zerahn, 1951) or with SH-groups binding reagents (Janáček, 1962), may be described as an increase in the shunt resistance R_y. Similarly, assuming time-variant properties of the resistance R_{Na}, transients in the electrical potential response to square waves of current of several minutes duration may be described (Janáček, 1963). Description using such an equivalent circuit, however, gives no information about the localization of the electrical potential gradients in the epithelial layers or about the mechanisms by which the active transport of ions may generate such gradients.

Measurements carried out by the technique of Steinbach (1933) suggested that the overall potential difference across frog skin is composed of at least two individual potential differences; in this technique two injury potentials are measured between the edge of the isolated skin and the two solutions bathing its surfaces, their sum being equal to the total potential difference across the skin and their changes being sometimes rather independent (Steinbach, 1933; Janáček, 1962). In a number of studies the potential profile across the epithelial layers of anurans was examined with microelectrodes. The technical difficulties encountered were rather great and the results obtained were not always comparable. Thus, in the frog skin one single step of the potential (Ottosen et al., 1953), two steps inside the epidermis (Engbaek and Hoshiko, 1957), two steps with only one of them inside the epidermis (Scheer and Mumbach, 1960), two or more steps, all of them inside the epidermis (Ussing and Windhager, 1964; Cereijido and Curran, 1965), as well as a smooth distribution of the potential without steps (Chowdhurry and Snell, 1965, 1966) have been found. Two steps of the potential difference (Frazier, 1962) as well as a smooth distribution of the potential (Chowdhurry and Snell, 1965, 1966) have been found in toad bladder. Finally, two steps of the potential in the bladder of the frog have been described (Janáček et al., 1968). It may be seen that in the majority of the studies two or more steps are found in the epidermal layers of the skins, containing several cell layers, and that two steps of the potential were found in the anuran bladders, containing a single layer of the mucosal epithelium. Hence it appears that the overall potential difference across the transporting epithelia is composed of several potential differences across cell membranes. Unless the concentration of sodium ions at the outer surface of the anuran membrane is very low, each of these steps increases the potential in the inward direction. Thus the polarity of the membrane potentials at the inner border of the epithelial layers is that found in most cells, with the cell interior negative with respect to the adjacent medium, whereas the polarity of the potentials nearer to the outer surface is reversed; the cell interior here is electrically positive with respect to the external medium. The situation is shown schematically in Fig. 22.2 for the histologically most simple anuran membrane, the bladder. Fine structure of the toad bladder was described by Choi (1963) and the unicellular character of its epithelial layer was demonstrated by Di Bona and co-workers (1969). The situation in the histologically more complex skins is essentially similar, many of the epidermal cells appearing to be connected by intercellular bridges (Ottosen *et al.*, 1953; Ussing and Windhager, 1964; Farquhar and

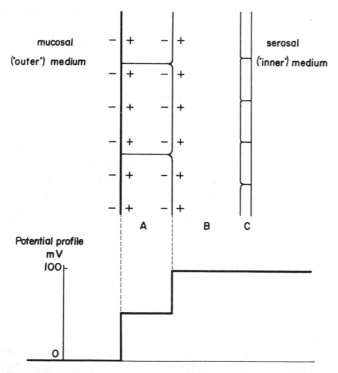

Fig. 22.2. Potential profile across the anuran bladder. **A** Layer of epithelial cells at the mucosal (urinary) side, **B** connective tissue with capillaries and smooth muscle fibers, **C** layer of serosal cells.

Palade, 1963*a,b*, 1964) by which greater complexes of cytoplasm at the same electrical potential may be formed.

Taking for granted that the potential differences across epithelial membranes are composed of membrane potentials across the membranes of the epithelial cells the origin of these membrane potentials may now be discussed. Some features of this system are explained by the well-known model of Koefoed–Johnsen and Ussing (1958). Some frog skins show a very low permeability to sulfate and their permeability to chloride can be considerably reduced by treatment with cupric ions. The potential difference across these skins (in sulfate media, if the skins were not treated with copper) was shown by Koefoed–Johnsen and Ussing to depend logarithmically on the concentration of sodium ions in the medium at the outer surface of the skin and on the concentration of potassium ions at the inner surface, with the slope close to the theoretical value for an ideally selective membrane (58 mV per tenfold change at 20°C). The potential difference rises when the sodium concentration at the outer surface is increased and drops with increasing potassium concentration at the inner surface. Hence, assuming that the intracellular concentration of sodium ions in the epithelium is rather low and that of potassium high (a well-founded assumption, as will be seen in section 22.3) Koefoed–Johnsen and Ussing explain the potential difference across the epithelial layer as a sum of two diffusion potentials across two membranes, the outer being preferentially sodium-permeable, the inner preferentially potassium-permeable. A nonelectrogenic sodium-potassium-coupled pump localized at the inner membrane was assumed to take care both of the homeostasis of the epithelial cells themselves and of the transepithelial sodium transport, as is shown schematically in Fig. 22.3. In view of more recent results it appears, however, that the concentration gradient of the potassium ions is not the only source of the potential difference across the inner membrane of epithelial layers, i.e., the sodium pump localized there is, in fact, electrogenic. The idea of a tightly coupled sodium-potassium pump originated mostly from the observation that in the absence of potassium ions in the solution bathing the inner side of the skin or the serosal side of the bladder, transcellular sodium transport is inhibited. However, it has been demonstrated that under these conditions the inhibition results from a decrease in the sodium permeability of the mucosal membrane of the toad bladder which follows a drop in the intracellular potassium concentration and an increase in the intracellular sodium concentration (Essig and Leaf, 1963). Moreover, if a potassium-free choline solution rather than potassium-free sodium solution is used at the serosal surface of the bladder, the tissue content of electrolytes is less influenced

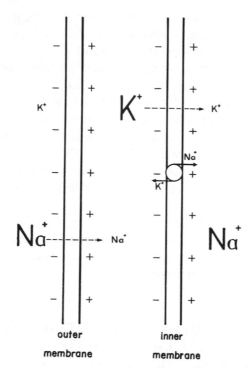

FIG. 22.3. Essential features of the origin of the potential difference across an epithelial layer according to Koefoed-Johnsen and Ussing (1958). Solid arrows: Active movements of the ions. Dashed arrows: Passive movements.

and the transcellular sodium transport much less inhibited (Essig, 1965). In the frog skin there is no correlation between potassium uptake and the transcellular transport and the two are not inhibited in the same degree by dinitrophenol and fluoroacetate, the potassium uptake being less sensitive to these metabolic inhibitors than the transcellular sodium transport (Curran and Cereijido, 1965). More direct evidence for the electrogenic character of the sodium pump comes from experiments with a high potassium concentration at the serosal surface of the toad bladder: Frazier and Leaf (1963) measured the serosal potential step under these conditions with micro-electrodes and demonstrated that none of the ions present can account for its size by its concentration gradient. Similarly, a consideration of the short-circuit current which can be drawn from the frog skin exposed to high potassium concentrations at the internal surface suggests that the situation in the frog skin is likely to be analogous (Bricker et al., 1963). Thus, whereas the diffusion of sodium ions across a membrane preferentially permeable to sodium ions may be considered as the main source of the potential step at the outer surface of the epithelial layers, two main generators of the electromotive force in parallel are to be considered at the inner surface of these epithelia: a concentration gradient of potassium ions and an electrogenic sodium pump. This view is supported by Herrera (1968) who demonstrated that under special conditions (choline Ringer solution at the serosal border) there is a marked fall in the potential difference across the toad bladder after inhibition with 10^{-3} M ouabain, without the sodium and potassium content of the tissue being appreciably affected. Thus, here again, the potential difference across the epithelium layer is rather independent of the concentration gradient of potassium ions across its serosal membrane.

For an erudite alternative view of the potential difference across epithelial layers, according to which the microelectrode technique is full of pitfalls and no completely satisfactory explanation of this potential difference can be given at this time, the article by Winn and co-workers (1966) should be consulted.

22.3. INTRACELLULAR CONCENTRATIONS OF IONS IN EPITHELIAL LAYERS. ACTIVE TRANSPORT AT THE INDIVIDUAL MEMBRANES

Adopting the view that the ion-transporting epithelial layers of anurans may be represented by a system of two membranes in series, enclosing a single cell layer in bladders and greater complexes of cells connected by intracellular bridges in skins, the mode of transport across each of the two membranes may be considered, but some information on the intracellular concentrations of ions is first required.

A number of attempts were undertaken to determine the concentration of sodium ions in the epithelial cells. Thus Andersen and Zerahn (1963), extrapolating from curves representing the disappearance of radiosodium from skins of *Rana arvalis* and *Rana temporaria* with sulfate Ringer solution on both sides (115 mM Na$^+$), found the intracellular concentration of free sodium to lie between 5.4 and 24.3 mM. When reducing the concentration of sodium in the solution bathing the outer surface of the skin below 1 mM, the intracellular concentration of free sodium was found to drop to values of the same order of magnitude. In short-circuited skin of *Rana pipiens* with chloride Ringer solution on both sides the average value of 39 mM Na$^+$ was calculated from isotopic experiments by Curran and co-workers (1963). In the bladder of *Rana temporaria* with Ringer solution on both sides the total sodium concentration in the space nonaccessible to inulin from the serosal side was found to be 33.0 ± 14.3 mM (Natochin *et al.*, 1965). Using the same tissue and cation-sensitive glass microelectrodes Janáček and co-workers found activities of sodium ions corresponding to concentrations of free sodium ions between 9.8 and 55 mM (average 28.4 mM). Finally, Cereijido and Rotunno (1967) measured the sodium concentrations inside the epithelial cells of the skin of *Leptodactylus ocellatus* under short-circuit conditions with low concentration of sodium (2.5–5 mM), to reduce the error in the extracellular sodium determination, on both sides. They found that only 37% of the total sodium content is exchangeable, thus giving the value for the concentration of free sodium about 26.7 mM. By analyzing thin slices of the epithelial layer, the authors found that both the free and the bound sodium are uniformly distributed across the thickness of the epithelial layer. Practically the same proportion of free and bound sodium in the skin of *Leptodactylus ocellatus* was found from a measurement of the nuclear spin resonance (Rotunno *et al.*, 1967). Judging from the localization of sodium pyroantimonate precipitates under an electron microscope it appears that the bound sodium in the frog skin may be trapped in cell nuclei (Zadunaisky *et al.*, 1968).

In a short-circuited skin the entry of sodium ions into the epithelial cells across the outer membrane is supported by a negative step of the potential difference. In *Rana pipiens* this step was reported to be about -18 mV (Cereijido and Curran, 1965), in the toad about -24 mV (Whittembury, 1964), in *Leptodactylus ocellatus* the values were between -12 and -40 mV (Cereijido and Rotunno, 1967). The bladders displayed lower, but still negative, values, the value of -5.5 mV being reported for the toad bladder (Frazier, 1962) and the value of -4.6 mV for the frog bladder (Janáček *et al.*, 1968). Combining these data with the values of the concentrations given above it may be seen that the sodium transport across the outer membrane was never found to proceed against an electrochemical gradient, which would prove its active character. Thus the entry of sodium ions into the epithelial cells may be a passive process, not directly coupled to metabolic reactions. It was shown, however, that it is not a simple electrodiffusion, but that it exhibits saturation kinetics: Frazier *et al.* (1962) demonstrated that the transport sodium pool in the toad bladder (the content of the labeled sodium in the bladder originating from the mucosal medium, the serosal medium being choline Ringer solution) depends on the sodium concentration in the mucosal medium in the same way as does the active transport of sodium across the bladder (i.e., there is a saturation at about 60 mM Na$^+$ in the mucosal medium). Hence, the active transport of sodium is approximately proportional to the amount (and thus, probably, to the concentration) of sodium in the transport pool and the whole process is limited by a saturable (possibly carrier-mediated) entry of sodium into the cells. Similar situation was found also in frog skin (Cereijido et al., 1964).

Whereas the entry of sodium ions into the epithelial layers of anurans may be a carrier-mediated electrodiffusion or similar passive process, there is no doubt about the active character of the extrusion of sodium ions across the inner membrane, the process being able to proceed and, unless unnatural electrical fields and concentrations are applied externally, always proceeding against a considerable electrochemical potential gradient. From the observations by Frazier and co-workers (1962) discussed above, it appears that the sodium pump would be saturated only at considerably higher concentrations than are those permitted by the saturable entry into the cells.

Several determinations of the intracellular potassium concentrations were performed. In slicing experiments on frog skin (Hansen and Zerahn, 1964) the intracellular potassium concentration of epithelial cells was found to lie between 100 and 140 mM (Andersen and Zerahn, 1963). In the non-inulin space of the bladder of Rana temporaria the total concentration of 208.3 $\pm$ 21.7 mM K$^+$ was reported (Natochin et al., 1965), whereas with cation-sensitive glass microelectrodes the activities of potassium ions corresponding to the concentrations of free ions between 71.1 and 119.1 mM (average 88.6 mM) were found in the same tissue (Janáček et al., 1968). Thus, if both these figures are correct, some potassium may be bound inside some bladder cells.

Whereas the outer membrane of the anuran epithelial layers appears to be little permeable toward potassium ions and the fluxes of potassium across it are of little importance (in the skin of Rana pipiens the fluxes of potassium ions from the outside solution are lower by a factor of approximately 30 than those from the inside solution—Curran and Cereijido, 1965), the permeability of the inner membrane is high. Although there is no rigid coupling between the active extrusion of sodium ions and the uptake of potassium ions (section 22.2), the uptake of potassium ions across the inner membrane is, nevertheless, active. In frog skin it was recognized on the basis of osmotic experiments in which the thickness of the epithelial layer was measured microscopically that the potassium ion is not in equilibrium across the inner membrane and is, therefore, subject to active transport (MacRobbie and Ussing, 1961). This conclusion was supported by Curran and Cereijido (1965) who demonstrated that the flux of potassium into the skin follows saturation kinetics and is markedly inhibited by metabolic inhibitors (2.4-dinitrophenol, sodium fluoroacetate) and by ouabain. Finally, in the frog bladder ho potential step was found at the serosal border as would correspond to a passive equilibrium for even the lowest concentration of free potassium ions found in the same preparation with cation-sensitive glass microelectrodes (Janáček et al., 1968).

Little is known about the intracellular concentrations of chloride anions under various conditions. In anuran bladders there appears to be no active chloride transport (as can be concluded from the equality of the short-circuit current and the net sodium flux; see section 22.1) and, hence, the chloride anion should be in thermodynamic equilibrium under short-circuit conditions when there is no net chloride flux across the preparation. The intracellular potentials in the cells of short-circuited bladders are, as shown above, very low and therefore the intracellular concentrations in the anuran bladders are, at least under these conditions, very high, almost the same as in the surrounding Ringer solutions. In anuran skins the intraepithelial potentials under short-circuiting are higher and, moreover, in some of them, the chloride anions are transported actively (section 22.1) so that their intracellular concentrations are uncertain. Whereas it may not yet be possible to decide about the character of the chloride anion transport across the individual membranes, some interesting dependences of permeability to chlorides on the presence of sodium ions were already described. Thus the permeability of the outer surface of the frog skin to chloride ions is reduced in the absence of sodium in the medium outside, the transepithelial potential difference becoming less dependent on chloride concentration under these conditions (Macey and Meares, 1963). On the other hand, the short-circuit current due to sodium flux is reduced when the chloride ions in the solution bathing the skin are replaced with sulfate or gluconate anions (Ferreira, 1968). The mechanism underlying this permeability change may be connected with the shrinkage due to loss of potassium chloride from the cells described by Ussing (1965) who found that in practically every instance shrinkage leads to inhibition of active sodium transport, possibly due to the fact that pores may be closing in the formerly stretched membranes of the shrinking cells and the viscosity of the cytoplasm, determining the diffusion rates of sodium transport and of metabolites used in active transport, may be reduced. In the skin of *Leptodactylus ocellatus*, however, in which the chloride ions are actively transported (see section 22.1), more complicated relations between chloride and sodium fluxes were described (Fischbarg et al., 1967), interpreted by the authors as a coupling superimposed on a competition between the two ions at the level of the ionic pumps on the inner membrane.

22.4. SOME PECULIARITIES RELATED TO THE TRANSPORT PROPERTIES OF ANURAN EPITHELIAL MEMBRANES

Before proceeding to the relations of the active transport in the anuran epithelial membranes to metabolism and to various chemical factors influencing its rate, some peculiar properties of these membranes, obviously related to their transport mechanisms, may be mentioned. First, some interesting phenomena resulting, as will be seen, from the coupling between flows of various substances will be discussed and the excitability phenomena on the anuran epithelial membranes described.

When measuring the transepithelial potential difference or the short-circuit current across frog skin in a split chamber, rather conspicuous changes in these parameters may be observed when even a minute hydrostatic pressure bulging the skin is applied to one or the other bathing medium. In a careful study in which bulging of the skins of *Rana temporaria* was achieved both by hydrostatic pressure gradients and by pushing the skins mechanically Nutbourne (1968) demonstrated that "bulging the skin in the absence of hydrostatic pressure gradients had no effect on sodium transport but that pressure gradients of less than 5 mm H_2O had a marked effect, increasing transport when the pressure was higher on the outside of the skin, and decreasing it when the pressure was higher on the inside" and concluded that "increasing surface area does not influence sodium transport, whereas small hydrostatic pressure gradients have a marked effect." According to Nutbourne the hydrostatic pressure may act in this way by changing directly or indirectly the bulk-flow rate down interspaces in the skin, the step which presumably follows the active extrusion of sodium ions from the epithelial cells.

In frog skin also the interesting phenomenon of nonosmotic water flow may be observed, explained by House (1964) in a two-membrane model, as described in section 4.2.

Another interesting phenomenon is the coupling between the fluxes of various solutes across the anuran skins. The phenomenon was first described by Ussing (1966) who observed that when the solution bathing the outer surface of the frog skin is made hypertonic with urea, creatinine, or even sodium chloride a pronounced asymmetry between the fluxes of sulfate or sucrose follows, the influx of these substances exceeding their efflux markedly. (Skins become rather permeable to these solutes when in contact with hypertonic medium outside.) A driving force is thus seen to act on the solutes, opposite to solvent drag due to osmotic water flow across the skin.

In further observations by Franz and co-workers (1968) the asymmetric movement of mannitol across the frog skin due to urea hyperosmolarity at the outer surface was demonstrated to be maintained when sodium transport across the skin is inhibited. In toad skin, the net flux of mannitol was shown by Biber and Curran (1968) to be reversed when the urea gradient was reversed. The solute flux asymmetry is thus seen to be due to the solute drag, i.e., due to a coupling between the fluxes of different solutes.

Finally, let us mention several excitability properties of anuran membranes. Lithium-induced oscillations of potential and resistance were described by Finkelstein (1961a). Oscillations in these parameters, with a period of 3–15 minutes, may be observed when the morphological outside of the skin is exposed to lithium solutions. "Even after the oscillations damped out, the system remained excitable, responding to a step of direct current or hydrostatic pressure with an oscillatory train" (Finkelstein, 1961a). But also an action potential of about 200 mV and 10 msec duration may be observed in frog skin and toad bladder with sodium or lithium solution at the outer surface (Finkelstein 1961b, 1964) as a response to a short current pulse of sufficient intensity which was interpreted as a consequence of current flow through a time-variant resistance element (Finkelstein, 1964). Using microelectrodes, Lindeman and Thorns (1967) demonstrated that in the frog skin the spike is generated at the outermost boundary of the surface cells, where the major resistive barrier of the skin is localized. In a detailed quantitative study Fishman and Macey (1969) demonstrated that the excitability of the frog skin is connected with the phenomenon of negative resistance and described its ionic requirements: Calcium, sodium, and chloride must be present in the outside solution, potassium and chloride ions in the inside solution. Unbinding and subsequent rebinding of calcium ions to membrane sites is suggested by the authors as the mechanism underlying the loss of excitability during the refractory state and the subsequent recovery.

22.5. ION TRANSPORT AND METABOLISM IN ANURAN EPITHELIAL MEMBRANES

A considerable part of oxygen consumption of the anuran epithelial membranes is directly related to the active sodium transport and it was actually in the experiments with anuran skins that the oxygen requirements of the sodium pump were established. As shown by Zerahn (1956) and by Leaf and Renshaw (1957a) only about 1/18 of an oxygen molecule corresponds to one sodium ion transported. This ratio was demonstrated by

Zerahn (1958) to be a true stoichiometric relation independent of all the factors tested, such as sodium concentration in the solution bathing the outside of the skin, potential across the skin, oxygen tension in the bathing solution, time of the year, species of the anuran, temperature, presence of neurohypophyseal hormone, and method by which the oxygen consumption in the absence of sodium transport was measured. A theoretical consideration of this ratio suggests that the efficiency of the sodium pump is extremely high (Ussing, 1967).

Some 20–40% of the sodium transport across the skin of *Rana temporaria* was reported to continue under anaerobic conditions with a simultaneous increase in lactic acid formation (Leaf and Renshaw, 1957b). On the other hand, in the experiments of Zerahn (1958) much lower values of transport were found with the skins of the same species. When traces of oxygen were removed with cysteine, sodium transport became zero. It appears that similar metabolic differences may be found between the animals of the same anuran species but from different geographical sources; such a difference exists between the bladders of the toads *Bufo marinus* from the Dominican Republic and from Colombia. In the absence of aldosterone, the addition of pyruvate stimulates the sodium transport in bladders from Colombian toads, whereas no increase in the transport occurs in bladders from Dominican toads (Davies *et al.*, 1968). Only traces of glycogen were found in the frog skin (Ussing, 1967), whereas in the toad bladder glycogenolysis is more important (Leaf *et al.*, 1958), its rate being regulated by the rate of the sodium transport (Handler *et al.*, 1968). The use of an exogenous substrate in the metabolism of an anuran membrane was demonstrated for the first time with acetate added to the medium bathing the inside of the frog skin (Van Bruggen and Zerahn, 1960; Zerahn, 1961); when acetate is present in 1 mM concentration, an appreciable part of the sodium transported derives the necessary energy from exogenous acetate. Exogenous palmitate may serve as a source of energy for the sodium transport in the toad bladder (Ferguson *et al.*, 1968).

The role of the magnesium-dependent sodium–potassium-activated adenosine triphosphatase in active sodium transport in the toad bladder was shown by Bonting and Canady (1964); both the enzyme activity and the sodium pump are inhibited by ouabain and erythrophleine.

For some metabolic studies the isolated cells of anuran bladders may be useful. It was shown that the cells of the toad bladder disaggregated with EDTA or trypsin have their active transport mechanisms damaged, whereas those disaggregated with collagenase or hyaluronidase appear to function normally (Gatzy and Berndt, 1968).

22.6. HORMONAL CONTROL OF THE TRANSPORT PHENOMENA IN THE ANURAN EPITHELIAL MEMBRANES

Finally, some of the great number of findings about the hormonal control of the transport phenomena will be summarized. Various hormones were found to be effective in this respect, the *neurohypophyseal hormones* being perhaps the most important among them. In the presence of these hormones in the media bathing the inner surfaces of the anuran epithelial membranes, active sodium transport was found to be enhanced both across the isolated frog skin (Ussing and Zerahn, 1951; Fuhrman and Ussing, 1951) and in the toad bladder (Leaf *et al.*, 1958). Permeability toward water is increased under these conditions in the frog skin (Fuhrman and Ussing, 1951) as well as in the toad bladder (Bentley, 1958). Permeability of the toad bladder toward urea and similar low-molecular solutes is also increased (Leaf and Hays, 1962). It is important that the effects on the active sodium transport and on the water permeability can be dissociated. Thus Bourguet and Maetz (1961) demonstrated that in the skin and bladder of the frog *Rana esculenta*, oxytocin and its two analogues, arginine-8-oxytocin (arginine vasotocin) and lysine-8-oxytocin (lysine vasotocin) may produce the same increase of water permeability, while stimulating in quite different ways sodium transport, and *vice versa*. An analogous dissociation of the effects by a perfected technique was described more recently by Bourguet and Morel (1967). Moreover, as shown by Petersen and Edelman (1964), at low concentrations of vasopressin, calcium ions have an inhibitory effect on the vasopression action on the water permeability, whereas the stimulating effect on sodium transport remains unaffected. Thus there are at least two different effects of application of a neorohypophyseal hormone to the anuran epithelial layers; an increase in water permeability and in the sodium transport is not achieved by the same mechanism. Let us turn our attention to the stimulating effect on active sodium transport.

There is little doubt that the entry of sodium into the epithelial layers across their outer surfaces (across the mucosal membranes in the bladders) is facilitated by application of the neurohypophyseal hormones. The increase of sodium content in these layers due to the presence of the hormone was demonstrated both in the short-circuited toad bladder (Frazier *et al.*, 1962) and in the short-circuited frog skin (Curran *et al.*, 1963). The simplest explanation of the hormone effect on the transepithelial sodium transport is that sodium transport is stimulated only due to the increased intracellular sodium concentration when the sodium pump is more saturated and, hence, expels more sodium out of the cells. Morel and Bastide (1965) demonstrated that the normal direction of the short-circuit current across the frog skin with a very dilute solution of sodium sulfate at the epidermal surface is recovered by oxytocin and it would seem that such an effect requires a direct stimulation of the sodium pump by the hormone, rather than a mere increase of the permeability of the outer permeability barrier. However, as shown by the same authors (Morel and Bastide, 1967), another explanation is possible: If there is a large intercellular leak of sodium across the frog skin, there may still be a net sodium entry into the cells, even if the short-circuit current is reversed due to a low concentration of sodium in the outside medium. Then again an increase of the passive permeability of the outer barrier would be sufficient to explain the change in the direction of the short-circuit current. The presence of a heterogeneous leak of the postulated nature in the toad bladder is suggested by the most instructive experiments of Civan and co-workers (1966) in which the magnitude of the electromotive force of the sodium pump was also determined and the independence of this value of the addition of the hormone was demonstrated. The experimental procedure was the following: The sodium chloride solution bathing the mucosal surface of the isolated toad bladder was diluted until the electrochemical potential difference across the tissue, measured with a pair of silver–silver chloride electrodes, was reduced to zero. The net movement of chloride ions across the bladder was then apparently also zero, the chloride ion being not actively transported across the toad bladder. But the net movement of actively transported sodium ions also vanished under these conditions, the movement of a cation unaccompanied by an anion being not permissible in the absence of an external circuit. Vasopressin added under these conditions increases the electrical potential difference across the bladder, the magnitude of the response being inversely dependent on the initial electrical potential. Extrapolation from the experimental data shows that for the initial transbladder potential of about 85 mV there would be no vasopressin effect on the potential. In each of the two solutions the concentration of the sodium ions was the same as that of the chloride ions; hence, if the electrochemical potential difference for the chloride ions was zero, the electrochemical potential difference for the oppositely charged sodium ions was to be twice the electrical potential difference across the bladder. The value of 170 mV is, therefore, to be considered as the electromotive force of the sodium pump and the results show that this value can be approached, but never exceeded, due to addition of a neurohypophyseal hormone. Civan and co-workers (1966) concluded from this that the electromotive force of the sodium pump is not influenced by the hormone.

To demonstrate that the electromotive force of the sodium pump is not influenced by the hormone is, however, not the same as to demonstrate that the sodium pump is not influenced by the hormone; indeed, in view of the ability of the hormone to increase the permeability of the mucosal membrane toward the possibly carrier-mediated and passive movement of sodium ions, it seems more probable that if the pump is itself stimulated at all one would predict a reduction of its internal resistance rather than an increase of the free energy of the reactions responsible for its electromotive force. A direct stimulation of the pump is suggested by experiments with a nonpolarized preparation of the frog bladder (vesicles made from paired half-bladders filled with liquid paraffin) in which no transport appears to take place across the mucosal membranes; in this preparation the sodium content is—highly significantly—decreased by oxytocin (Janáček and Rybová, 1967).

It was shown by Civan and Frazier (1968) that 98% of the total resistance in the toad bladder after vasopressin may be localized at the mucosal membrane and this finding appears to contradict the hypothesis about the reduction of the internal resistance by the hormone. However, it should be remembered that the electromotive force of the sodium pump is much higher than the potential difference across the serosal membrane and hence the internal resistance of the sodium pump is to be considerably higher than the parallel resistances of other paths across the serosal membrane. Now, the resistance of a parallel combination of a high and a low resistance is not very sensitive to small variations of the high resistance. The effect of the hormone on the sodium pump may be small and unimportant physiologically, but it may suggest that the sodium entry across the mucosal membrane and the sodium extrusion by the pump may proceed by a similar mechanism, say, by similar carriers, the mobility or the number of which is increased by the hormone.

The effect of the neurohypophyseal hormones on the water permeability is characterized by much greater increase in the osmotic water permeability than in the diffusional water permeability, which, according to the ideas discussed in section 4.2 is considered as evidence for a pore mechanism of water permeation with the dimensions of the pore increased in the presence of the hormones (Koefoed–Johnsen and Ussing, 1953; Andersen and Ussing, 1957; Hays and Leaf, 1962). According to the hypothesis pronounced by Ginetzinsky (1958) the increase of the water permeability observed on epithelial structures in the presence of the neurohypophyseal hormones is due to an increased secretion of hyaluronidase or a similar mucolytic enzyme by the cells, the enzyme depolymerizing the mucopolysaccharides in the interstitial tissue and increasing in this way the *intercellular* permeability to water. The hypothesis was recently strongly supported by the experiments of Ivanova and Natochin (1968), who demonstrated that hyaluronidase at a suitable *p*H, corresponding to the type of the enzyme used in the experiment, produces similar water permeability changes in the frog bladder as the neurohypophyseal hormones, and, moreover, is inhibited in the same way by higher concentrations of calcium ions. The observation by MacRobbie and Ussing (1961) that frog skin epithelium swells when a hypotonic solution is present at the outer surface and neurohypophyseal hormone at the inner surface does not appear to contradict this mechanism of the hormone action; the cells are likely to swell if the hypotonic medium flows through the intercellular spaces. It is interesting that the effects of the hormone on water permeability are preserved in preparations of the frog bladders fixed by formaldehyde and glutaraldehyde (Jard *et al.*, 1966).

To explain the fact that after water permeability was increased by the hormones the anuran epithelial membranes still remain little permeable to small solute molecules a thin dense permeability barrier was postulated at the surface of the epithelia, in series with the porous barrier limiting the bulk flow of water, both in the frog skin (Andersen and Ussing, 1957) and the toad bladder (Leaf and Hays, 1962). Amphotericin B, a polyene antibiotic, appears to increase greatly the permeability of the dense barrier in the toad bladder to sodium and to small solutes with little effect on the osmotic water permeability (Lichtenstein and Leaf, 1965). Amphotericin B thus appears to increase the permeability of one of the two permeability barriers acted upon by neurohypophyseal hormones (but in a less selective manner with respect to small solutes) and does not influence the other, limiting the bulk flow of water. This interpretation was, however, refuted by Mendoza and co-workers (1967) observing different results of amphotericin B action on the toad bladder; in their experiments, an osmotic permeability change was found and the increase in the sodium permeability had another time course than the increase in permeability to water and urea.

Adenosine 3′,5′-phosphate (cyclic AMP) was shown to be the intracellular mediator of the action of neurohypophyseal hormones (Orloff *et al.*, 1962); cyclic AMP and theophylline inhibiting, like other methyl xanthines, the inactivation of cyclic AMP to 5′-AMP, mimic the effects of neurohypophyseal hormones. A considerable amount of evidence for this hypothesis was obtained and reviewed together with much other information on the anuran epithelial transport by Orloff and Handler (1964, 1967).

The relation between structure and activity of more than thirty synthetic analogues of neurohypophyseal hormones was studied and discussed by Morel and Bastide (1964).

Another important group of hormones affecting the sodium transport in the anuran epithelial membranes are the corticosteroids and notably aldosterone, the most active of these. Its effects on the toad bladder *in vitro* were discovered by Crabbé (1961) and a great number of studies concerned with the mechanism of its action was reviewed by Sharp and Leaf (1966). Aldosterone increases sodium transport after a latent period of 40–120 min and its effects are mediated through protein synthesis.

Of the other hormones influencing the sodium transport in anuran epithelial layers mention should be made of thyroxine which increases not only the active sodium transport but also the general respiratory metabolism of the cells (Green and Matty, 1963) and of insulin, which appears to stimulate directly the sodium pump, since the transport increases with a simultaneous tendency of the sodium pool within the tissue to decrease (Herrera, 1965, 1968). Finally, noradrenalin was shown by Bastide and Jard (1968) to stimulate the active sodium transport resulting from an increase of the sodium influx, without the sodium efflux or the chloride fluxes being affected and to increase the osmotic water permeability in the frog skin in concentrations of 10^{-8} to 10^{-6} M. At much higher concentrations a large increase in the permeability of the skin to the two ions was found.

23. INTESTINE

23.1. INTRODUCTION

The intestinal epithelium represents a fine object for transport studies because of its possessing highly specific mechanisms for transporting various substances from the intestinal lumen into the blood stream. The morphology of the mucosal layer provides an enormous enlargement (by a factor of 600 relative to a cylinder; Wilson, 1962) of the absorptive surface area of the mucosal border, the following structures being responsible (see Fig. 23.1):

1. Folds of Kerkring (human small intestine).
2. Villi, finger-like structures lined with epithelial cells on the surface; the core of villi (lamina propria) consists of connective tissue with blood and lymphatic capillaries.
3. Microvillus structure of the brush-border membrane designated often as luminal or mucosal membrane of the columnar epithelial cells (Fig. 23.2).

The transport process takes place across the whole layer of columnar

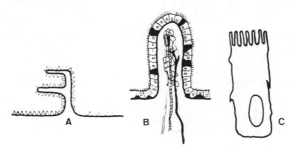

Fig. 23.1. Three mechanisms for increasing the surface area of the small intestine. A Fold of Kerkring, B villus, C microvilli. (According to Wilson, 1962.)

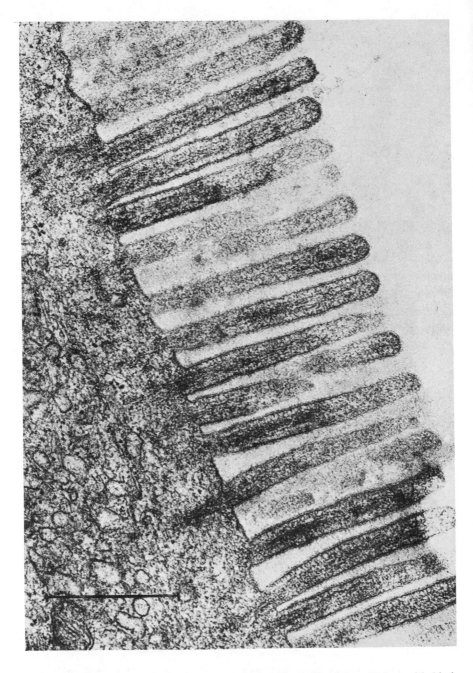

FIG. 23.2. Brush-border region in mouse jejunum. Scale line 0.5 μ. (Taken with kind permission from Sjöstrand, 1963a.)

cells which separates two fluids of different composition and includes at least two barriers arranged in series, the brush-border (mucosal) and the basal (serosal) membranes of different structural and functional activities. The brush-border membrane has been extensively studied in the last few years. Intestinal brush borders of various species and using different methods were isolated (reviewed by Porteous, 1969) and their chemical and enzymatic composition was studied. They represent a digestive–absorptive surface (Crane, 1966a,b) in which the elements responsible for digestion and absorption are arranged in a mosaic-like macromolecular complex.

Intestinal cells transport a wide range of substances, both lipid and water soluble, presumably with a high degree of selectivity. Although active processes are largely responsible for the transport of most substances (sugars, amino acids, bile salts, pyrimidines) other transport mechanisms are involved. (a) Pinocytosis is apparently highly useful for some newborn animals in the absorption of proteins, especially of intact antibodies, to obtain passive immunity (Clark, 1959). The pinocytosis hypothesis was applied to the absorption of lipid particles about 500 Å in diameter as studied by electron microscopy (Palay and Karlin, 1959). (b) Simple diffusion, a mechanism for the absorption of lipid-soluble substances, water-soluble vitamins (see Matthews, 1967), some nucleic acid derivatives, weak electrolytes, and highly ionized compounds. (c) Carrier-mediated diffusion of compounds which are not actively transported but share the common carrier with actively transported compounds (such as 6-deoxy-1,5-anhydro-D-glucitol; Bihler *et al.*, 1962).

In discussing the accumulation process in the mucosal cell layer, the vectorial component, derived from the structural arrangement of enzymatic systems and metabolically dependent asymmetries, will be considered. Two systems of transport (or possibly a combination of the two) are coming into play here.

(a) The transport system located at the luminal membrane is responsible for the transport of substances from a lower concentration in the lumen across the brush-border membrane to a higher concentration in the cell if metabolic energy is available. The substance leaves the cell along the concentration difference across the basal membrane (e.g., sugar transport).

(b) The substance crosses the luminal membrane along the concentration difference (from a higher concentration or electrochemical potential at the luminal side to a lower one in the cell), then it is pumped out of the cell by a system located at the basal membrane (e.g., sodium pump).

(c) The transport of a substance across both membranes is a metabolically dependent (or carrier-mediated) process, the difference in the efficiencies of both pumps determining whether the substance is maintained in the cell at a higher or at a lower concentration than outside the cell.

Intestinal transport has been studied by both *in vivo* and *in vitro* methods (reviewed by Wilson, 1962; Levin, 1967). *In vivo* fluxes from the lumen to blood (insorption) and from blood to the lumen (exsorption) are estimated. In studies carried out *in vitro* it is easy to separate the mucosal and serosal fluids. The substance has to pass through submucosal and smooth muscle layers to the serosal side. A ratio of serosal/mucosal concentrations greater than 1 is taken as an indicator of active transport. When incubating intestinal strips, mucosal sheets, villi (Crane and Mandelstam, 1960), or isolated cells (Huang, 1965) the accumulation of substrate can be measured and compared with its medium concentration, taking a correction for extracellular space. Using these methods it is not possible to separate cleanly the processes occurring at either pole of the cell. However, measurements of unidirectional fluxes across the mucosal border into the cell, as applied by Curran's group (Schultz *et al.*, 1967) confirmed the sodium dependence of sugar and amino acid entry at this pole of the cell.

It is important to note that different activities are differently distributed along the length of the small intestine; thus disaccharidases are more abundant in the jejunum than in the ileum, being practically absent in the duodenum (see Semenza, 1968), monosaccharide absorption is also most effective in the jejunum (Crane and Mandelstam, 1960), and active transport of bile salts is a unique function of the ileum (Lack and Weiner, 1961).

For transport studies it is also important to consider species differences. Thus, glucose inhibits amino acid transport in the hamster intestine due to competition for a site on the carrier (Alvarado, 1966a) but stimulates amino acid transport in the rat intestine either by serving as a fuel (Newey and Smyth, 1964) or due to an increased water transport, "dragging" the amino acid with it (Munck, 1968). Likewise, the mode of activation of sucrase and sugar transport by sodium is different in different species. The Lineweaver–Burk plot shows that sodium affects the apparent K_m of both sucrase and the sugar carrier in the rat and the hamster, with the maximum velocity not appreciably affected. The pattern in man and rabbit is different, since Na^+ does not affect significantly the apparent K_m but increases the maximum velocity (Semenza, 1968).

In recent years, investigation of intestinal transport has been considerably extended by modern *in vitro* methods, histochemical methods, and electron microscopy. The reader is referred to three important monographs in the field (Wilson, 1962; Smyth, 1967; Code, 1968).

23.2. SUGAR ABSORPTION

23.2.1. Location of Disaccharidases

The carbohydrate diet of humans and of most animals consists mainly of polysaccharides, disaccharides, and phosphate sugar esters. The intestinal mucosal cells are impermeable to polysaccharides and to products of their hydrolysis by pancreatic amylase as well as to disaccharides. For this reason, disaccharides are split by disaccharidases prior to absorption; monosaccharides then enter the cell interior. It is now well established that both disaccharide hydrolysis and monosaccharide transport are located in the brush-border membrane. In spite of the divergent views it has been shown using different techniques that a number of disaccharidases, including maltase, sucrase, isomaltase, lactase, and trehalase, as well as alkaline phosphatase, are associated almost exclusively with intestinal brush borders (Eichholz, 1967, 1969). Disaccharidases were demonstrated in the brush borders histochemically by Lojda (1965) and others (Jos *et al.*, 1967). Fractionation and density gradient techniques permitted isolation of a fraction of brush borders identified under the electron microscope (Overton *et al.*, 1965) as microvillus membranes (Eichholz and Crane, 1965), hollow finger-like structures having mucopolysaccharide "fuzz"—glycocalyx—on their outer surface. These membranes contain all the above-mentioned disaccharidases (4× purified over the whole brush borders). Sucrase, maltase, and isomaltase activities have been identified with particles 45–60 Å in diameter, fixed in the outer protein coat of the microvillus membranes and buried in the glycocalyx (Johnson, 1967, 1969). These particles (see Fig. 23.3) markedly resemble the papain-solubilized and purified (Kolínská and Semenza, 1967) sucrase–isomaltase complex forming particles of the same size (Nishi *et al.*, 1968). By a short exposure to papain, lactase and β-glucosidase (phlorizin hydrolase) can also be released, while trehalase and alkaline phosphatase remain associated with the trilaminar unit membrane of the microvilli (Eichholz, 1968). Actually, there does not seem to be an appreciable diffusion barrier between sucrase and the medium as the K_m for sucrose in both intact rings of the hamster intestine and the homogenate are the same (Semenza, 1968). For a scheme of the digestive absorptive surface see Fig. 23.4.

The hypothesis proposed by Crane (1965) suggests that disaccharidases are located in the brush borders externally to the monosaccharide transport site since phlorizin prevents glucose entry, irrespective of whether glucose was added in the medium or was formed as a result of hydrolytic activity

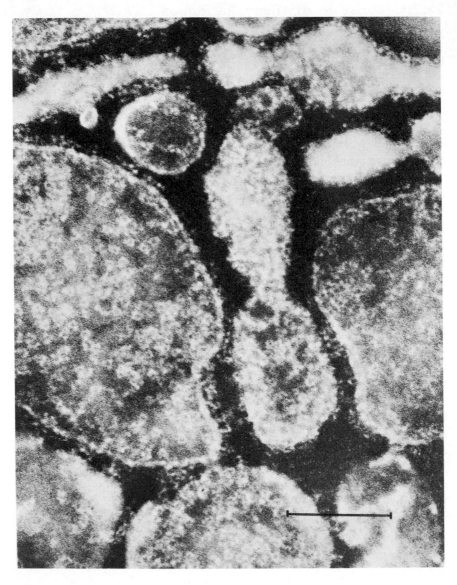

FIG. 23.3. Microvillus fragments treated with trypsin which does not solubilize sucrase. Scale line 0.1 μ. (Taken with kind permission from Nishi *et al.*, 1968.)

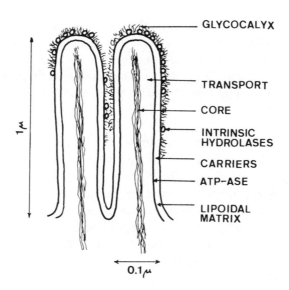

GLYCOCALYX

TRANSPORT

CORE

INTRINSIC
HYDROLASES

CARRIERS

ATP–ASE

LIPOIDAL
MATRIX

FIG. 23.4. Schematic representation of the digestive absorptive surface. (According to Crane, 1969.)

of sucrase (Miller and Crane, 1961) and lactase and trehalase (Malathi and Crane, 1968) or maltase (Newey *et al.*, 1963). Also the absence of Na^+ which is required for monosaccharide entry into the cells, does not stop disaccharidase activity (Crane, 1962). Under such conditions, the products of hydrolysis must appear at the membrane and then diffuse into the medium as shown by Ugolev and co-workers (1964).

Other *in vitro* data of Miller and Crane (1961) indicate that the intestinal uptake of glucose formed hydrolytically from disaccharide has a kinetic advantage over the free glucose for entry into cells; this entry is more rapid then the liberation of monosaccharide into the medium. Thus, the accumulation of glucose derived from sucrose is a function of sucrose and not of glucose concentration in the medium, glucose oxidase added to the incubation media not affecting glucose accumulation.

23.2.2. Monosaccharide Transport

McDougle and co-workers (1960) showed by microdissection that *in vitro* preparations of hamster intestine accumulated monosaccharides in highest concentrations inside the epithelial cells. The results were confirmed by autoradiographic studies (Kinter, 1961; Kinter and Wilson, 1965) of layers of intestinal mucosa. Csáky and Fernald (1961) obtained the same results with frog intestine. An analysis of the separated mucosal and serosal halves after incubation showed that sugars were accumulated only from the mucosal side (Bihler and Crane, 1962). The data suggest that we are dealing here with a transport system located at the luminal brush-border pole of the mucosal cells.

Specificity of Monosaccharide Transport. There appear to be two separate paths for sugar entry into epithelial cells, a Na^+-dependent and a Na^+-independent one. The Na^+-dependent sugar transport has a specific requirement for the pyranose ring structure and a hydroxyl group at C_2 oriented equatorially as in the D-glucose in the *C1* chair conformation. The known substrates for this Na^+-dependent process are D-glucose, 3-methyl-glucose, D-xylose (Csáky and Lassen, 1964; Alvarado, 1964), 6-deoxy-D-glucose, 6-deoxy-D-galactose, and other related compounds (Crane, 1960*a,b*) as well as L-glucose (Caspary and Crane, 1968; Neale and Wiseman, 1968) with a similarly oriented group in position 4 of the *1C* chair conformation. These sugars mutually compete for the same carrier system. It is possible to demonstrate here counterflow phenomena (Alvarado, 1966*b*, 1967*a*; Caspary and Crane, 1968) by which the existence of a mobile carrier is suggested.

Since glucose is liberated from sucrose as the α-anomer the question arose as to whether the conformation of C_1 plays a role in intestinal sugar transport. The data presented by Semenza (1969*a*) showed no kinetic preference for either α-glucose or β-glucose by the sugar carrier. Thus, the kinetic advantage of the glucose liberated from sucrose over free glucose cannot be explained by the difference in transport of the α- and β-anomeric forms. The glycoside phlorizin inhibits the Na^+-dependent sugar transport as a fully competitive inhibitor (Alvarado and Crane, 1962) without entering in appreciable amounts into the cells (Stirling, 1967). The lack of penetration is presumably due to the attachment of its aglycone to a membrane site (Diedrich, 1961; Alvarado, 1967*b*). The finding of β-glucosidase activity in the brush-border membrane (Malathi and Crane, 1969) which splits phlorizin into the aglycone phloretin and glucose, brought up the question of whether phloretin formed hydrolytically within the brush-border membrane may not be the inhibitor of sugar transport rather than phlorizin (Diedrich, 1968). Free phloretin, however, has been found to be some 1000 times less effective than phlorizin in the intestine (Alvarado and Crane, 1962) and it inhibits sugar transport in a noncompetitive manner (Diedrich, 1966*b*).

The Na^+-independent process does not lead to uphill accumulation of sugars. Fructose, 2-deoxy-glucose, mannose, and sorbitol are transported by this system which is distinct from the Na^+-dependent one. It seems to be relatively nonspecific as shown recently for 2-deoxyglucose (Goldner et al., 1969a).

Mechanism of Na^+-Dependent Sugar Transport. Two steps of Na^+-dependent accumulation of monosaccharides in the small intestinal mucosa have been considered (Bihler et al., 1962). (1) The carrier-mediated entry of sugars into the cells which follows Michaelis–Menten kinetics and is independent of metabolic energy. It results in a quick equilibration between the extracellular and the intracellular compartments and can be distinguished from the accumulation against a concentration difference by uncoupling the energy sources of the cell with dinitrophenol or anaerobic conditions. Of particular interest is the sodium dependence of the entry step. In the absence of Na^+ (replaced with K^+) under anaerobic conditions sugar entry is greatly inhibited and mutual inhibition between different sugars abolished. In Li^+ media, however, as shown by Bihler and Adamič (1967), mutual inhibition as well as phlorizin inhibition of sugar transport persists, indicating that Li^+ can partially substitute for Na^+ at the level of the equilibration system. (2) The second component is a Na^+-dependent energy-dependent movement against the concentration difference. Anaerobiosis and reduction of tissue energy supplies prevent sugar accumulation. Sugar accumulation is abolished also by ouabain which is effective at the serosal side only. Na^+ is required at the mucosal side, serosal Na^+ has no effect on sugar movement (Csáky and Thale, 1960). The effect of Na^+ at the mucosal side is very specific and cannot be produced by another cation of the first group. Actually, K^+, Li^+, Cs^+, Rb^+, and NH_4^+ suppress the accumulation below the Na^+-free control (Bosáčková and Crane, 1965a).

According to Smyth and his co-workers (Sanford et al., 1965) at least two metabolic pathways are functional which can supply energy for sugar transport in the rat intestine, the citric acid cycle and aerobic glycolysis. Fluoroacetate inhibits sugar transfer provided for by energy from endogeneous substrates but does not affect the glucose-stimulated transfer of other sugars. On the other hand, fluoride inhibits the latter and not the former pathway. Uranyl nitrate appears to inhibit glucose metabolism which can supply energy for galactose and 3-methyl-glucose transport (Newey et al., 1965, 1966). Anaerobic glycolysis seems to provide energy for active sugar transport in human foetal intestine (Jirsová et al., 1966).

23.3. PROTEIN ABSORPTION

Prior to absorption, proteins must be split in the lumen of the small intestine of adult mammals. According to Fisher (1967) the time course of peptide bond hydrolysis in the intestinal lumen by trypsin, chymotrypsin, and erepsin follows a logarithmic curve. Thus the rate of splitting of peptide bonds is directly proportional to the number of bonds which remain to be split. Peptides have been shown in both *in vivo* and *in vitro* experiments to be absorbed very poorly (Agar *et al.*, 1953; Newey and Smyth, 1959). The terminal stage of peptide bond hydrolysis is associated with the brush-border membrane and it is known that leucyl-naphthyl amidase and leucyl-glycine hydrolase are building blocks of the brush-border membrane (Rhodes *et al.*, 1967). Amino acids as end products of protein digestion are readily absorbed in the small intestine.

Some new-born mammals can absorb even intact proteins by pinocytosis (Wilson, 1962). The intestinal pinocytotic function of the suckling rat was clearly demonstrated by Clark (1959).

23.3.1. Amino Acid Transport

Amino acids are transported in the intestinal mucosa against a concentration difference. Their transport is inhibited by dinitrophenol and lack of oxygen (Fridhandler and Quastel, 1955) and by cyanide (Agar *et al.*, 1954). As with sugars, the amino acid transport has a special requirement for Na^+. Amino acids increase the short-circuit current and this stimulation is inhibited by ouabain when added to the serosal solution (Schultz and Zalusky, 1965). Amino acid transport has been extensively studied since the first observation of Elsden and co-workers (1950) and Wiseman's (1951) demonstration of an accumulation of L-isomers of the neutral amino acids against a concentration difference in the rat small intestine. Transport of basic amino acids using hamster intestine was first shown by Hagihira and co-workers (1961).

The available data suggested the existence of four separate systems for amino acid transport: (a) neutral amino acids, (b) basic amino acids, L-cystine, (c) L-proline, L-hydroxyproline and sarcosine, dimethylglycine, and betaine, (d) dicarboxylic amino acids (transaminated extensively; Wiseman, 1953).

Within each group, the amino acids mutually compete for the transport site in the brush-border membrane (Nathans *et al.*, 1960; Robinson, 1966; Finch and Hird, 1960). Investigating this further, Alvarado (1966a) found this mutual inhibition between neutral amino acids to be fully competitive (*cf*. Thorn's plot on p. 236).

Incubation studies suggest that L-amino acids share the same transport system with their D-enantiomorphs, the affinity of the membrane site for the latter being relatively low (Lin *et al.*, 1962). Nevertheless, the transport of D-amino acids has been demonstrated by Jervis and Smyth (1960) and Lerner and Taylor (1967) for D-methionine and by Randall and Evered (1964) for D-serine and D-norvaline. Wilson (1962) has pointed out the importance of the nonpolar character of the side chain of the neutral amino acids for the affinity of the carrier system. The charge in the side chain decreases the affinity of the carrier (Huang, 1961; Lin *et al.*, 1962). The degree of solubility of the side chain in the lipid phase of the membrane is suggested to be important for the attachment of amino and carboxyl groups and of α-hydrogen to the active sites of the carrier. Daniels and co-workers (1969) emphasized that the lipid–water partition coefficient determines to some extent the chemical specificity of the neutral amino acid. Whereas larger, more lipid-soluble neutral amino acids have a greater affinity for the methionine transport system, small water-soluble neutral amino acids are more inhibitory for the sarcosine transport system.

The specificities of the amino acid transporting systems overlap somewhat so that the system for imino acids in the rat intestine is used also by the neutral amino acids leucine, glycine, and alanine (Munck, 1966). Inhibition of the glycine transport pathway via the system for neutral amino acids by methionine permitted to show the affinity of glycine for the amino acid transport system. Likewise, according to Spencer and Brody (1964), L-phenylalanine inhibits the transport of L-proline in hamster intestine and *vice versa*.

Studies of the interaction between neutral cyclo-leucine and basic L-arginine (Alvarado, 1966*a*) showed that L-arginine is a partially competitive inhibitor of active transport of neutral amino acids. In this case, Thorn's plot gives a hyperbolic curve instead of the straight line expected from the equation for a fully competitive inhibitor. The evidence that basic amino acids elicit counterflow of neutral amino acid supports further the assumption that interactions between these two groups of amino acids occur at the membrane carrier.

The stimulation of L-arginine uptake by L-methionine and L-leucine (Robinson, 1968) was explained in terms of an inhibition of L-arginine efflux by intracellular L-methionine or L-leucine. This phenomenon was explained as being due to an inhibition of the back flux of a basic amino acid across the brush-border membrane by intracellular neutral amino

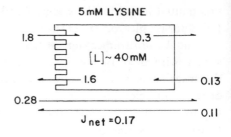

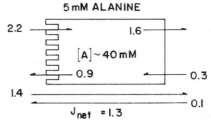

FIG. 23.5. Comparison of unidirectional fluxes of lysine and alanine across the mucosal and serosal membranes in the presence of 5 mM amino acid and 140 mM Na$^+$. All values are expressed in (μmol/ h)/cm^2. (According to Munck and Schultz, 1969.)

acids. Under control conditions the efflux of L-lysine across the brush-border membrane is very high and almost equals the L-lysine influx across the brush-border membrane (Munck and Schultz, 1969). Figure 23.5 shows the unidirectional fluxes of neutral and basic amino acids and it can be seen that the influx of both neutral and basic amino acid in the presence of Na$^+$, as well as their intracellular concentrations, are quantitatively similar. However, the net flux of lysine is much lower than that of alanine. The reason for the low net flux of L-lysine is seen in a very high efflux of L-lysine back across the brush-border membrane. The primary reason as viewed by the authors is a low maximal rate of efflux of L-lysine out of the cells across the serosal membrane. This work showed that net flux data can lead to misinterpretation unless phenomena occurring at the serosal and the mucosal membranes are analyzed separately.

23.3.2. Interaction between Amino Acid and Sugar Transport

On the basis of the present knowledge three possibilities for amino acid and sugar interaction have been proposed and all three appear to be justified, considering species differences. (1) Competition for a common carrier in the microvillus membrane was proposed by Alvarado (1966*a*, 1968). He found that galactose and other actively transported sugars are partially competitive inhibitors of neutral amino acid transport in the hamster small intestine. In kinetic theory of the enzymes, the partially competitive inhibitor is viewed as a substance which binds to a site different from but close to the substrate binding site. Such inhibition is properly classified as allotopic and is reflected in a change of K_m to a higher value without a change in V_{max}. (2) Competition for the source of energy produced by the cell. According to this hypothesis (Newey and Smyth, 1964; Bingham *et al.*, 1966a,b); nonmetabolizable actively transported sugars compete with amino acids for the source of energy which is not sufficient to allow both transport processes to operate at maximum capacity. Glucose, on the other hand, can itself provide an energy supply for amino acid transport and stimulate the transfer of amino acids across the intestinal wall. Glucose overcomes galactose inhibition as enough energy is produced for transfer of both hexoses and amino acids. This was observed in the rat intestine where the glycolytic activity is high. It does not apply to rabbit and hamster intestine. Here glucose inhibits transport of amino acids as well as of actively transported sugars. (3) By increasing the net water flux (Munck, 1968) from the mucosal to the serosal fluid, glucose increases the serosal compartment available for equilibration of the accumulated amino acid. Using the test tube method of Crane and Wilson (1958) Munck measured the effect of glucose on the transport of L-valine and L-proline across the intestinal wall of the rat. Glucose caused a 90% increase in the serosal volume and, due to the fluid movement, a rapid disappearance of amino acids from the mucosal fluid. The concentration of the amino acid in the serosal fluid was the same as in the control with no glucose present.

23.4. TRANSPORT OF IONS AND WATER

23.4.1. Monovalent Cations

In the intestinal transport of solutes, a major role is played by sodium ions, not only for the general principle of regulating the concentration of ions and of the cell volume by the sodium pump but also for their characteristic translocation which is important for other transport processes. In contrast to sugar accumulation inside cells the intracellular concentration of Na^+ remains lower than in the extracellular fluid (Bosáčková and Crane, 1965b; Schultz *et al.*, 1966). Intracellular potential recordings also showed that the cell interior is negative, relative to both the mucosal and the serosal fluid (Gilles-Baillien and Schoffeniels, 1965, 1967; Wright, 1966), the potential step being larger—and opposite in sign—across the serosal membrane than across the mucosal membrane. Transmural potentials across the intestinal wall of some 5–7 mV were recorded with serosa positive towards mucosa (Baillien and Schoffeniels, 1961; Barry *et al.*, 1961). When the short-circuit current technique is applied there should be no movement of passively transported ions as the potential difference is reduced to zero. However, an appreciable transfer of Na^+ still takes place. This shows that the transmural potential is a result of active sodium transport against the electrochemical gradient (Curran, 1965). Cardiac glycosides (like ouabain), when applied to the serosal side, inhibit this net Na^+ flux (Schultz and Za-lusky, 1964*a,b*). These findings may be explained by the sodium pump being localized at the serosal membrane of the epithelial cells.

Divergent findings have been reported on the relation between the short-circuit current and the net Na^+ transport. (a) Equality between the short-circuit current and the net Na^+ transport, (b) short-circuit current greater than the net sodium flux, indicating an active Cl^- transport mechanism directed from serosa to mucosa, (c) short-circuit current smaller than the sodium flux, indicating a nonelectrogenic sodium transport directed from mucosa to serosa.

The effect of hexoses on the short-circuit current showed that glucose, galactose, and 3-methylglucose increase the transmural potential and the short-circuit current of rabbit and rat ileum which corresponds to a net Na^+ transfer from the mucosal to the serosal side (Schultz and Zalusky, 1964*a*; Clarkson and Toole, 1964).

However, Barry and co-workers (1965) reported that the net sodium transport in rat jejunum accounts only for some 30% of the short-circuit current when galactose or α-methylglucoside were present. Under the same conditions, glucose brought about an equal increase in the short-circuit current and the net Na^+ flux. These two different findings arose obviously from a difference in response of the jejunum and the ileum to the actively transported sugars. Recently Taylor and associates (1968) studied the relation between Na^+ transport and short-circuit current and the sugar effect in both the jejunum and the ileum of rat and rabbit. They confirmed the previous findings that in the ileum of both rat and rabbit the short-circuit current equals the net Na^+ transfer in the presence of both glucose and galactose. In the rat jejunum, the difference between the short-circuit current and the Na^+ flux in the presence of galactose was found to be due to the active secretion of chloride in the opposite direction. A similar phenomenon was described by Parsons (1967) who found that the rates of chloride and sodium absorption differ in the rat jejunum: Na^+ was absorbed faster than chloride. It was pointed out that chloride may be transported by an independent process involving exchange with bicarbonate ions which are absorbed in the rat jejunum. The difference between the jejunum and the ileum may be attributed to a metabolic limitation in the jejunum in the absence of glucose. Other substrates, such as pyruvate and citrate, could not replace glucose in sustaining the ion and water transport across the rat jejunum (Smyth and Taylor, 1957; Gilman and Koelle, 1960).

Occasionally, more Na^+ is transported than would correspond to the short-circuit current. Thus, sugars which are not transported against a concentration difference but are metabolized (fructose, mannose) stimulate Na^+ transport accompanied by volume flow. However, the net Na^+ movement is greater than the short-circuit current. According to Smyth (1966), there exist two transfer mechanisms for Na^+, an electrogenic one, involving hexose transfer and producing the short-circuit current, and a neutral sodium pump depending on hexose metabolism. The latter system does not cause a potential difference and is associated with transfer of chloride, bicarbonate, and lactate in the same direction as sodium.

23.4.2. Divalent Cations

The place of calcium absorption in the gut has been localized in the proximal part of the small intestine (Harrison and Harrison, 1951; Nicolaysen, 1951; Moore and Tyler, 1955), the transport process being at least in part an active one, calcium passing from the intestinal lumen through the mucosal cell surface to the serosal surface of the intestine against its concentration and electrical potential gradient (Schachter, 1963; Wiseman, 1964). The process is inhibited by anaerobic conditions, 2,4-dinitrophenol, fluoride, cyanide, iodoacetate, citrate, oxaloacetate, and by low temperature. The capacity of active transport of calcium by the small intestine of young rabbits was found to diminish substantially on passing from the duodenum to the upper jejunum (Schachter and Rosen, 1959).

The mechanism of calcium transfer is not fully understood but it is known that calcium interacts with a number of ligands in the intestinal lumen, such as citrate, fructose, or lactose (Dupuis and Fournier, 1963; Wiseman, 1964; Helbock et al., 1966). The effect of phosphate on calcium transfer was demonstrated by Helbock and co-workers (1966), who used the short-circuit technique to follow the movement of calcium through the wall of small intestine sections of rats. They found that the calcium transfer in the absence of phosphate is a passive process while in its presence calcium appears to be actively transported together with the phosphate anion. If the phosphate level is high, absorption of calcium is inhibited by formation of insoluble salts. The authors demonstrated, however, that in the presence of suitable chelates (fructose, citrate, EDTA) the ion is solubilized and its movement across the intestinal wall enhanced.

Vitamin D plays an important function in calcium absorption as it increases the efficiency and the rate of calcium absorption from the intestinal lumen (Schachter, 1963; Wiseman, 1964).

Transport of strontium and magnesium in the gut has some features in common with calcium transfer and there may exist a common pathway for their absorption (Hendrix *et al.*, 1963; Lengeman, 1963; Wiseman, 1964). Vitamin D influences positively the transfer both of strontium and magnesium (Schachter, 1963; Wiseman, 1964) even if this stimulation is not as pronounced as in the case of calcium.

A calcium-transporting protein has been isolated from the chick duodenum (see chapter 8).

23.4.3. Water

Curran and co-workers (see Curran, 1965), using rat ileum *in vitro*, obtained a close relationship between the rate of water absorption and the NaCl concentration in the mucosal fluid. A quantitative determination shows (Curran, 1960) that volume flow follows net solute flow. At zero solute flow there was no significant volume change and they thus concluded that water movement is a passive process. This observation was confirmed by Clarkson and Rothstein (1960). When the mucosal solution was made hypertonic with mannitol water absorption was stopped but net Na^+ absorption remained unaltered (Green *et al.*, 1962). This again supports Curran's view on the predominant active Na^+ transport when water movement may be explained without postulating a specific water pump.

23.5. INTERACTION BETWEEN THE TRANSPORT OF NONELECTROLYTES AND OF SODIUM

As mentioned above, the transport of sugars and amino acids against a concentration difference requires the presence of Na^+ at the mucosal side and is abolished by ouabain (Crane *et al.*, 1961; Csáky *et al.*, 1961) which exerts its effect when added to the serosal side (Schultz and Zalusky, 1964*b*). The location of the effect of ouabain, an inhibitor of the (Na^+-K^+)-stimulated ATPase, indicates rather strongly the presence of the sodium pump at the basal membrane. How then can one explain the effect of Na^+ on nonelectrolyte transport if the sugar and amino acid carriers are located in the brush-border membrane while the sodium pump is at the opposite pole of the cell?

Two hypotheses (Crane, 1962, 1965; Csáky, 1963, 1965) have been proposed to explain the Na^+ requirement of nonelectrolyte transport. The major differences between Crane's and Csáky's hypotheses are the following: (1) Crane proposed a direct interaction of sodium with the sugar carrier, under formation of a ternary sugar–carrier–sodium complex in the brush-border membrane; Csáky's model postulates an independent entry of sugar or amino acid and Na^+ at the brush-border membrane. (2) According to Crane's model the uptake of sugar is not directly linked with metabolic energy; the maintenance of Na^+ downhill gradient across the brush-border membrane by the ouabain-sensitive sodium pump provides a driving force for the accumulation of the sugar. Csáky thinks that sugar and amino acid uptakes are directly coupled with Na^+-dependent utilization of metabolic energy; inhibition of the $(Na^+$–$K^+)$-stimulated ATPase by ouabain thus brings about an inhibition of nonelectrolyte accumulation.

It had been difficult for some years to design experiments for distinguishing between these two hypotheses. However, several observations strongly support Crane's hypothesis. (a) Under anaerobic conditions, when no uphill sugar transport takes place, the entry of a sugar is still sodium-dependent (Bihler et al., 1962). Thus, the equilibrium system was affected by Na^+ without an energy-yielding process being involved. (b) Sugar moves out of cells against its concentration gradient when the Na^+ gradient is reversed. The tissue was first equilibrated with the sugar at a high Na^+ concentration in the medium and then placed into a sodium-free medium (Crane, 1964). (c) Interaction of Na^+ with the sugar carrier was confirmed by short-circuit current measurements indicating an increased rate of net Na^+ flux on adding of the sugar (Clarkson and Rothstein, 1960; Barry et al., 1964; Schultz and Zalusky, 1963, 1964a). Ouabain did not inhibit the net Na^+ flux stimulated by the sugar at the mucosal side but it was inhibitory at the serosal side. Thus it appears that the sugar–Na^+ interaction at the brush-border membrane is not directly affected by ouabain. (d) Curran's group (Schultz et al., 1967) developed a method for direct and simultaneous measurement of unidirectional fluxes of Na^+ and amino acids from the mucosal solution into the cell of rabbit illeum. Their observation is consistent with Crane's model of a common entry mechanism for Na^+ and amino acid or sugar. Briefly, both amino acid and sugar influxes displaying Michaelis–Menten kinetics are dependent upon Na^+ in the mucosal solution (Li^+ and K^+ are inhibitory) and involve a simultaneous entry of Na^+ (so-called amino acid or sugar-dependent Na^+ influx) into the cell. It was demonstrated that a reduction of the intracellular Na^+ pool does not influence alanine influx (Schultz *et al.*, 1967). Na^+-dependent alanine entry is not significantly affected by dinitrophenol, iodoacetate, and ouabain (Chez *et al.*, 1967).

Two models for amino acid (Curran *et al.*, 1967 *a*) and sugar transport (Goldner *et al.*, 1969*b*) were developed to account for differences between amino acid and sugar transport in rabbit ileum. (a) Na^+ affects the V_{max} of sugar influx with little change of the apparent K_m; an opposite pattern holds for amino acid influx. (b) In Na^+-free media very little of the sugar influx seemed to be mediated, whereas amino acid influx is mediated. (c) The ratio of sugar-dependent Na^+ influx and sugar influx is constant at different Na^+ concentrations in the mucosal solution. The ratio of alanine-dependent Na^+ influx and alanine influx varies with Na^+ concentration.

According to Crane's hypothesis, the accumulation of the sugar across the brush-border membrane is a result of three asymmetric phenomena. (1) Downhill movement of Na^+ from the mucosal solution into the cell. The sugar carrier is loaded by Na^+ from the mucosal solution and thus increases the entry of the sugar into the intracellular compartment. (2) Downhill movement of K^+ from the cell into the mucosal fluid. K^+ is at a high concentration inside the cell and displaces Na^+ from the binding site at the inner face of the brush-border membrane. Sugar carrier becomes ineffective for the back movement of the sugar into the mucosal fluid. (3) The sugar transport system is affected by Na^+ so as to increase the affinity of the carrier for the sugar (hamster and rat). The apparent K_m for the sugar increases up to about 200 times within the range of Na^+ concentrations 145 mM to zero. K^+ brings about a further decrease of the affinity of the carrier for the sugar (Crane *et al.*, 1965). At this particular point it is important to point out that in the rabbit intestine the mode of action of Na^+ is different from that mentioned above for hamster and rat intestine. The increase of the V_{max} with no change of the apparent K_m was reported for the rabbit intestine by Semenza (1968) and Goldner and co-workers (1969*b*). Theoretically, both patterns are possible for the interaction of an ion with an enzyme system (Dixon and Webb, 1965).

Crane's model thus predicts that the effect of Na^+ on the influx of sugar is a matter of the equilibration system in the brush-border membrane, not itself vectorial, which is not directly coupled with the vectorial Na^+ pump. All the asymmetries mentioned depend on the efficiency of the Na^+ pump which provides for Na^+ translocation and the Na^+ gradient between the cell and its environment. The location of the Na^+ pump is then of little relevance.

Some recent observations suggest that the transport of other organic substrates, such as bile salts (Holt, 1964) and pyrimidines (Csáky, 1961), is coupled with Na^+ transport. Sodium ions stimulate the activity of intestinal sucrase (Semenza *et al.*, 1964; Kolínská and Semenza, 1967) while cations of the first group and NH_4^+ competitively inhibit it with respect to Na^+. Similarity of the ratio of the K^+ inhibition constant and the Na^+ activation constant of intestinal sucrase and of that for sugar transport system indicates that sucrase, which is located close to the sugar carrier in the brush-border membrane (Miller and Crane, 1961), may share a common Na^+-binding site with the sugar transport system. Moreover, purified sucrase has a glucose binding site different from the substrate binding site (sucrose, isomaltose) (Semenza, 1969*b*). ^{14}C-Glucose fixation was found to be Na^+-dependent and inhibited by most actively transported sugars.

Alvarado (1966*a*) put forward the possibility of a polyfunctional carrier system in the membrane which contains binding sites for different groups of organic substrates and Na^+. Allotopic interactions between these binding sites may bring about inhibition or activation of the individual transport processes.

23.6. ABSORPTION OF LIPIDS

23.6.1. Pinocytosis

A variety of authors published evidence of particulate absorption of fats (Ashworth *et al.*, 1960; Strauss, 1963). Palay and Karlin (1959) demonstrated lipid droplets 50 mμ in diameter at the base of the microvilli. These droplets were engulfed in the form of pinocytotic vesicles in the membrane but the authors admit that the number of pits and signs of active pinocytosis seemed insufficient to account for the rapid accumulation of lipids in the smooth endoplasmic reticulum during the early stage of lipid absorption. Recently Porter (1969) presented evidence indicating an independence between fat absorption and pinocytosis. Different lipid markers, like colloidal suspension of metallic silver in corn oil, were never found in the pinocytotic vesicles nor in the fat globules of the smooth endoplasmic reticulum where the fat droplets occur shortly after entering the cells. It was also seen that the membrane enclosing the lipid globules in the endoplasmic reticulum is thin and symmetric and differs very much from the unit membrane of pinocytotic vesicles which is thick and asymmetric.

23.6.2. Diffusion

It appears that simple diffusion of fatty acids and monoglycerides is the major route of lipid absorption. Two hypotheses have been put forward to account for monoglyceride and fatty acid entry across the brush-border membrane, one suggesting entry in the intact micellar form, the other in a molecular form.

(a) If the intact bile salt–fatty acid–monoglyceride micelle crosses the membrane by diffusion two aspects are of importance—the size of the micelle and its ionic nature. It has been thought that such mixed micelles (Ashworth and Lawrence, 1966) about 27–30 Å across as described by Feldman and Borgström (1966a) and by Borgström (1965) could penetrate through the pores of the epithelial membrane of the small intestine. The ionic nature of the micelles is very important for their movement across the membrane as is affected by the electrochemical gradient and possibly by solvent drag (Gordon and Kern, 1968). The degree of ionization of the bile salt–fatty acid–monoglyceride micelle is pH-dependent, the pK_a of fatty acids in the micelle having been found to be 6.4 (Hoffman and Small, 1967). It is probable that at the physiological pH of 7.4 these mixed micelles carry a net negative charge on their surface, formed by acid groups of the bile salts. Such micelles would bind counterions, most likely Na^+ and K^+, but little information exists on the binding of these cations in the intestinal lumen. The presence of fatty acids and of lecithin may also be of importance in the counterion binding. Recently, Lyon (1968) showed that triolein is taken up by the rat ileum in the form of a mixed micelle with taurocholate and glycerol mono-oleate and that this uptake is Na^+-dependent. From his short- and open-circuit measurements he could not conclude that the mixed micellar lipid crosses the brush-border membrane as a counterion to Na^+. Gordon and Kern (1968) support the hypothesis of intact micellar diffusion in the hamster jejunum. They demonstrated an identical rate of absorption of oleic acid and taurodeoxycholate and a stoichiometric relationship between the concentration of these compounds in the tissue.

(b) Both *in vivo* and *in vitro* experiments indicate that molecular diffusion of fatty acids and monoglycerides and bile salts takes place. Pessoa and co-workers (1953) found an appreciable absorption of fatty acids in experiments with the bile fistula. Fox (1965) who worked with turtle intestine found an absorption of fatty acids *in vitro* from media containing albumin. At chain length of C_{12} fatty acids were absorbed maximally. No evidence of accumulation on the serosal side was obtained. According to Johnston and Borgström (1964) and Feldman and Borgström (1966b) the entry is an energy-independent process unaffected by metabolic inhibitors. Hogben (1960) proposed that the fatty acids enter by virtue of their solubility in the lipid portion of the membrane. According to Hogben, short-chain fatty acids enter the mucosal membrane faster in the unionized form and thus a pH difference across the intestinal wall would increase the gradient of fatty acids.

There are indications that some lipids may enter the membrane by mediated diffusion in a Na^+-dependent process (Lyon, 1968).

Bile salts are transported against their concentration difference as shown for ten species of animals (Glasser et al., 1965). Generally, bile salts are absorbed in the ileum with a requirement for Na^+ (Playoust and Isselbacher, 1964; Holt, 1964). Their entry is a saturable function of their concentration in the medium, it is inhibited by metabolic inhibitors and ouabain, and requires a negative charge on the molecule. Conjugation of bile salts has a pronounced effect on transport (Lack and Weiner, 1966). The pattern of mutual inhibition between bile salts for entry is not a simple process. It has been found to be competitive only at low concentrations of the inhibitor. The inhibitory potency observed in mutual inhibition was in the order of triketo < trihydroxy < dihydroxy bile salts. It should be pointed out that bile salts inhibit also unrelated systems as sugar and amino acid transport (Faust, 1964; Nunn et al., 1963; Parkinson and Olson, 1963).

23.6.3. Chylomicrons

In the endoplasmic reticulum the products of lipid degradation are reformed into triglycerides, fatty acids being first activated to form acyl-CoA. Triglycerides are released into the intercellular space in the form of chylomicrons, the formation of which requires the presence of phospholipids, free or esterified cholesterol and protein which provides the wrapping of the chylomicrons.

24. KIDNEY

24.1. MORPHOLOGY

An enormous amount of research work in the field of biological transport has been devoted to the study of the specialized excretory organ, the kidney. According to Morel (1967) this investigation may be divided into roughly four classes: (1) biophysics of the epithelial cells; (2) physiology of the nephron; (3) function of the kidney as a whole; (4) homeostasis of the body. Only the first group belongs properly to the subject matter of this book but it appears that due to the great morphological and functional complexity of the kidney this kind of information is, in fact, derived mostly from studies belonging to the other groups. Only a short survey of some aspects of kidney function will be attempted here; for a more complete information, specialized literature should be consulted.

The kidney contains a number of functional units called *nephrons*, each of them consisting in its turn of the *glomerulus* and of the *urinary tubule*. The glomerulus is composed of a number of capillaries formed by branching of the *afferent arteriole* and joining again into the *efferent arteriole*. The diameter of the efferent arteriole being smaller than that of the afferent one, high hydrostatic pressure is produced in the capillaries: the glomerulus is the site of ultrafiltration. The ultrafiltrate of the blood plasma enters *Bowman's capsule* which is the extended beginning of the tubule. The main parts of the kidney tubule are the *proximal tubule*, the *loop of Henle*, and the *distal tubule*. From the distal tubules, the luminal fluid passes into the system of the collecting ducts. The glomeruli and the proximal and distal tubules are situated in the outer layer of the kidney, called the *cortex*, whereas the U-shaped loops of Henle with a descending and an ascending part project into the inner part, called the *medulla*. The efferent arteriole

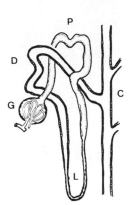

FIG. 24.1. Structure of the nephron. **G**
Glomerulus, **P** proximal tubule, **L** loop of
Henle, **D** distal tubule, **C** collecting duct.

forms once more a network of capillaries, this time around the tubules and
Henle's loops. Into these capillaries proceed the substances reabsorbed by
the epithelial cells of the tubule from the luminal fluid and from the capillaries arrive the substance secreted by the cells into the lumen (Fig. 24.1).

24.2. EXPERIMENTAL TECHNIQUES

Of the various experimental techniques by which the transport mechanisms in the kidney have been studied the following may be mentioned.

1. Measurement of *clearance*, in which the concentration of a given
substance is determined both in the blood plasma shortly before ultrafiltration and in the final urine. The results are expressed as the volume of the
plasma containing the amount of the substance excreted by the kidney
per minute, i.e., as the volume of plasma cleared of the given substance
during one minute. The physiological conditions of this measurement are
of advantage, but, on the other hand, no information about the localization
and character of the processes is obtained.

2. The *stop-flow* technique offers some information on the probable
localization of the transport processes but the experimental conditions are
less physiological. By stopping the outflow of the urine from the kidney
the pressure inside the tubules prevents further ultrafiltration and the composition of the luminal fluid is modified, depending on the function of this
part of the nephron with which it is in contact. Relieving the flow and
analyzing the successively collected samples, the first ones coming from the
most distal parts of the nephrons, the localization of some transport processes can be attempted.

3. The *micropuncture* techniques are of greatest interest for the local-

ization of the individual processes. With sharp micropipettes, a few microns in diameter, one can not only withdraw samples of the luminal fluid but also inject experimental solutions, this procedure being followed by an analysis of the final urine or of the luminal fluid after a subsequent withdrawal, etc. A shortcoming of this technique is that only the superficial nephrons, not necessarily identical with those situated deeper, are accessible to the micropuncture.

4. Finally, *in vitro techniques* are used with aim of obtaining a better insight into the biochemical processes connected with the transport mechanisms. Cortex tissue slices have been used most but it has been criticized that the luminal membranes are only poorly accessible in the slices (e.g., Bojessen and Leyssac, 1965). On the other hand, it can be demonstrated that glucose accumulation, localized on these membranes, occurs even in slices (Kleinzeller *et al.*, 1967 *a*, *b*). Isolated cells and even isolated tubules have been used by some authors, as will be seen later.

24.3. TRANSPORT OF SODIUM IONS AND OF WATER

A substantial part of sodium ions in the glomerular ultrafiltrate are reabsorbed already in the proximal tubule of the nephron. The rate of their reabsorbtion in the proximal tubule was measured by the shrinking droplet method of Gertz (1963) in which isotonic sodium chloride solution was microinjected between two oil droplets in the lumen of rat proximal tubule and its volume and sodium chloride concentration changes with time were examined. This so-called isotonic reabsorption represents a considerable flux of sodium ions of about 9×10^{-9} equiv$\cdot$cm$^{-2}\cdot$sec^{-1} (Ullrich, 1967) and its value is independent of sodium concentration in the plasma (Hierholzer *et al.*, 1965). The extent to which the sodium concentration in the luminal fluid of the proximal tubule can be reduced by the transport mechanism may be estimated by injecting into the tubule a solution of a poorly permeating substance (e.g., raffinose) which prevents osmotically the rapid reabsorption of water (Kashgarian *et al.*, 1963). The "steady-state sodium concentration at zero net flux" thus obtained in the proximal tubule of the rat changes in parallel with plasma sodium concentration, being always by 35 mM lower than the sodium concentration in the plasma (Ullrich, 1967). The transport of sodium ions is thus seen to proceed against a concentration gradient. Sodium reabsorption (efflux from the lumen) is not assisted by electrical forces: The lumen was reported to be electrically negative by some 20 mV with respect to the peritubular fluid both in the proximal tubule of *Necturus* (Giebisch, 1958, 1961; Whittembury, 1963) and of the rat (Kashgarian *et al.*, 1963; Marsh and Solomon, 1964). However, newer measurements with a better control of the tip position suggest that the potential difference across the proximal tubular wall of the rat is actually zero (Frömter und Hegel, 1966). In no case was a gradient of electrical potential favoring the reabsorption of cations from the proximal tubule found. Together with the evidence that sodium transport is not due to interaction with flows of other substances this is a sufficient criterion of the active character of sodium transport in the proximal tubule. It was suggested that a considerable part of the sodium transport may be due to solvent drag (Rector *et al.*, 1966) but this conclusion was contradicted by demonstrating that higher reflection coefficients for NaCl (and, hence, a less important solvent drag) may be found in the proximal tubule (Ullrich, 1967). The localization of the sodium pump in the proximal tubule follows from the consideration of the low intracellular sodium concentration and negative electrical potential in the tubular epithelial cells. The extrusion of sodium ions from the cells into the peritubular space is necessarily active but, as shown by Marsh and co-workers (1963), during stationary microperfusion experiments the sodium ion is far from equilibrium also across the luminal cell membrane and, hence, also here an active mechanism extruding sodium from the cells (but less important than the passive entry at the same place) must be postulated. The sodium-linked water transport in the proximal tubule by which a considerable part of the glomerular filtrate (representing, as with sodium, as much as some 80% in mammals) is reabsorbed was recently analyzed in great detail by Whittembury (1967) and the feasibility of the intercellular pathway for water movement was demonstrated.

In the Henle loop of the mammalian kidney the high passive permeability toward sodium ions, together with the active transport of sodium, form the basis of the concentrating mechanism of the mammalian kidney, the so-called *countercurrent system*. The high permeability of the Henle loop to sodium was demonstrated by the tracer microinjection technique. Whereas an immobilization of the injected fluid containing both radioactive sodium and labeled inulin in the loop prevents any excretion of labeled inulin in the final urine, neither the fraction of the sodium excreted (6.8%) nor the time course of its excretion are changed. Thus, a rapid exchange of sodium takes place between adjacent Henle loops in the medulla (De Rouffignac and Morel, 1967; Morel, 1967). Due to the high permeability to sodium, the sodium ions transported actively out of the ascending limb of the Henle loop permeate into the descending limb, proceed at a higher concentration to the ascending one and are there subjected to a more extensive transport and so on until a steady state is achieved in which the medulla is hyperosmotic and thus capable of concentrating urine by an osmotic withdrawal of water from the collecting ducts. This appears to be the basis of the countercurrent concentrating mechanism in the mammalian kidney as reviewed, e.g., by Ullrich and collaborators (1961). Stimulation of the active reabsorption of sodium from the ascending limb by neurohypophyseal hormones may explain both the build-up of the medullary gradient of osmotic pressure during antidiuresis and the natriuretic effects of the hormones (Morel, 1964).

In the distal tubule then net sodium efflux (reabsorption) is considerably smaller than in the proximal tubule; the value of 1.6×10^{-9} equiv $\cdot$cm$^{-2}\cdot$sec^{-1} was given for the isotonic efflux from the distal tubule of the rat (Ullrich, 1967). The net efflux is the difference between the unidirectional efflux, considered to be almost entirely active, the lumen of the distal tubule being electrically negative with respect to the peritubular fluid (Kashgarian *et al.*, 1963) and a passive influx. The steady-state concentration of sodium ions in the distal tubule of the rat at zero net flux is between 30 and 50 mM (Ullrich, 1967).

Whereas the water permeability of the proximal tubule is not influenced by the neurohypophyseal hormones, the osmotic water permeability of the distal tubule is markedly increased by the hormones, resembling in this respect the amphibian bladders (see chapter 22 for a discussion of the mechanism of this effect). In the absence of hormones (the condition known as *diabetes insipidus* in which enormous amounts of hypotonic urine are being excreted by the kidney and which follows after lesions of the hypothalamus) the permeability of the distal tubule of the rat may be as much as 10 times lower than under antidiuretic conditions (Ullrich *et al.*, 1964). A similar effect of the neurohypophyseal hormones on the walls of the collecting ducts lead to osmotic equilibration between the fluid in the ducts and the fluid in the Henle loops whereby urine is concentrated (Ullrich *et al.*, 1961).

24.4. TRANSPORT OF OTHER MONOVALENT IONS

According to the ideas developed by Berliner and his associates (Berliner, 1960) most of the *potassium* ions from the ultrafiltrate are reabsorbed in the nephron, the potassium excreted in the urine originating from the secretion of potassium by the tubular cells in the distal part of the nephron. It was argued that the potassium reabsorption in the proximal tubule is an active process, assuming the lumen to be electrically negative (Marsh *et al.*,

1963). Newer results, however, show in agreement with the finding by Frömter and Hegel (1966) of the zero transtubular potential difference in the proximal tubule, that the distribution ratio of potassium across the tubular wall, consistently close to 1, corresponds to a passive equilibration of potassium ions (Bennet et al., 1967). Also the secretion of potassium ions in the distal tubule appears to be a passive process as shown by Morel and Boudiak (1962) using the stop-flow technique.

Chloride ions appear to be transported passively both in the proximal tubule (Ullrich, 1967) where the chloride concentrations in the lumen and in the interstitium are equal under steady-state conditions of zero net flux (Kashgarian et al., 1963) and in the distal tubule across the wall of which the electrochemical potential difference for chloride ions is not significantly different from zero (Kashgarian et al., 1963).

The mechanism of bicarbonate reabsorption in the proximal tubule of kidney is rather interesting. Using the ordinary and pH-sensitive glass microelectrodes and carbonic anhydrase inhibitors it has been shown that bicarbonate reabsorption is mediated by an active hydrogen ion secretion into the tubule, the H_2CO_3 being immediately decomposed to highly diffusible CO_2 and H_2O by carbonic anhydrase located in the luminal membrane of the proximal tubule (Rector et al., 1965). In the distal tubule, the bicarbonate reabsorption is also accomplished by the (probably active) hydrogen ion secretion. Carbonic anhydrase being not localized in the luminal membrane of the distal tubular cells, the pH in the distal tubule is somewhat lower than would correspond to the equilibrium value (Rector et al., 1965).

24.5. TRANSPORT OF CALCIUM

Practically all the calcium from the glomerular filtrate is reabsorbed in renal tubules, the likely site of the reabsorption being near the glomerulus perhaps in the proximal tubule (Bronner and Thompson, 1961). In dogs the amounts of calcium reabsorbed are greater than 99% (Chen and Neuman, 1955; Freeman and Jacobsen, 1957; Poulos, 1957). Even with this high percentage of reabsorbed calcium it is not easy to define the electrochemical gradient against or along which calcium ions move from the luminal to the peritubular space. According to the transtubular potential difference (Giebisch, 1961) it appears that calcium ions are transported against a high electrochemical gradient but recent investigations show that the electrical potential difference across the proximal tubule of the rat kidney is near zero (Frömter and Hegel, 1966).

Frick and co-workers (1965) found on the basis of *in vivo* experiments a similarity between reabsorption of calcium and sodium. As discussed above, sodium ions appear to cross the luminal membrane of tubular cells passively, following their electrochemical gradient, their passage across the basal membrane being probably an active energy-dependent process (Cort and Kleinzeller, 1958; Whittembury, 1960; Giebisch, 1961). However, Höfer and Kleinzeller (1963a,b,c) and Janda (1970a) postulate that membranes of tubular cells are impermeable to calcium ions and that calcium is transported from lumen into cells actively by a specific carrier in the form of a diffusible electrically neutral complex so that its transport is not determined by a difference of the electrical potential. It may be released from cells without any metabolic energy requirement through the basal membrane into the peritubular liquid.

The transport of calcium is coupled with that of inorganic phosphate, both ions using apparently the same carrier (Janda, 1970a).

There may exist two different mechanisms for the transport of magnesium and calcium, as indicated by *in vivo* (Walser and Robinson, 1963) as well as *in vitro* (Höfer and Kleinzeller, 1963a, b, c; Janda, 1970b) experiments. Transport of strontium, on the other hand is, very probably effected by the same mechanism as that of calcium.

24.6. TRANSPORT OF UREA

Whereas uric acid is the end product of nitrogen metabolism in reptiles and birds, urea is the main catabolite in amphibians and mammals. However, the amphibian kidney differs in handling the urea from the mammalian one. In the former, urea is actively secreted by the tubular cells into the tubular lumen, in the latter urea enters the proximal tubule by glomerular filtration (Marshall and Crane, 1924).

The secreted urea in the amphibian kidney is derived either from the blood or formed in the kidney cells from arginine and uric acid (Carlisky *et al.*, 1966). It has been suggested that arginase is involved in the reaction, catalyzing the formation of arginine at the basal side and the hydrolysis of arginine at the luminal border of the cells (Robinson and Schmidt-Nielsen, 1964). In common with other active processes, urea secretion is inhibited by dinitrophenol and convallatoxin (inhibitor of sodium transport), the latter effect suggesting the sodium dependence of the process (Vogel and Kürten, 1967). By lowering the sodium concentration in the perfusing fluid in the kidney of *Rana ridibunda*, secretion of urea was significantly decreased; the ratio of Na^+ reabsorbed/urea secreted remained constant.

In this way, the Na^+-requiring urea secretion resembles the secretion of *p*-aminohippurate (Vogel, 1966).

In mammals, the clearence of urea is lower than that of inulin, indicating a backflow of urea from the glomerular filtrate into blood, the process being considered as passive. The data of Ullrich and co-workers (1967) suggest that on a low-protein diet the transport of urea across the collecting duct wall of the rat is active in contrast to that in the normal animal, the concentration of urea in the collecting duct being lower than that in the *vasa recta* plasma. Active reabsorption of urea from the collecting duct had been previously observed by Lassiter and co-workers (1966).

24.7. REABSORPTION OF SUGARS

D-Glucose. Reabsorption of glucose by the kidney is a carrier-mediated active process coupled with the energy metabolism of the tubular cells in the first half of the proximal tubule. Such mechanism prevents glucose from being lost to the glomerular filtrate into urine and thus represents a potential regulation of the glucose level in the blood. Perfusion techniques have been widely used for studying glucose reabsorption in the kidney. Perfusion of the whole kidney used to be carried out as a standard method but later it was superseded by micropuncture and microperfusion of single tubular segments. Microperfusion was used by Walker and Hudson (1937) for the study of glucose reabsorption in the renal tubule of amphibia and by Deetjen and Boylan (1968) in the renal tubule of homoiotherm animals. The advantage of microperfusion over perfusion of the whole kidney lies in the possibility of following the dependence of glucose reabsorption on its intratubular concentration. Thus Loeschke and Baumann (1969) demonstrated Michaelis–Menten kinetics of reabsorption of glucose from the convoluted proximal tubule with the local maximum rate $V = 6 \times 10^{-10}$ mole·cm^{-2}·sec^{-1} and a $K_m = 0.6$ mM. These authors found that at 5.5 mM glucose in the plasma the glucose carrier was 90% saturated. From this and many other observations on dog, cat, and rat kidney (Shannon *et al.*, 1941; Eggleton and Shuster, 1953; Robson *et al.*, 1968) it follows that at normal glucose concentrations in the plasma, glucose is completely reabsorbed and that at normal glomerular filtration the plasma glucose concentration can be appreciably increased (above 17 mM or even 30 mM; Loeschke *et al.*, 1969) before glucose appears in the urine.

It ought to be pointed out that in view of the location of the sugar transport system in the brush-border membrane of the proximal tubule cell one would expect an accumulation of sugar against its concentration difference in the cells before its outflow through the basal membrane. This has been actually found for glucose, galactose, fructose, and xylose which are reabsorbed in the kidney (see Smith, 1951) and accumulate in kidney cortex slices (Krane and Crane, 1959; Kleinzeller et al., 1967a,b) as well as isolated kidney tubules (Kleinzeller et al., 1967a). With regard to glucose transport it has been shown by the latter authors that a s_{II}/s_I ratio higher than 1 was established only at low external glucose concentrations (below 0.5 mM) whereas for higher s_I the intracellular glucose concentration fell below that in medium. On the other hand, the results for galactose show s_{II}/s_I ratios approaching unity with increasing s_I concentrations. It was supposed that the steady-state s_{II}/s_I ratios for glucose may be due to its intracellular metabolism or marked differences in the maximum rate of transport into and out of the cell. Another factor involved in glucose intracellular accumulation is gluconeogenesis from added substrates, such as glycerol and fumarate (see Krebs et al., 1963).

An important characteristic of D-glucose reabsorption is the two-component transport system. The major component is an active transport and the minor one is a passive diffusion (Loeschke et al., 1969; Brückner, 1951). These two components seem to be independent and additive and can be separated from each other by phlorizin, an inhibitor of the sugar transport system. At high transtubular concentration difference of glucose the passive component may compensate for active transport, but under normal conditions the passive component plays practically no role in D-glucose reabsorption.

Phlorizin. The apple-root glycoside phlorizin inhibits the reabsorption of glucose in the tubular lumen at very low administration rates (0.6–10 (μg/kg)/min; Lotspeich and Woronkow, 1958) without affecting the metabolism of renal cells (see Lotspeich, 1961). 10^{-5}–10^{-7} M phlorizin blocks glucose reabsorption in the cat kidney (Chan and Lotspeich, 1962) which agrees with its high affinity for the glucose carrier. The binding is dissociable when phlorizin infusion is stopped. The aglycone phloretin can also inhibit renal glucose transport but it is some ten times less effective than phlorizin. On the basis of ^{14}C-phlorizin determination in the kidney when a maximum depression of glucose reabsorption was observed, Diedrich (1966a) calculated the turnover rate of the carrier (1.39 μmoles glucose transported by 1 μmole carrier per min) and the number of glucose carriers (10^7 molecules per cell). Diedrich (1963) studied the influence of a number of phlorizin analogues on the renal tubule reabsorption of glucose and presented some conclusions on the critical structure of the phlorizin molecule for its inhibitory mechanism. In the configuration of the glycosidic group an equatorial configuration of the OH group at C_4 and a free OH function at C_3 were shown to be indispensable for the receptor site. The activity of the aglycone was found to depend on the attachment by formation of a π–π bond between the aromatic B ring to the receptor surface and on a hydrogen bond formation between oxygen of the 4-hydroxyl group of the B ring, in the absence of which the inhibitory activity of phlorizin was pronouncedly depressed (Kotyk *et al.*, 1965).

Structural Requirements of Renal Sugar Transport. Kleinzeller and co-workers (1967*b*) showed that, in addition to D-glucose, other sugars are readily accumulated in the rabbit kidney cortex cells *in vitro*: D-galactose, α-methyl glucoside, and also D-fructose, in spite of its rapid conversion to glucose (Krebs and Lund, 1966). Transport of the poorly accumulated sugars D-xylose, 6-deoxy-D-glucose, 6-deoxy-D-galactose, and L-arabinose shows the essential features of the transport mechanism responsible for the uphill transport, such as phlorizin inhibition, mutual competition between sugars, and a sodium requirement.

The rapid accumulation of 2-deoxy-D-glucose and 2-deoxy-D-galactose against a high concentration difference without Na^+ being required is somewhat exceptional. It ought to be mentioned that ouabain does not affect 2-deoxy-D-galactose accumulation whereas D-galactose accumulation was appreciably inhibited by ouabain in experiments of Kleinzeller and Kotyk (1961). Recent observations (Kolínská, 1968) strongly indicate that also 2-deoxy-sugars share the common carrier system with other actively transported sugars (competitive inhibition between D-galactose and 2-deoxy-D-galactose found both in the presence and in the absence of Na^+). Thorn's plot shows that this inhibition is partially competitive, indicating that the two above-mentioned sugars bind in the brush-border membrane at different binding loci which are located close to each other.

L-Glucose shares a common carrier system with D-glucose (Woosley and Huang, 1967; Huang and Woosley, 1968; Baumann and Huang, 1969) but, still, the kidney seems to handle L-glucose differently from D-glucose. Whereas D-glucose is reabsorbed from the tubules, intravenously infused L-glucose is actively secreted from the cells into the lumen of the perfused tubule. The maximum rate of L-glucose secretion and D-glucose reabsorption are almost equal, the K_m for L-glucose being higher (3.1 mM) than that for D-glucose (0.6 mM). Phlorizin in the tubular lumen inhibits L-glucose secretion while D-glucose elicits L-glucose secretion, possibly due to a counterflow phenomenon.

Thus, the structural requirements of sugar transport in the kidney may be summarized as follows: (1) OH group on C_1 does not appear to be essential for transport since α-methyl-ᴅ-glucoside is readily accumulated. (2) OH group on C_2 plays an important role in sugar-carrier–Na$^+$ interaction since 2-deoxy-sugar transport does not require Na$^+$. In renal cells, similarly to rabbit small intestine, Na$^+$ affects V rather than K_m of glucose (Vogel *et al.*, 1965) and α-methyl glucoside (Kleinzeller *et al.*, 1967b) transport. (3) Free OH group on C_3 appears to be required for active accumulation of sugars, 3-O-methyl glucose poorly penetrating into the cells. In *Necturus* kidney, the last-named sugar is reabsorbed, the absorption rate being by an order of magnitude lower than that of ᴅ-glucose (Khuri *et al.*, 1966). (4) OH group on C_6 enhances accumulation, 6-deoxy-analogues of ᴅ-glucose and ᴅ-galactose being but poorly accumulated. From the work of Hauser (1969a,b) it may be inferred that myo-inositol uptake in the rat kidney slices may proceed by a similar mechanism as sugar transport since it is an energy- and sodium-requiring process.

24.8. REABSORPTION OF AMINO ACIDS

Whereas *p*-aminohippurate is normaly secreted by the tubular cells, passing from the capillaries through the extracellular space into the cells, amino acids are reabsorbed from the tubular lumen at the brush-border absorbing surface. Amino acids are accumulated *in vitro* both in kidney cortex slices (Rosenberg *et al.*, 1961) and in isolated kidney tubules (Hillman *et al.*, 1968). Amino acid reabsorption seems to be provided by five transport systems, each of which being shared by more or less structurally related amino acids: (1) neutral α-amino acids, aliphatic, aromatic, and heterocyclic (Scriver, 1965); (2) basic amino acids, related to cystine (Schwartzman *et al.*, 1966); (3) imino acids (proline and hydroxyproline) and glycine (Scriver and Goldman, 1966); (4) dicarboxylic amino acids (Webber, 1963); (5) β-amino acids (Scriver *et al.*, 1966). The transport function of the first three groups of amino acids may be lost by mutation, as in Hartnup's disease (Scriver, 1965) caused by a genetic modification of renal transport of some neutral α-amino acids, in cystinuria, leading to hyperexcretion of cystine and dibasic amino acids (Rosenberg *et al.*, 1966; Fox *et al.*, 1964), or in a mutation affecting the transport of imino acids and glycine (Ribierre *et al.*, 1958).

In this connection, the question arises whether for a given amino acid, more than one transport mechanism are available. Thus, Rosenberg and co-workers (1967) demonstrated two distinct transport systems for lysine

(α and β) which differ in affinity and capacity and are under separate genetic control. Only one of them (α) is shared also by cystine, arginine, and ornithine and is defective in cystinuria in both kidney and intestine. Similarly, evidence for a genetic control of two types of transport of imino acids and glycine was presented by Scriver and Wilson (1967). Type A has the highest affinity for proline and the lowest affinity for glycine, whereas the affinity of system B is reversed. Wilson and Scriver (1967) confirmed the multiple transport system for imino acids and glycine in rat kidney, as well as a transport system for basic amino acids separate from that for cystine.

The investigation of amino acid accumulation in kidney cortex slices indicated a specific role of Na^+ in amino acid transport (Fox *et al.*, 1964*b*; Thier *et al.*, 1967). Active accumulation of the neutral amino acids glycine and α-amino isobutyric acid was abolished in Na^+-free media, whereas the basic amino acid lysine differs from the neutral amino acids in that its accumulation is diminished but persists to a significant degree. Ouabain inhibits the active transport of amino acids but does not inhibit the Na^+-independent component of lysine accumulation. Another peculiarity of lysine transport is its resistance to anaerobiosis (Segal *et al.*, 1967). This implies two separate pathways for lysine transport, one Na^+-dependent, the other Na^+-independent. Whether these two suggested systems have any relation to those mentioned above is unknown.

Optimal transport of amino acids occurred at physiological K^+ concentrations while at higher concentrations it is diminished as compared with the control when Tris was used to replace Na^+.

In this connection, an observation on the renal papilla is of interest (Lowenstein *et al.*, 1968). It accumulates amino acids from hypertonic (439 m-equiv Na^+/liter) as well as oxygen-deficient media.

Compartmental analysis done on rat kidney slices by Rosenberg and co-workers (1963) with nonmetabolizable α-aminoisobutyric acid during temperature-dependent incubation indicates that the amino acid efflux from the intracellular space is a mediated process similar to that used for influx.

Both sugars (Segal *et al.*, 1962) and phlorizin (Segal *et al.*, 1963; Webber, 1965) have been found to affect renal amino acid transport. Evidence has been presented that neither phlorizin nor sugars (glucose, fructose, galactose) affect the affinity of the carrier site for amino acids, the effects being noncompetitive (Thier *et al.*, 1964). The suggestion of Segal and co-workers (1963) that phlorizin diminishes incorporation of amino acids into protein would indicate phlorizin entry into the renal cells. This may be justified as phlorizin excretion has been observed (Braun *et al.*, 1957).

25. MITOCHONDRIA

25.1. INTRODUCTORY REMARKS

Morphological studies of subcellular structure have shown that most actively transporting cells contain mitochondria and that these are in some cases located strikingly near the cell membrane across which the active transport takes place. This has raised the question whether the mitochondria participate in cellular transport.

Oxidative phosphorylation reactions furnish most of the energy for active ion transport in aerobic cells and tissues, either stored as ATP or taken directly from the energy-conserving chain during substrate oxidation. This mechanism, through which the mitochondria indirectly participate in cell ion transport, will not be dealt with in this section. The other form of mitochondrial participation in cell ion translocation results from their ability to accumulate certain ions from the suspending medium in a process tightly coupled with the reactions of oxidative phosphorylation. This latter process, which has recently attracted much attention of investigators, will now be considered in some detail.

Although mitochondria doubtless contain mechanisms for the uptake of various nonionic substances these have not been studied very extensively and we shall refrain from discussing them here.

25.2. UNSPECIFIC PERMEABLE SPACE OF MITOCHONDRIA

Since the quantity of ions found associated with mitochondrial particles depends upon the method of separation and the washing procedure, it is important to define the free mitochondrial permeable space. When mitochondria are spun down or filtered off, the pellet includes an extraparticulate fraction of water, which varies with the conditions. The fraction of freely accessible mitochondrial water measured with carboxypolyglucose or with sucrose corresponds to about 20 or 60% of the total water of the pellet, respectively (Werkheiser and Bartley, 1957).

From the fact that the concentration of K^+ and the sucrose-inaccessible water of mitochondria rise and fall together during induced swelling, changes in osmolarity and ageing, the conclusion was drawn that K^+ resides in an osmotically responsive space not penetrated by sucrose (see, for example, Harris and van Dam, 1968). Pfaff (1967) measured the space permeated by aspartate, along with either adenine nucleotides or nicotinamide adenine dinucleotide, as well as the sucrose permeability, using the centrifugal-filtration technique. He showed that, under constant conditions, the permeability of the mitochondrial outer membrane is exactly the same for all small molecules and that the mitochondrial water amounts to 2.5 ml/g protein.

The mitochondrial protein fraction, as determined by the biuret method, corresponds within 10% to the mitochondrial dry weight which, in turn, represents about 20% of the mitochondrial fresh weight. The average metal ion content of normal rat liver mitochondria freshly isolated from a sucrose homogenate and washed in metal-free 0.25 M sucrose is as follows: K^+ 130, Na^+ 6.3, Mg^{2+} 42.0, Ca^{2+} 5.6, Zn^{2+} 1.9, $Fe^{2+/3+}$ 7.5, Mn^{2+} 0.4 μmole/g protein (Lehninger, 1965).

25.3. TRANSPORT OF CATIONS

It has been shown that actively accumulated K^+ cannot be replaced with Na^+ from the medium and that the uptake of the two cations proceeds at different rates, the amounts of K^+ and Na^+ taken up being relatively independent of each other. The accumulation of Ca^{2+} is similarly specific and opposed to that of K^+.

The transfer of a cation across the mitochondrial membrane must be consistent with the maintenance of electroneutrality. This can be accomplished by the transfer of a counterion, e.g., K^+ for H^+, or by the simultaneous transfer of an anion. Consequently, the availability of a penetrating anion may be important in determining both the maximum uptake of the cation and the energy required.

25.3.1. Monovalent Cations

The study of mitochondrial monovalent cation transport has been in progress for some 15 years and has been greatly stimulated in the last few

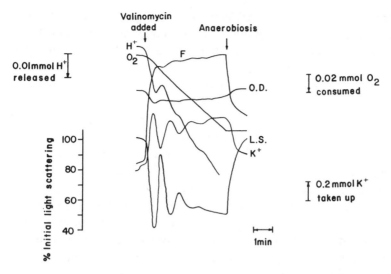

FIG. 25.1. Representative record obtained with a system of ion-specific electrodes, oxygen electrode, fluorescence and light-scattering sensors, connected to a multichannel recorder. Experimental conditions: rat liver mitochondria (about 5 mg protein/ml suspension), 10 mM KCl, 20 mM Tris-chloride, 5 mM Tris-phosphate, 250 mM sucrose, final pH 7.2, final volume 10 ml, temperature 24°C. The transport of potassium ions and the attendant changes were initiated by the addition of 0.5 μg valinomycin. Trace identification: K$^+$, potassium level in medium; H$^+$ pH of the medium; O$_2$, level of dissolved oxygen in the medium; O.D., rate of change of oxygen consumption; F, fluorescence of mitochondrial nicotinamide adenine dinucloetides; L.S., light scattering.

years by the introduction of cation-sensitive glass electrodes (p. 257). In Pressman's laboratory an apparatus has been developed enabling one to record simultaneously the K$^+$ concentration (Beckman K$^+$-specific electrode) the Na$^+$ concentration (Beckman Na$^+$-specific electrode), pH (A. H. Thomas pH-combination electrode, its Ag/AgCl electrode being used as the common reference), oxygen tension (radiometer Clark-type electrode), fluorescence of the nicotinamide-adenine dinucleotides (exciting light 336 nm, measured fluorescence at 450 nm), and mitochondrial volume changes (light scattering measured at 546 nm) (*cf.* Pressman, 1967). A representative record obtained with this assembly of sensors is shown in Fig. 25.1.

Transport of Potassium Ions. Metabolizing mitochondria or mitochondria supplemented with external ATP take up K$^+$ from a suspending medium which contains at least 1 mM K$^+$. Harris and associates (1966a) have shown that under favorable conditions such as are provided by high acetate concentration (40 mM) and 10 mM K$^+$, an uptake of 380 μequiv/g protein at an initial rate of 100 (μequiv/g)/min can be obtained. The uptakes are less in chloride media than in acetate. The possibility of potassium uptake is dependent on the presence of a penetrating anion.

An important feature of the mitochondrial K$^+$ uptake is that it can be highly stimulated by certain antibiotics such as valinomycin, macrolide actins, etc. Thus, addition of less than 0.1 μg/ml valinomycin to the mitochondrial suspension under the above conditions increases the uptake to 800 μequiv K$^+$/g protein with an initial rate of 1000 (μequiv/g)/min. The stimulation of K$^+$ uptake brings about an increased rate of oxygen consumption. The extra oxygen consumption is less when acetate is present than in its absence. This supports the idea that the energy demand reflects the ease of the movement of the two ions making up the potassium salt. A rapid rate of movement is associated with a high energy demand as shown either by the respiration rate or, in an ATP-energized system, by the adenosine triphosphatase activity.

When a high rate is induced in the absence of acetate, the supply of energy may become less than the demand. If a critical limit is exceeded, the activity resulting in K$^+$ accumulation vanishes and the accumulated ions are released. With carefully chosen concentrations, several successive cycles of uptake and loss can occur (Harris *et al.*, 1968). Simultaneous monitoring of H$^+$ liberated during K$^+$ uptake shows that the H$^+$ for K$^+$ exchange accounts for up to 40% of the K$^+$ moved, the rest of K$^+$ being presumably accompanied by anions. The availability of a penetrating anion alters the K$^+$/H$^+$ ratio and increases the amount of K$^+$ taken up even in the absence of a stimulating agent like valinomycin. Examples of readily entering anions are phosphate (or arsenate) and acetate, but substrate anions can also take part in the cation movement. When K$^+$ uptake is accompanied by movement of an anion there is a simultaneous uptake of water, as indicated by decreased light scattering as a measure of mitochondrial swelling.

Freshly prepared mitochondria, suspended in a medium without added K$^+$, retain their K$^+$ even without oxygenation or addition of exogenous substrate. Only very little K$^+$ leaks out, resulting in a K$^+$ concentration in the suspending medium of about 0.3 mM. In such cases, mitochondria presumably use an endogenous source of energy since they can take up added K$^+$, which is then maintained in mitochondria, even under anaerobiosis. The role of endogenous metabolism in maintaining ion gradients is of particular significance during preparation of mitochondria. In contrast, the massive K$^+$ uptake induced by valinomycin is immediately reversed when the system becomes anaerobic; K$^+$ is returned to the medium and

H^+ enters the mitochondria. Addition of dinitrophenol under such conditions causes further loss of K^+ and a gain of H^+.

Experiments with ^{42}K demonstrate the inhomogeneity of potassium content in the mitochondria. At 25°C, about 40% of the total K^+ is exchanged in about 60 sec without much further change in the next 60 min (Amoore, 1960). When additional K^+ is taken up, the unlabeled K^+ pool remains as before. Organic mercurials greatly increase the rate of the K^+ exchange reaction (Gamble, 1957). These agents depress the ability to maintain bound K^+ in the mitochondria, as does dinitrophenol which, however, inhibits the exchange reaction. Valinomycin promotes a K^+ turnover (up to 300–500 times with 50 $\mu g/g$ protein) in addition to causing a net uptake (Harris et al., 1967a).

Transport of Sodium Ions. Liver mitochondria incubated with ^{22}Na show only a partial exchange which is not appreciably affected even by administration of organic mercurials. An attempt to determine a Na^+–K^+ exchange in liver mitochondria (Harris et al., 1966b) resulted in no convincing evidence in favor of the exchange. However, a passive Na^+-in, K^+-out exchange can be forced by the use of dinitrophenol. The uptake of Na^+ can be stimulated by gramicidins. The effect of these antibiotics is, nonetheless, not specific and will be mentioned briefly below.

Effects of Transport-Enhancing Agents. A whole class of agents found among the toxic antibiotics (Pressman, 1965) increases, with various degrees of specificity, mitochondrial permeability to monovalent cations. The increase of cation permeability brings about a further release of K^+ from fresh water-washed mitochondria. On the other hand, in the presence of an energy source the increased permeability activates a process that not only replenishes the K^+ which has leaked out but also increases the total quantiy associated with the particles. Given a sufficient supply of K^+ and of energy, the addition of a permeability-increasing agent leads to a sustained increase of respiration or to an ion-dependent uncoupling. In the absence of an appropriate permeant cation these agents exert minimal effects on respiration.

Pressman and co-workers (1967) have divided the transport-influencing antibiotics into two classes. The first class, called after valinomycin, stimulates transport of K^+ and of other monovalent cations into mitochondria. All of the antibiotics are small molecules, lipid-soluble, and lack ionizable groups. The group includes valinomycin, enniatins, macrolide actins, gramicidins, and their analogues. The second class, called after nigericin, includes nigericin, dianemycin, and others and has been found to reverse the transport induced by valinomycin antibiotics. The members of this class differ from the former in containing an ionizable carboxyl group in their molecules.

The authors conclude that both the valinomycin and the nigericin groups of antibiotics induce alkali ion permeability by carrying ions across lipid barriers as lipid-soluble complexes. The major difference between the effects of the classes appears to consist of the fact that the nigericin compounds confer a H^+ permeability on the membrane and provide ion pathways across it at random loci while the valinomycin group interact preferentially at the mitochondrial ion pump assembly without increasing the H^+ permeability. Thus only the valinomycin group can promote the alkali ion movement against a concentration gradient while the nigericin group primarily facilitate the downhill dissipation of such concentration gradients.

Some members of the valinomycin class have been shown to stimulate, through the induced transport of K^+ into mitochondria, the phosphorylation of both endogenous and added adenine nucleotides. Thus, the addition of an appropriate amount of valinomycin to mitochondria in the presence of K^+ leads to an increased rate of phosphorylation with concomitant elevation of respiration, even beyond the rates obtainable with uncoupling agents, so that the P/O ratio is maintained at the conventional level (Höfer and Pressman, 1966). A similar effect can also be obtained with the macrolide actins but not with the gramicidins (Harris *et al.*, 1967*b*). Induced Na^+ transport does not substitute for induced K^+ transport.

The glycoside ouabain, which in low concentrations is known to inhibit Na^+ and K^+ transport across the plasma membrane, presumably by inhibiting the specific membrane ATPase activity, does not affect the active maintenance or uptake of K^+ by isolated mitochondria. This fact clearly differentiates between mitochondrial K^+ and Na^+ transport mechanisms and those functioning in the cell membrane.

Transport of Hydrogen Ions. The movement of any cation species across the mitochondrial membrane is totally (in the absence of a permeant anion), or at least partially (in other cases), counterbalanced by an opposite movement of H^+. Some authors believe it is the H^+-pump that provides the driving force for the transport of other cations by an exchange mechanism, possibly on the basis of electrostatic forces only. The question of H^+ translocation has become, because of Mitchell's hypothesis, intimately linked with that of the mechanism of oxidative phosphorylation. In terms of Mitchell's model (1966) the H^+-pump functions as the primary ion pump and is directly linked to the respiratory chain.

According to the traditional concepts of a chemical coupling of oxidative phosphorylation (Chance, 1963; Lehninger, 1965) the coupling between the respiratory chain and the phosphorylation of adenine nucleotides is mediated by a series of as yet unidentified, unphosphorylated, high-energy intermediates. The chemiosmotic hypothesis of Mitchell, on the other hand, is based on a coupling between electron flow and proton flow in an anisotropic membrane. The chemiosmotic system consists of four principal parts: (1) the proton-translocating reversible ATPase system; (2) the proton-translocating oxido-reduction chain; (3) the exchange diffusion system, coupling proton translocation to that of anions and cations; (4) the ion-impermeable coupling membrane, including systems (1), (2), and (3).

The operation of systems (1) and (2) in an ion-tight membrane creates both a pH difference and a membrane potential, together called a proton-motive force (pmf, by analogy with emf). The presence of the exchange diffusion system (3) regulates the internal pH by enhancing the membrane potential component of the pmf at the expense of the pH difference. Coupling between oxido-reduction and phosphorylation is described by a circulating proton current connecting the systems (1) and (2) at a pmf of some 250 mV. The two systems would tend to come into equilibrium and allow some reversibility of both systems (1) and (2), accounting for respiratory control and reversed electron transport.

25.3.2. Divalent Cations

Mitochondria isolated from a variety of tissues are capable of taking up all divalent cations. Certain amounts (about 40 μmol/g protein) can be taken up independently of metabolism and the presence of inhibitors. Such cation uptakes are accompanied by H^+ efflux with mutual competition between different ions (Judah et al., 1965). Metabolizing mitochondria take up additional amounts of divalent cations and these uptakes can be inhibited or reversed, unless the ions are deposited in the mitochondria as insoluble phosphates. The energy-dependent uptake is greatly enhanced by the presence of inorganic phosphate, clearly because of the formation of low-solubility products within mitochondria. The divalent cations possess an uncoupling activity, presumably because they release fatty acids from the mitochondria which act as a sink for high-energy phosphate groups. Addition of ATP counteracts the activity.

Transport of Calcium Ions. The accumulation of calcium can be observed under different sets of conditions (Carafoli and Lehninger, 1967).

(1) Mitochondria exposed to relatively high concentrations of Ca^{2+} (about 3 mM) accumulate up to 2.5 μmol Ca^{2+}/mg protein. Phosphate ions are required and also accumulated. This process, called massive loading, is stoichiometrically coupled with electron transport, although the excess Ca^{2+} brings about loss of respiratory control and of capacity for oxidative phosphorylation. (2) Mitochondria supplied with small amounts of Ca^{2+} (about 100 μM) take up Ca^{2+} in a process that does not require phosphate (however, if present it is also accumulated) and is referred to as limited loading. Under the latter set of conditions both the oxidative phosphorylation and the respiratory control are undamaged.

Massive accumulation of Ca^{2+} and phosphate can be energized either by substrate oxidation or by ATP hydrolysis. However, in the former case, ATP is also required to retain the accumulated calcium phosphate. For each energy-conserving site of the respiratory chain about 1.67 molecules of Ca^{2+} and 1.0 molecule of phosphate are accumulated. Thus, Ca^{2+} and phosphate are accumulated in the molar ratio of hydroxylapatite. Mg^{2+} is essential for the incubation medium and some of it is also accumulated.

On the other hand, under limited loading conditions usually not more than 1% of the added Ca^{2+} is left in the suspending medium. H^+ is exchanged for Ca^{2+}. Mg^{2+} and ATP are not required in the absence of phosphate. When phosphate is added, it is also accumulated if the system contains ATP and Mg^{2+} or the ATPase activity is blocked by oligomycin. Arsenate will replace phosphate.

The Ca^{2+}/O accumulation ratio (Ca^{2+} accumulated per extra oxygen uptake stimulated by the addition of Ca^{2+} as determined by the oxygen electrode trace) depends on the respiratory substrate used. It is approximately 5–6 for NAD-dehydrogenase-linked substrates and 3.6–4.0 for succinate. Two calcium ions produce the same affect on oxygen consumption as one molecule of ADP. However, in the presence of oligomycin, ADP no longer stimulates the oxygen uptake while Ca^{2+} still does (cf. Fig. 25.2).

In the presence of a permeant anion, such as acetate, the addition of Ca^{2+} leads to swelling of the particles. The (Ca + acetate) swelling is reversed by the administration of phosphate; phosphate also leads to loss of acetate. When starting with mitochondria loaded with K^+, the addition of Ca^{2+} can lead to contraction of the particles because of losing both K^+ and H^+ (Harris et al., 1966a).

The univalent cation/Ca^{2+} ratios have been found to vary from 2 to 0.2. A H^+/Ca^{2+} ratio of 1 is obtained when limited amounts of Ca^{2+} are added to mitochondria incubated in isosmotic sucrose or NaCl media buffered with Tris-HCl at pH 7.0–7.4. Permeant anions decrease the H^+/Ca^{2+}

ratio to as low as 0.2. An univalent cation/Ca^{2+} ratio of 2 is observed under three sets of experimental conditions: when Ca^{2+} is added to valinomycin-treated mitochondria which have accumulated K^+ against a gradient; when buffer is omitted from the medium; and when aerobic uptake of Ca^{2+} is initiated by addition of succinate or of mitochondria (Azzone et al., 1967).

Transport of Other Divalent Cations. Rat liver mitochondria, first depleted of Mg^{2+} and K^+ by a 20-min incubation at 30°C in Mg^{2+}-free medium, take up considerable amounts of Mg^{2+} during subsequent incubation with Mg^{2+} and substrate, whether or not phosphate is present. This uptake does not occur with fresh mitochondria, so that it appears that Mg^{2+} becomes attached to the groups vacated by K^+ and that Mg^{2+} competes with K^+ for these groups (Judah et al., 1965). In the presence of phosphate, however, fresh mitochondria accumulate massive amounts of Mg^{2+} and phosphate.

In addition to an uptake independent of metabolism, Mn^{2+} is accumulated by mitochondria in an energy-dependent process that stimulates oxygen consumption (Mn^{2+}/O about 6) and that is subject to inhibition by metabolic inhibitors. Both H^+ and K^+ are evolved during Mn^{2+} uptake. The addition of dinitrophenol brings about loss of Mn^{2+} although in the presence of phosphate mitochondria maintain the Mn^{2+} taken up.

Sr^{2+} behaves in many respects like Ca^{2+}. Its uptake includes an initial rapid phase which, unlike adsorption of other divalent cations, apparently requires respiration, followed by a slower phase also dependent upon a supply of metabolic energy.

25.4. TRANSPORT OF ANIONS

Studies of the permeability properties of mitochondrial membranes to anions are based primarily on the swelling of the particles in iso-osmotic solutions of the ammonium salts of the anion under investigation. Mitochondria from various sources show a rapid change in volume when in media of varying tonicity, obeying the relationship valid for a perfect osmometer: $V = k/c + V_c$, where V is the volume at an external concentration of solute c, k is a constant, and V_c is the volume at infinite concentration of external solute.

25.4.1. Inorganic Anions

Chappell and Crofts (1966) have shown that the mitochondrial membrane is impermeable to K^+ and Cl^-. The authors followed the swelling of mitochondria in iso-osmotic solutions of KCl, K-phosphate, NH_4Cl, NH_4-phosphate, and NH_4-acetate. Swelling occurred only when NH_4^+, together with phosphate or acetate, composed the suspending medium. It was suggested that actually NH_3 and CH_3COOH are the species which penetrate the membrane while the penetration of phosphate ions is mediated by an exchange-diffusion carrier permitting $H_2PO_4^-$ to enter in exchange for OH^-. Arsenate appears to be able to enter mitochondria on the same carrier. The facilitated-diffusion carrier for phosphate is specifically inhibited by organic mercurials, as pointed out by Tyler (1969) for sodium mersalyl (and formaldehyde) and Fonyo and Bessman (1968) for p-mercuribenzoate.

25.4.2. Organic Anions

Mitochondrial membrane permeability to various substrate anions was studied, using the method of mitochondrial swelling in iso-osmotic solutions of NH_4^+ salts of tricarboxylic acid cycle intermediates and related compounds (Chappell and Haarhoff, 1967). It was found that formate, propionate, and butyrate, as well as acetate, enter the particles rapidly. Little or no swelling occurred in NH_4-succinate until NH_4-phosphate (or arsenate) was added. The phosphate-dependent swelling was also observed when succinate was replaced with D-malate, L-malate, malonate, or meso-tartrate. For the entry of citrate and cis-aconitate and D- and L-tartrate, not only phosphate (or arsenate) but also an addition of L-malate was necessary before swelling occurred. Swelling was not observed with trans-aconitate, fumarate, or maleate.

In view of these results two vectorial transporting systems were suggested to be involved in allowing di- and tricarboxylic acids to penetrate the mitochondrial membrane. Carrier I does not accept a double bond between the two carbon atoms bearing the carboxyl groups, which must be able to take up mutual *cis*-configuration. Carrier II accepts molecules with the carboxyl groups in *trans*-configuration, and specifically requires L-malate for its action.

The role of phosphate is important since phosphate is required for swelling to occur in NH_4^+ salts. Nevertheless, phosphate is not required for the oxidation of those substrates by mitochondria. It has been suggested that phosphate enters as $H_2PO_4^-$, then dissociates to form a proton and HPO_4^{2-} which then exchanges with the dicarboxylic acid anions. Succinate may be considered as an example to illustrate the results when the mitochondria are oxidizing the substrate in the absence of phosphate. The succinate which enters is exchanged for the malate formed inside through succinate oxidation; hence phosphate is not required.

The ease of entry for substrate anions seems to play an important role in the maximum respiratory capacity of mitochondria. If for any reason the permeation of substrate fails to keep up with its metabolic demand (e.g., in the presence of excess uncoupler) respiration is inhibited (Harris *et al.*, 1967c). If, on the other hand, the permeation of substrate is enhanced (e.g., through induced K^+ uptake) respiration is also stimulated (Harris *et al.*, 1967b).

Translocation of Adenine Nucleotides. Studies of the effect of atractyloside on mitochondrial oxidative phosphorylation have revealed the probable existence in the mitochondrial membrane of a transporting system for ATP–ADP exchange. Klingenberg, Vignais, and their co-workers investigated the exchange of endogenous and exogenous adenine nucleotides in isolated mitochondria. They concluded that there is an exchange-diffusion carrier which permits the exchange of ADP and ATP across the mitochondrial inner membrane and that this carrier is the site of action for atractyloside (Klingenberg and Pfaff, 1966).

25.4.3. Uncoupler Ions

The problem of anion transport into mitochondria seems to be associated with that of a possible mechanism of action of uncouplers of oxidative phosphorylation. According to the theory of chemical coupling, uncoupling activity is accounted for by hydrolysis of a hypothetical high-energy intermediate prior to the entry of phosphate into the sequence of energy-conserving reactions. While according to the chemi-osmotic theory the uncouplers simply increase the membrane permeability to H^+, thus discharging the proton gradient across the membrane.

Van Dam and Slater (1967) suggested a possible mechanism for the action of uncouplers, based on a cyclic energy-dependent transport of uncoupler anions into mitochondria. According to their proposal, substrate anions are transported along with a cation into the particles at the cost of energy. Actually, it might be a cation carrier that draws the anions with the cations. The product of biological oxidation might then be released to the outside through the same transporter system, thus regenerating the energy utilized to bring the substrate inside. An uncoupler (in its anionic form) would be transported exactly as a substrate anion. Once inside, the uncoupler anion would take up a proton and diffuse freely as an uncharged molecule through the membrane to the outside, where it again dissociates into H^+ and an anion; another cycle may then start.

The well-known inhibition of respiration by high uncoupler concentrations could be accounted for by competition between uncoupler anion and substrate anion for entry into mitochondria (see also Harris et al., 1967b).

25.5. STOICHIOMETRY OF TRANSPORT

The efficiency ratios cation/$\sim$ P differ for various cations under different conditions. Cockrell and co-workers (1966) reported values greater than 7 K^+/$\sim$ P for ATP-driven valinomycin-induced potassium accumulation while that supported by oxidizable substrate showed an efficiency of 3.2 K^+/$\sim$ P. The former value exceeds those obtained for divalent cations, even after dividing by two to allow for the difference in charges. (Many authors found the divalent cation/$\sim$ P ratios to be near 2.) However, the findings of Carafoli and co-workers (1965) indicate values of Ca^{2+}/$\sim$ P as high as 10. The corresponding figures for Mn^{2+} and Sr^{2+} are 1.8 and 1.8–2.1, respectively.

25.6. EFFECT OF INHIBITORS

The inhibitors of mitochondrial ion transport may be divided into three groups according to their effect on mitochondrial metabolism, viz., blockers of electron transfer (e.g., amytal, rotenone, antimycin A, cyanide), blockers of oxidative phosphorylation (e.g., oligomycin, atractyloside, aurovertin), and uncouplers of oxidative phosphorylation (e.g., dinitrophenol, carbonyl cyanide phenylhydrazones, dicoumarol, etc.).

Addition of antimycin A to a respiring mitochondrial suspension blocks ion transport. It may be reactivated either if ATP is added or if tetramethyl p-phenylenediamine and ascorbate serve as oxidizable substrates (oxidation of ascorbate proceeds with only a single phosphorylation between cytochrome c and oxygen). Ion transport blocked by rotenone or amytal can be further energized by oxidation of succinate or ascorbate. (The two inhibitors block electron transfer between NAD^+ and flavoprotein.) Cyanide inhibits ion movement supported by oxidation of ascorbate. None of the electron transfer inhibitors affects the ATP-driven ion translocation.

The movement of ions supported by electron transfer is unaffected by oligomycin, this showing that the energy does not come from ATP. If ATP

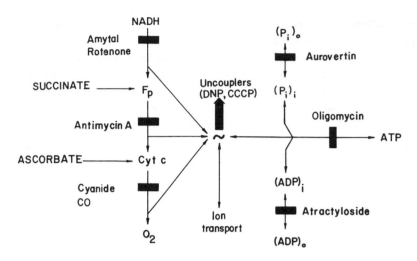

FIG. 25.2. Suggested sites of action of various inhibitors on mitochondrial oxidative phosphorylation.

is added as the donor of energy oligomycin becomes a potent inhibitor of mitochondrial transport.

Uncouplers prevent the use of respiratory energy for creating and maintaining ionic gradients so that after their addition ions are released from mitochondria. The effects of all the three groups of inhibitors are schematically shown in Fig. 25.2.

The outlined pattern of inhibition suggests that mitochondrial ion transport is not directly linked to the splitting of ATP or to electron transfer but to a high-energy intermediate of oxidative phosphorylation located between the locus of action of oligomycin and the energized electron transport carrier(s) (Chance, 1963; Lehninger, 1965). Thus, all three systems, the electron transport chain, ATP synthesis, and the ion transport, appear to be linked through a common component allowing mutual interconversion of energy shifts. This conclusion has recently found support in that induced transport of K^+ ceases during phosphorylation of added ADP and resumes only after respiration returns to the original (state 4) level (Harris et al., 1967b). When ADP and Ca^{2+} are added together, Ca^{2+} is accumulated first and phosphorylation of ADP follows, both processes taking place stoichiometrically (Rossi et al., 1966). If the direction of the net potassium movement is reversed (during K^+ efflux induced by valinomycin) a net synthesis of ATP by mitochondria in the absence of respiration can be observed (Cockrell et al., 1967). All these observations support the conclusion that intact mitochondria possess a highly efficient cation transport system tightly coupled with ATP synthesis.

It is interesting to note that phosphoprotein–phosphate turnover is not inhibited by oligomycin (Ahmed and Judah, 1963) and that the energy for ion transport thus may be derived from protein-bound phosphate. The addition of valinomycin to mitochondria in the presence of K^+ increases the incorporation of ^{32}P-labeled phosphate into protein (Pressman, 1964). Addition of Ca^{2+} causes the appearance of an unidentified protein-bound phosphate fraction (Norman et al., 1965).

26. TUMOR CELLS

Among the various tumor cells that can be grown in a liquid culture, the Ehrlich ascites cells have been investigated by far the most thoroughly with respect to their transport properties. The two types of substances examined in detail are amino acids and monosaccharides.

26.1. SUGARS

The transport of monosaccharides in Ehrlich cells was studied first by Crane and co-workers (1957), Nirenberg and Hogg (1958) and recently by Kolber and LeFevre (1967). It was found to be identifiable with facilitated diffusion. The kinetic parameters are shown in Table 26.1. The uptake of sugars is apparently mediated by a mobile carrier (evidence from counter-transport) and is inhibited by both phloretin and phlorizin.

Table 26.1. Kinetic Constants of Monosaccharide Uptake by Ehrlich Ascites Cells *

Sugar	K_T, mM	V, (μmol/ml)/min
D-Glucose	6.6	—
D-Galactose	5.5–26	6.1
D-Xylose	2–15	55
D-Arabinose	300	—
D-Ribose	125–300	3
L-Arabinose	65	—
3-Methylglucose	2.1	3.6

* Data of Crane *et al.* (1957); Kolber and LeFevre (1967); and Saha and Coe (1967).

The maximum accumulation ratio does not exceed unity but the temperature quotient Q_{10} is rather high (3.8 for D-ribose between 20 and 30°C).

As in many other systems, glucose behaves somewhat anomalously in that it can apparently enter the Ehrlich cells by two different paths and that it shows a low Q_{10} of 1.4 (Saha and Coe, 1967).

26.2. AMINO ACIDS

The work with amino acids is mostly due to the laboratories of Heinz (e.g., Heinz and Walsh, 1958) and Christensen (e.g., Christensen and Riggs, 1952; Inui and Christensen, 1966), later extended by Johnstone and Scholefield (1961), Eddy (1968), and others.

The Ehrlich cells appear to contain several distinct transport systems for amino acids with overlapping specificities (analogy to yeast and bacterial cells) which are capable of transporting amino acids against a concentration gradient. The role of Na^+ in some of the systems is a fine example of coupled uphill transport.

There are at least two distinct carriers for neutral amino acids, the L-site (for leucine-preferring) and the A-site (for alanine-preferring). The two systems differ in a number of respects, the L-site showing affinity for leucine, isoleucine, valine, phenylalanine, and methionine; the A-site for alanine, glycine, serine, threonine, proline, asparagine, glutamic acid, and methionine, there being some overlapping in specificities. The L-site is little affected by pH, the A-site rather pronouncedly. The L-site is involved in exchange diffusion (maximum s_{II}/s_I ratios of 1–5) while the A-site in uphill transport (maximum s_{II}/s_I ratios of 10–20). The L-site transport is depressed by increasing the K^+/Na^+ ratio much more than is the A-site transport. Table 26.2 summarizes the kinetic parameters of the uptake of some of the amino acids (Oxender and Christensen, 1963).

In addition to the above transport systems, there is apparently a distinct mechanism for the uptake of glutamate (K_T 0.2 M, V 70 (μmol/ml)/min) (Heinz et al., 1965) and one for diamino acids like lysine (Christensen and Liang, 1966).

The various systems enumerated interact to a greater or lesser degree, apparently due to their adjacent localization on the cell membrane.

It should be noted that, using the experimental evidence of amino acid transport in Ehrlich carcinoma cells, Heinz and co-workers developed the concept of pump-and-leak of nonionic compounds (p. 145) and that here too, the stimulating effect of preloading on subsequent inward flux was

Table 26.2. Uptake Parameters of Neutral Amino Acids by the Ehrlich Ascites Cells

Amino acid	K_T, mM	V, (μmol/ml)/min
Glycine	3–4	5–6
Alanine	0.5	5–6
Serine	0.6	3–5
α-Aminoiso-butyric acid	1–2	4–6
Valine	3–4	4–6
Leucine	0.5	2
Methionine	1	2

established (p. 76). The transport fluxes of amino acids but not the exchange fluxes at saturation conditions are inhibited by cyanide and by 2-deoxy-D-glucose. This remarkable property is probably due to the fact that energy increases the rate of outward movement of the free carrier or else that the dissociation of the carrier–substrate complex at the inner phase of the membrane is increased (Jacquez and Sherman, 1965).

One of the striking features of amino acid transport in Ehrlich cells is its interconnection with the movement of alkaline metal ions, particularly of sodium. The uptake of neutral amino acids is obligatorily coupled with the movement of Na^+ and K^+ in a stoichiometric relationship, reminding one of the ATPase transport of alkaline cations in various cell types, viz., 0.9 ± 0.11 sodium and 0.62 ± 0.11 potassium per glycine (Eddy, 1968). It is likely that glycine can cross the membrane as C-Gly, C-Na-Gly, C-K-Gly, while Na^+ can cross as C-Na-Gly or C-Na and K^+ can cross as C-K-Gly or C-K. Schafer and Jacquez (1967) showed that actually 1 Na^+ is bound to an A-site carrier while no Na^+ is required by the L-site carrier. Much remains to be elucidated on the actual molecular mechanism of the ion-coupled amino acid transport, but the hypothesis is likely that the uphill transport of neutral amino acids in these cells is actually run by the Na^+–K^+ exchange pump. However, the dependence on Na^+ transport is not a yes-or-no question with all amino acids as there appear to be systems insensitive to sodium ions in parallel with those requiring it for operation.

There is good evidence that the transport systems for diamino acids can function as the Na^+-requiring system for neutral amino acids, the position of one of the NH_3^+-groups being then taken by the sodium ion (Christensen and Handlogten, 1969).

26.3. IONS

The transmembrane movement of ions does not differ in any significant respect from that in erythrocytes. There appear to be two components of Na^+ flux, a mediated passive influx and an active efflux (Aull and Hempling, 1963). The active influx of K^+ is linearly related to the amount of ATP present (Weinstein and Hempling, 1964), both the active fluxes occurring at the expense of endogenous sources of energy but glucose having a stimulating effect.

Calcium is transported actively outward, the amount bound intracellularly being rather high (70% in Ehrlich cells, 55% in HeLa cells) (Levinson, 1967; Borle, 1969).

BIBLIOGRAPHY

BOOKS OF GENERAL INTEREST TO MEMBRANE TRANSPORT INVESTIGATORS

Comprehensive Biochemistry (ed. by Florkin, M. and Stotz, M.). Elsevier Publ. Co., Amsterdam, 1962–1969.

Methods in Enzymology (ed. originally by Colowick, S. P. and Kaplan, N. O.). Academic Press, New York, 1955–1969.

Methods of Biochemical Analysis (ed. by Glick, D.). Interscience Publ., New York, 1954–1969.

Physical Techniques in Biological Research (ed. originally by Oster, G. and Pollister, A. W.). Academic Press, New York, 1955–1969.

Enzymes (by Dixon, M. and Webb, E. C.). Longmans, Green & Co., London, 1965.

Transport and Accumulation in Biological Systems (by Harris, E. J.). Butterworths, London, 1960.

Membrane Transport and Metabolism (ed. by Kleinzeller, A. and Kotyk, A.). Academic Press, New York, 1961.

Biological Transport (by Christensen, N. H.). W. A. Benjamin, New York, 1962.

The Cellular Functions of Membrane Transport (ed. by Hoffman, J. F.). Prentice–Hall, Inc., Englewood Cliffs, 1964.

Problems of Cell Permeability (by Troshin, A. S.). Pergamon Press, Oxford, 1966.

The Movement of Molecules across Cell Membranes (by Stein, W. D.). Academic Press, New York, 1967.

Biological Membranes (ed. by Dowben, R. M.). Little, Brown & Co., Boston, 1969.

REFERENCES

Abbott, B. C., Hill, A. V., and Howarth, J. V. (1958). *Proc. Roy. Soc.* **B 148**, 149.

Adrian, R. H. (1956). *J. Physiol.* **133**, 631.

Adrian, R. H. (1960). *J. Physiol.* **151**, 154.

Adrian, R. H. (1961). *J. Physiol.* **156**, 623.

Adrian, R. H. and Freygang, W. H. (1962*a*). *J. Physiol.* **163**, 61.

Adrian, R. H. and Freygang, W. H. (1962*b*). *J. Physiol.* **163**, 104.

Adrian, R. H. and Slayman, C. L. (1966). *J. Physiol.* **184**, 970.

Agar, W. T., Hird, F. J. R., and Sidhu, G. S. (1953). *J. Physiol.* **121**, 255.

Agar, W. T., Hird, F. J. R., and Sidhu, G. S. (1954). *Biochim. Biophys. Acta* **14**, 80.

Agin, D. and Holtzman, D. (1966). *Nature* **211**, 1194.

Ahmed, K. and Judah, J. D. (1963). *Biochim. Biophys. Acta* **71**, 295.

Alvarado, F. (1964). *Experientia* **20**, 302.

Alvarado, F. (1966*a*). *Science* **151**, 1010.

Alvarado, F. (1966*b*). *Biochim. Biophys. Acta* **112**, 292.

Alvarado, F. (1967*a*). *Comp. Biochem. Physiol.* **20**, 461.

Alvarado, F. (1967*b*). *Biochim. Biophys. Acta* **135**, 483.

Alvarado, F. (1968). *Nature* **219**, 276.

Alvarado, F. and Crane, R. K. (1962). *Biochim. Biophys. Acta* **56**, 170.

Akedo, H. and Christensen, H. N. (1962). *J. Biol. Chem.* **237**, 118.

Ames, G. F. (1964). *Arch. Biochem. Biophys.* **104**, 1.

Amoore, J. E. (1960). *Biochem. J.* **76**, 438.

Andersen, B. and Ussing, H. H. (1957). *Acta Physiol. Scand.* **39**, 228.

Andersen, B. and Zerahn, K. (1963). *Acta Physiol. Scand.* **59**, 319.

Anderson, B., Kundig, W., Simoni, R., and Roseman, S. (1968). *Fed. Proc.* **27**, 643.

Andreoli, T. E., Bangham, J. A., and Tosteson, D. C. (1967*a*). *J. Gen. Physiol.* **50**, 1729.

Andreoli, T. E., Tieffenberg, M., and Tosteson, D. C. (1967*b*). *J. Gen. Physiol.* **50**, 2527.

Andrianov, B. K., Kurella, G. A. and Litvin, F. F. (1969). *Biofizika* **14**, 78 (in Russian).

Anraku, Y. (1967). *J. Biol. Chem.* **242**, 793.

Anraku, Y. (1968*a*). *J. Biol. Chem.* **243**, 3116.

Anraku, Y. (1968*b*). *J. Biol. Chem.* **243**, 3123.

Anraku, Y. (1968*c*). *J. Biol. Chem.* **243**, 3128.

Arisz, W. H. (1953). *Acta Bot. Neerl.* **1**, 506.

Arisz, W. H. (1956). *Protoplasma* **46**, 5.

Armstrong, W. McD. and Rothstein, A. (1964). *J. Gen. Physiol.* **48**, 61.

Armstrong, W. McD. and Rothstein, A. (1967). *J. Gen. Physiol.* **50**, 967.

Ashworth, C. T. and Lawrence, J. F. (1966). *J. Lipid Res.* **7**, 465.

Ashworth, C. T., Steinbridge, V. A., and Sanders, E. (1960). *Amer. J. Physiol.* **148**, 1326.

Aull, F. and Hempling, H. G. (1963). *Amer. J. Physiol.* **204**, 789.

Autilio, L. A., Norton, W. T., and Terry, R. D. (1964). *J. Neurochem.* **11**, 17.

Äyräpää, T. (1950). *Physiol. Plant.* **3**, 402.

Azam, F. and Kotyk, A. (1969). *FEBS Letters* **2**, 333.

Azzone, G. F., Azzi, A., and Rossi, C. S. (1967). In: *Mitochondrial Structure and Compart-mentation* (ed. by Quagliariello, E., Papa, S., Slater, E. C., and Tager, J. M.), p. 234. Adriatica Editrice, Bari.

Bader, H., Sen, A. K., and Post, R. L. (1966). *Biochim. Biophys. Acta* **118**, 106.

Baillien, M. and Schoffeniels, E. (1961). *Biochim. Biophys. Acta* **53**, 537.

Baker, B. R. (1967). *Design of Active-Site-Directed Irreversible Enzyme Inhibitors.* John Wiley and Sons, New York.

Baker, P. F., Hodgkin, A. L., and Shaw, T. I. (1961). *Nature,* **190**, 885.

Baker, P. F., Hodgkin, A. L., and Shaw, T. I. (1962). *J. Physiol.* **164**, 355.

Barber, J. (1968a). *Biochim. Biophys. Acta* **150**, 618.

Barber, J. (1968b). *Biochim. Biophys. Acta* **150**, 730.

Barber, J. (1968c). *Biochim. Biophys. Acta* **163**, 531.

Barnett, J. A. (1968). *J. Gen. Microbiol.* **52**, 131.

Barr, C. E. (1965). *J. Gen. Physiol.* **49**, 181.

Barr, C. E. and Broyer, T. C. (1964). *Plant Physiol.* **39**, 48.

Barry, R. J. C. (1967). *Brit. Med. Bull.* **23**, 266.

Barry, R. J. C., Dikstein, S., Matthews, J., and Smyth, D. H. (1961). *J. Physiol.* **155**, 17P.

Barry, R. J. C., Dikstein, S., Matthews, J., Smyth, D. H., and Wright, E. M. (1964). *J. Physiol.* **171**, 316.

Barry, R. J. C., Smyth, D. H., and Wright, E. M. (1965). *J. Physiol.* **181**, 410.

Bastide, F. and Jard, S. (1968). *Biochim. Biophys. Acta* **150**, 113.

Battaglia, F. C., Manchester, K. L. and Randle, P. J. (1960). *Biochim. Biophys. Acta* **43**, 50.

Battaglia, F. C. and Randle, P. J. (1960). *Biochem. J.* **75**, 408.

Baumann, K. and Huang, K. C. (1969). *Pflügers Arch.* **305**, 155.

Ben-Amotz, A. and Ginzburg, B. Z. (1969). *Biochim. Biophys. Acta* **183**, 144.

Benko, P. V. and Segel, I. H. (1968). *Fed. Proc.* **27**, 831.

Bennet, C. M., Clapp, J. R., and Berliner, R. W. (1967). *Amer. J. Physiol.* **213**, 1254.

Bentley, P. J. (1958). *J. Endocrin.* **17**, 201.

Berliner, R. W. (1960). *Circulation* **21**, 892.

Bernstein, J. (1902). *Pflügers Arch. ges Physiol.* **92**, 521.

Bernstein, R. E. (1954). *Science* **120**, 459.

Bhattacharya, G. (1961). *Biochem. J.* **79**, 369.

Bibb, W. R. and Straughn, W. R. (1964). *J. Bacter.* **87**, 815.

Biber, T. U. L. and Curran, P. F. (1968). *J. Gen. Physiol.* **51**, 606.

Bihler, I. (1968). *Biochim. Biophys. Acta* **163**, 401.

Bihler, I. and Adamič, Š. (1967). *Biochim. Biophys. Acta* **135**, 466.

Bihler, I. and Crane, R. K. (1962). *Biochim. Biophys. Acta* **59**, 18.

Bihler, I., Hawkins, K. A., and Crane, R. K. (1962). *Biochim. Biophys. Acta* **59**, 94.

Bingham, J. K., Newey, H., and Smyth, D. H. (1966a). *Biochim. Biophys. Acta* **120**, 314.

Bingham, J. K., Newey, H., and Smyth, D. H. (1966b). *Biochim. Biophys. Acta* **130**, 281.

Blei, M. (1967). *Acta Cient. Venezolana, Suppl.* **3**, 106.

Blinks, R. L. (1951). In: *Manual of Phycology* (ed. by Smith, G. M.). Chronica Botanica, Waltham, Massachusetts.

Blinks, R. L. (1955). In: *Electrochemistry and Biology in Medicine* (ed. by Shedlovsky, T.), p. 187. John Wiley and Sons, Inc., New York.

Blostein, R. (1966). *Biochem. Biophys. Res. Comm.* **24**, 598.

Blount, R. W. and Levedahl, B. H. (1960). *Acta Physiol. Scand.* **49**, 1.

Blumenthal, R. and Katchalsky, A. (1969). *Biochim. Biophys. Acta* **173**, 357.

Bobinski, H. and Stein, W. D. (1966). *Nature* **211**, 1366.

Bojessen, E. and Leyssac, P. (1965). *Acta Physiol. Scand.* **65**, 20.

Bonting, S. L. and Canady, M. R. (1964). *Amer. J. Physiol.* **207**, 1005.

Bonting, S. L., Simon, K. A., and Hawkins, N. M. (1961). *Arch. Biochem. Biophys.* **95**, 416.

Booij, H. L. (1962). In: *Conference on Permeability*, p.l. Tjeenk Willink, Zwolle.

Borgström, B. (1965). *Biochim. Biophys. Acta* **106**, 171.

Borle, A. B. (1969). *J. Gen. Physiol.* **53**, 43, 57.

Borst Pauwels, G. W. F. H., Peter, J. K., Jager, S., and Wijffels, C. C. B. M. (1965). *Biochim. Biophys. Acta* **94**, 312.

Bosáčková, J. (1963). *Biochim. Biophys. Acta* **71**, 345.

Bosáčková, J. and Crane, R. K. (1965a). *Biochim. Biophys. Acta* **102**, 423.

Bosáčková, J. and Crane, R. K. (1965b). *Biochim. Biophys. Acta* **102**, 436.

Bourguet, J. and Jard, S. (1964). *Biochim. Biophys. Acta* **88**, 442.

Bourguet, J. and Maetz, J. (1961). *Biochim. Biophys. Acta* **52**, 552.

Bourguet, J. and Morel, F. (1967). *Biochim. Biophys. Acta* **135**, 693.

Boyle, P. J. and Conway, E. J. (1941). *J. Physiol.* **100**, 1.

Bozler, E. (1948). *Experientia* **4**, 213.

Bozler, E., Calvin, M. E., and Watson, D. W. (1958). *Amer. J. Physiol.* **195**, 38.

Bozler, E. and Lavine, D. (1958). *Amer. J. Physiol.* **195**, 45.

Bradley, J. and Williams, E. J. (1967a). *J. Exp. Bot.* **18**, 241.

Bradley, J. and Williams, E. J. (1967b). *Biochim. Biophys. Acta* **135**, 1078.

Braun, W., Whittaker, V. P., and Lotspeich, W. D. (1957). *Amer. J. Physiol.* **190**, 563.

Brauner, L. and Diemer, R. (1967). *Planta* **77**, 1.

Bricker, N. S., Biber, T., and Ussing, H. H. (1963). *J. Clin. Invest.* **42**, 88.

Briggs, G. E., Hope, A. B., and Robertson, R. N. (1961). *Electrolytes and Plant Cells.* Blackwell Scientific Publications, Oxford.

Britten, R. J. and McClure, F. T. (1962). *Bacter. Rev.* **26**, 292.

Britton, H. G. (1966). *J. Theoret. Biol.* **10**, 28.

Brod, J. (1962). *Ledviny. Fysiologie, klinická fysiologie a klinika.* (In Czech.) SZN, Prague.

Bronner, F. and Thompson, D. D. (1961). *J. Physiol.* **157**, 232.

Brower, R. (1965). *Ann. Rev. Plant. Physiol.* **16**, 241.

Brückner, J. (1951). *Helv. Physiol. Pharmacol. Acta* **9**, 259.

Burnstock, G., Holman, M. E., and Prosser, C. L. (1963). *Physiol. Rev.* **43**, 482.

Burnstock, G. and Prosser, C. L. (1960). *Amer. J. Physiol.* **199**, 553.

Caldwell, P. C. (1954). *J. Physiol.* **126**, 169.

Caldwell, P. C. (1958). *J. Physiol.* **142**, 22.

Caldwell, P. C. (1968). *Physiol. Rev.* **48**, 1.

Caldwell, P. C., Hodgkin, A. L., Keynes, R. D., and Shaw, T. L. (1964). *J. Physiol.* **171**, 119.

Caldwell, P. C. and Keynes, R. D. (1957). *J Physiol* **137**, 13P.

Caldwell, P. C. and Keynes, R. D. (1959). *J. Physiol.* **148**, 8P.

Caldwell, P. C. and Lowe, A. G. (1966). *J. Physiol.* **186**, 24P.

Carafoli, E., Gamble, J. L., and Lehninger, A. L. (1965). *Biochem. Biophys. Res. Commun.* **21**, 215.

Carafoli, E. and Lehninger, A. L. (1967). In: *Methods in Enzymology*, Vol. X (ed. by Colowick, S. P. and Kaplan, N. O.), p. 745. Academic Press, New York–London.

Carlisky, N. J., Jard, S., and Morel, F. (1966). *Amer. J. Physiol.* **211**, 593.

Caspary, W. F. and Crane, R. K. (1968). *Biochim. Biophys. Acta* **163**, 395.

Cereijido, M. and Curran, P. F. (1965). *J. Gen. Physiol.* **48**, 543.

Cereijido, M., Herrera, F. C., Flanigan, W. J., and Curran, P. F. (1964). *J. Gen. Physiol.* **47**, 879.

Cereijido, M. and Rotunno, C. A. (1967). *J. Physiol.* **190**, 481.

Chan, S. S. and Lotspeich, W. D. (1962). *Amer. J. Physiol.* **203**, 975.

Chance, B. (1963). *Energy-Linked Functions of Mitochondria.* Academic Press, New York.

Chappell, J. B. and Crofts, A. R. (1966). In: *Regulation of Metabolic Processes in Mito-chondria* (ed. by Tager, J. M., Papa, S., Quagliariello, E., and Slater, E. C.), p. 293 Elsevier Publishing Company, Amsterdam–London–New York.

Chappell, J. B. and Haarhoff, K. N. (1967). In: *Biochemistry of Mitochondria* (ed. by Slater, E. C., Kaniuga, Z., and Wojtczak, L.), p. 75. Academic Press, London-New York.

Chen, P. S. and Neuman, W. F. (1955). *Amer. J. Physiol.* **180**, 623.

Chez, R. A., Palmer, R. R., Schultz, S. G., and Curran, P. F. (1967). *J. Gen. Physiol.* **50**, 2357.

Choi, J. K. (1963). *J. Cell Biol.* **16**, 53.

Chowdhury, T. K., and Snell, F. M. (1965). *Biochim. Biophys. Acta* **94**, 461.

Chowdhury, T. K. and Snell, F. M. (1966). *Biochim. Biophys. Acta* **112**, 581.

Christensen, H. N. and Handlongten, M. E. (1969). *FEBS Letters* **3**, 14.

Christensen, H. N. and Liang, M. (1966). *J. Biol. Chem.* **241**, 5542.

Christensen, H. N. and Riggs, T. R. (1952). *J. Biol. Chem.* **194**, 57.

Cirillo, V. P. (1967). In: *Symposium über Hefe-Protoplasten* (ed. by Müller, R.), p. 153. Akademie-Verlag, Berlin.

Cirillo, V. P. (1968). *J. Bacteriol.* **95**, 603.

Cirillo, W. P., Wilkins, P. O. and Anton, J. (1963). *J. Bacteriol.* **86**, 1259.

Civan, M. M. and Frazier, H. S. (1968). *J. Gen. Physiol.* **51**, 589.

Civan, M. M., Kedem, O., and Leaf, A. (1966). *Amer. J. Physiol.* **211**, 569.

Clark, S. L., Jr. (1959). *J. Biophys. Biochem. Cytol.* **5**, 41.

Clarkson, T. W. and Rothstein, A. (1960). *Amer. J. Physiol.* **199**, 898.

Clarkson, T. W. and Toole, S. R. (1964). *Amer. J. Physiol.* **206**, 658.

Cochrane, V. W. and Tull, D. L. W. (1958). *Phytopathology* **48**, 623.

Cockrell, R. S., Harris, E. J., and Pressman, B. C. (1966). *Biochemistry* **5**, 2326.

Cockrell, R. S., Harris, E. J., and Pressman, B. C. (1967). *Nature* **215**, 1487.

Code, C. F. (1968). In: *Handbook of Physiology* (ed. by Code, C. F.). Amer. Physiol. Soc., Washington.

Cohen, G. N. and Rickenberg, H. V. (1956). *Ann. Inst. Pasteur.* **91**, 693.

Cohen, L. B., Keynes, R. D., and Hille, B. (1968). *Nature* **218**, 438.

Cole, K. S. (1949). *Arch. Sci. Physiol.* **3**, 253.

Cole, K. S. and Curtis, H. J. (1939). *J. Gen. Physiol.* **22**, 649.

Collander, R. (1936). *Protoplasma* **25**, 201.

Collander, R. (1949). *Physiol. Plantarum* **2**, 300.

Collander, R. (1957). *Ann. Rev. Plant Physiol.* **8**, 335.

Conway, E. J. (1951). *Science* **113**, 270.

Conway, E. J. (1953). *Internat. Rev. Cytol.* **2**, 419.

Conway, E. J. (1954). *Symp. Soc. Exptl. Biol.* **8**, 297.

Conway, E. J. (1957). *Physiol. Rev.* **37**, 84.

Conway, E. J. (1960*a*). *J. Gen. Physiol.* **43**, Part 2, 17.

Conway, E. J. (1960*b*). *Nature* **187**, 394.

Conway, E. J. (1964). *Fed. Proc.* **23**, 680.

Conway, E. J. and Duggan, F. (1958). *Biochem. J.* **69**, 265.

Cook, G. M. W. (1968). *Brit. Med. Bull.* **24**, 118.

Cope, F. W. (1965). *Proc. Nat. Acad. Sci.* **54**, 225.

Copeland, F. (1964). *J. Cell Biol.* **23**, 253.

Cort, J. H. and Kleinzeller, A. (1958). *J. Physiol.* **142**, 208.

Coster, H. G. (1965). *Biophys. J.* **5**, 669.

Coster, H. G. L. (1966). *Austr. J. Biol. Sci.* **19**, 545.

Coster, H. G. L. and Hope, A. B. (1968). *Austr. J. Biol. Sci.* **21**, 243.

Crabbé, J. (1961). *J. Clin. Invest.* **40**, 2103.

Crane, R. K. (1960a). *Biochim. Biophys. Acta* **45**, 477.

Crane, R. K. (1960b). *Physiol. Rev.* **40**, 789.

Crane, R. K. (1962). *Fed. Proc.* **21**, 891.

Crane, R. K. (1964). *Biochem. Biophys. Res. Comm.* **17**, 481.

Crane, R. K. (1965). *Fed. Proc.* **24**, 1000.

Crane, R. K. (1966a). *Gastroenterology* **50**, 254.

Crane, R. K. (1966b). In: *Intracellular Transport* (ed. by Warren, K. B.), p. 7. Academic Press, New York–London.

Crane, R. K. (1969). *Fed. Proc.* **28**, 5.

Crane, R. K., Field, R. A., and Cori, C. F. (1957). *J. Biol. Chem.* **224**, 649.

Crane, R. K., Forstner, G., and Eichholz, A. (1965). *Biochim. Biophys. Acta* **109**, 467.

Crane, R. K. and Mandelstam, P. (1960). *Biochim. Biophys. Acta* **45**, 460.

Crane, R. K., Miller, D., and Bihler, I. (1961). In: *Membrane Transport and Metabolism* (ed. by Kleinzeller, A. and Kotyk, A.), p. 439. Academic Press, New York–London.

Crane, R. K. and Wilson, T. H. (1958). *J. Appl. Physiol.* **12**, 145.

Crank, J. (1956). *Mathematics of Diffusion.* Oxford University Press.

Crocken, B. and Tatum, E. L. (1967). *Biochim. Biophys. Acta* **115**, 100.

Cross, S. B., Keynes, R. D., and Rybová, R. (1965). *J. Physiol.* **181**, 865.

Csáky, T. Z. (1961). *Biochem. Pharmacol.* **8**, 38.

Csáky, T. Z. (1963). *Fed. Proc.* **22**, 3.

Csáky, T. Z. (1965). *Ann. Rev. Physiol.* **27**, 415.

Csáky, T. Z. and Fernald, G. W. (1961). *Nature* **191**, 709.

Csáky, T. Z., Hartzog, H. G., and Fernald, G. W. (1961). *Amer. J. Physiol.* **200**, 459.

Csáky, T. Z. and Lassen, U. A. (1964). *Biochim. Biophys. Acta* **82**, 215.

Csáky, T. Z. and Thale, M. (1960). *J. Physiol.* **151**, 59.

Cummins, J. T., Strand, J. A., and Vaughan, B. E. (1969). *Biochim. Biophys. Acta* **173**, 198.

Curran, P. F. (1960). *J. Gen. Physiol.* **43**, 1137.

Curran, P. F. (1965). *Fed. Proc.* **24**, 993.

Curran, P. F. and Cereijido, M. (1965). *J. Gen. Physiol.* **48**, 1011.

Curran, P. F., Herrera, F. C., and Flanigan, W. J. (1963). *J. Gen. Physiol.* **46**, 1011.

Curran, P. F. and McIntosh, J. R. (1962). *Nature* **193**, 347.

Curran, P. F., Schultz, S. G., Chez, R. A., and Fuisz, R. E. (1967a). *J. Gen. Physiol.* **50**, 1261.

Curran, P. F., Taylor, A. E., and Solomon, A. K. (1967b). *Biophys. J.* **7**, 879.

Curtis, H. J. and Cole, K. S. (1942). *J. Cell. Comp. Physiol.* **19**, 135.

Dainty, J. (1960). *Symp. Soc. Exptl. Biol.* **14**, 140.

Dainty, J. (1962). *Ann. Rev. Plant. Physiol.* **13**, 379.

Dainty, J. (1963). *Adv. Bot. Res.* **1**, 279.

Dainty, J., Croghan, P. C., and Fensom, D. S. (1963). *Can. J. Bot.* **41**, 953.

Dainty, J. and Ginzburg, B. Z. (1963). *J. Theoret. Biol.* **5**, 256.

Dainty, J. and Ginzburg, B. Z. (1964a). *Biochim. Biophys. Acta* **79**, 102.

Dainty, J. and Ginzburg, B. Z. (1964b). *Biochim. Biophys. Acta* **79**, 129.

Dainty, J. and Hope, A. B. (1959a). *Austr., J. Biol. Sci.* **12**, 395.

Dainty, J. and Hope, A. B. (1959b). *Austr. J. Biol. Sci.* **12**, 136.

Dainty, J. and Hope, A. B. (1961). *Austr. J. Biol. Sci.* **14**, 541.

Dainty, J., Hope, A. B., and Denby, C. (1960). *Austr. J. Biol. Sci.* **13**, 267.

Dainty, J. and House, C. R. (1966a). *J. Physiol.* **182**, 66.

Dainty, J. and House, C. R. (1966b). *J. Physiol.* **185**, 172.

Damadian, R. (1966). TR 66-19 of the USAF School of Aerospace Medicine.

Damadian, R. (1967). *Biochim. Biophys. Acta* **135**, 378.

Danielli, J. F. (1943, 1952). In: *The Permeability of Natural Membranes.* Cambridge University Press, 1943, 2nd Edition 1952.

Danielli, J. F. (1954). *Symp. Soc. Exptl. Biol.* **8**, 502.

Danielli, J. F. and Davson, H. (1935). *J. Cell. Comp. Physiol.* **5**, 495.

Daniels, V. G., Dawson, A. G., Newey, H., and Smyth, D. H. (1969). *Biochim. Biophys. Acta* **173**, 575.

Daniels, V. G., Dawson, A. G., Newey, H., and Smyth, D. H. (1969). *Biochim. Biophys. Acta* **173**, 575.

Davies, H. E. F., Martin, D. G., and Sharp, G. W. G. (1968). *Biochim. Biophys. Acta* **150**, 315.

Deák, T. and Kotyk, A. (1968). *Folia microbiol.* **13**, 205.

Dean, R. D. (1941). *Biol. Symp.* **3**, 331.

Dee, E. and Kernan, R. P. (1963). *J. Physiol.* **165**, 550.

Deetjen, P. and Boylan, J. W. (1968). *Pflügers Arch. ges. Physiol.* **299**, 19.

de Gier, J. and van Deenen, L. L. M. (1961). *Biochem. Biophys. Acta* **49**, 286.

Deierkauf, F. A. and Booij, H. L. (1968). *Biochim. Biophys. Acta* **150**, 214.

de la Fuente, D. and Sols, A. (1962). *Biochim. Biophys. Acta* **56**, 49.

Demis, D. J., Rothstein, A. and Meier, R. (1954). *Arch. Biochem. Biophys.* **48**, 55.

Denbigh, K. G. (1958). *The Thermodynamics of the Steady State.* Methuen & Co., Ltd., London; John Wiley & Sons., Inc., New York.

de Robertis, E. D. P., Nowinski, W. W., and Saez, F. A. (1965). *Cell Biology.* W. B. Saunders Company, Philadelphia–London.

DeRouffignac, C. and Morel, F. (1967). *Nephron* **4**, 92.

Desmedt, J. E. (1953). *J. Physiol.* **121**, 191.

Diamond, J. M. and Solomon, A. K. (1959). *J. Gen. Physiol.* **42**, 1105.

Diamond, J. M. (1964). *J. Gen. Physiol.* **48**, 15.

DiBona, D. R., Civan, M. M., and Leaf, A. (1969). *J. Cell. Biol.* **40**, 1.

Dick, D. A. T. (1964). *J. Theoret. Biol.* **7**, 504.

Dick, D. A. T. (1966). *Cell Water.* Butterworths, London.

Diedrich, D. F. (1961). *Biochim. Biophys. Acta* **47**, 618.

Diedrich, D. F. (1963). *Biochim. Biophys. Acta* **71**, 688.

Diedrich, D. F. (1966a). *Amer. J. Physiol.* **211**, 581.

Diedrich, D. F. (1966b). *Arch. Biochem. Biophys.* **117**, 248.

Diedrich, D. F. (1968). *Arch. Biochem. Biophys.* **127**, 803.

Dietz, G., Jr., Anraku, Y. and Heppel, L. A. (1968). *Fed. Proc.* **27**, 831.

Dilley, R. A. and Vernon, L. P. (1965). *Arch. Biochem. Biophys.* **111**, 365.

Dixon, M. and Webb, E. G. (1965). *Enzymes*, p. 429. Longmans, Green and Co., Ltd., London.

Dodd, W. A., Pitman, M. G., and West, K. R. (1966). *Austr. J. Biol. Sci.* **19**, 341.

Dodge, J. T., Mitchell, C. D., and Hanahan, D. J. (1963). *Arch. Biochem. Biophys.* **100**, 119.

Donaldson, P. E. K. (1958). *Electronic Apparatus for Biological Research.* Butterworths Scientific Publications, London.

Dourmashkin, R. R., Dougherty, R. M., and Harris, R. J. C. (1962). *Nature* **194**, 1116.

Dowd, J. E. and Riggs, D. S. (1965). *J. Biol. Chem.* **240**, 863.

Dudel, J., Peper, K., Rüdel, R., and Trautwein, W. (1967). *Pflügers Arch. ges. Physiol.* **295**, 197.

Duerdoth, J. K., Newey, H., Sanford, P. A., and Smyth, D. H. (1965). *J. Physiol.* **176**, 23P.

Dupuis, Y. and Fournier, P. (1963). In: *The Transfer of Calcium and Strontium across Biological Membranes* (ed. by Wasserman, R. H.), p. 277. Academic Press, New York–London.

Durbin, R. P. (1961). *J. Gen. Physiol.* **44**, 315.

Durbin, R. P., Frank, H., and Solomon, A. K. (1956). *J. Gen. Physiol.* **39**, 535.

Dydynska, M. and Harris, E. J. (1966). *J. Physiol.* **182**, 92.

Eadie, G. S. (1942). *J. Biol. Chem.* **146**, 85.

Eavenson, E. and Christensen, H. N. (1967). *J. Biol. Chem.* **242**, 5386.

Eccles, J. C. (1964). *The Physiology of Synapses.* Springer–Verlag, Berlin–Göttingen–Heidelberg.

Eddy, A. A. (1963). *Biochem. J.* **88**, 34P.

Eddy, A. A. (1968). *Biochem. J.* **108**, 195.

Eddy, A. A. and Indge, K. J. (1962). *Biochem. J.* **85**, 35P.

Edsall, J. T. and Wyman, J. (1958). *Biophysical Chemistry, Vol. I*, p. 618. Academic Press Inc., New York.

Egan, J. B. and Morse, M. L. (1965a). *Biochim. Biophys. Acta* **97**, 310.

Egan, J. B. and Morse, M. L. (1965b). *Biochim. Biophys. Acta* **109**, 172.

Egan, J. B. and Morse, M. L. (1966). *Biochim. Biophys. Acta* **112**, 63.

Eggleton, M. G. and Shuster, S. (1953). *J. Physiol.* **22**, 54P.

Eichholz, A. (1967). *Biochim. Biophys. Acta* **135**, 475.

Eichholz, A. (1968). *Biochim. Biophys. Acta* **163**, 101.

Eichholz, A. (1969). *Fed. Proc.* **28**, 30.

Eichholz, A. and Crane, R. K. (1965). *J. Cell Biol.* **26**, 687.

Einstein, A. (1908). *Z. Elektrochem.* **14**, 235. Also in: Einstein, A. (1956). *Investigation on the Theory of the Brownian Movement*, p. 68–85. Dover Publications Inc.

Eisenman, G. (1961). In: *Membrane Transport and Metabolism* (eds. Kleinzeller, A. and Kotyk, A.), p. 163. Academic Press, New York-London.

Eisenman, G. (1962). *Biophys. J.* **2**, Suppl., 259.

Eisenman, G. (1968). *Fed. Proc.* **27**, 1249.

Elsden, S. R., Gibson, Q. H., and Wiseman, G. (1950). *J. Physiol.* **111**, 56P.

Emmelot, P., Bos, C. J., Benedetti, E. L., and Rümke, P. L. (1964). *Biochim. Biophys. Acta* **90**, 126.

Engbaek, L. and Hoshiko, T. (1957). *Acta Physiol. Scand.* **39**, 348.

Engelman, D. M. and Morowitz, H. J. (1968). *Biochim. Biophys. Acta* **150**, 385.

Eppley, R. W. (1958). *J. Gen. Physiol.* **41**, 901.

Epstein, E. (1956). *Ann. Rev. Plant Physiol.* **7**, 1.

Epstein, E. and Hagen, C. E. (1956). *Plant Physiol.* **27**, 457.

Epstein, W. and Schultz, S. G. (1966). *J. Gen. Physiol.* **49**, 469.

Ernst, E. (1963). *Biophysics of the Striated Muscle.* Publishing House Hung. Acad. Sci., Budapest.

Essig, A. (1965). *Amer. J. Physiol.* **208**, 401.
Essig, A., Frazier, H. S., and Leaf, A. (1963). *Nature* **197**, 701.
Essig, A. and Leaf, A. (1963). *J. Gen. Physiol.* **46**, 505.
Etherton, B. (1963). *Plant Physiol.* **38**, 581.
Evans, D. R., White, B. C., and Brown, R. K. (1967). *Biochem. Biophys. Res. Comm.* **28**, 699.
Eyring, H. (1936). *J. Chem. Phys.* **4**, 283.

Fairclough, G. F. and Fruton, J. S. (1966). *Biochemistry* **5**, 673.
Falk, G. and Fatt, P. (1964). *Proc. Roy. Soc. B*, **160**, 69.
Farquhar, M. G. and Palade, G. E. (1963*a*). *J. Cell Biol.* **19**, 22A.
Farquhar, M. G. and Palade, G. E. (1963*b*). *J. Cell. Biol.* **17**, 375.
Farquhar, M. G. and Palade, G. E. (1964). *Proc. Nat. Acad. Sci.* **51**, 569.
Faust, R. G. (1964). *J. Cell. Comp. Physiol.* **63**, 55.
Fawcett, D. W. (1962). *Circulation* **26**, 1105.
Feldman, E. B. and Borgström, B. (1966*a*). *Biochim. Biophys. Acta* **125**, 136.
Feldman, E. B. and Borgström, B. (1966*b*). *Biochim. Biophys. Acta* **125**, 148.
Fensom, D. S. and Dainty, J. (1963). *Can. J. Bot.* **41**, 685.
Fensom, D. S., Meylan, S., and Pilet, P. (1965). *Can. J. Bot.* **43**, 453.
Fensom, D. S. and Wanles, I. R. (1967). *J. Exp. Bot.* **18**, 563.
Ferguson, D. R., Handler, J. S., and Orloff, J. (1968). *Biochim. Biophys. Acta* **163**, 150.
Fernández–Morán, H. (1962). In: *Symposia of the International Society for Cell Biology* Vol. *I*, p. 411. Academic Press, New York.
Fernández–Morán, H., Oda, T., Blair, P. V. and Green, D. E. (1964). *J. Cell. Biol.* **22**,63.
Ferreira, K. T. G. (1968). *Biochim. Biophys. Acta* **150**, 587.
Finch, L. R. and Hird, F. J. R. (1960). *Biochim. Biophys. Acta* **43**, 278.
Findlay, G. P. and Hope, A. B. (1964*a*). *Austr. J. Biol. Sci.* **17**, 62.
Findlay, G. P. and Hope, A. B. (1964*b*). *Austr. J. Biol. Sci.* **17**, 400.
Finean, J. B. (1958). *Exptl. Cell Res. Suppl.* **5**, 18.
Finkelstein, A. (1961*a*). *J. Gen. Physiol.* **44**, 1165.
Finkelstein, A. (1961*b*). *Nature* **190**, 1119.
Finkelstein, A. (1964). *J. Gen. Physiol.* **47**, 545.
Finkelstein, A. and Mauro, A. (1963). *Biophys. J.* **3**, 215.
Fischbarg, J., Zadunaisky, J. A., and DeFisch, F. W. (1967). *Amer. J. Physiol.* **213**, 963.
Fisher, R. B. (1967). *Brit. Med. Bull.* **23**, 241.
Fisher, R. B. and Lindsay, D. B. (1956). *J. Physiol.* **131**, 526.
Fishman, H. M. and Macey, R. I. (1969). *Biophys. J.* **9**, 151.
Folch, J. and Lees, M. (1951). *J. Biol. Chem.* **191**, 807.
Fonyo, A., and Bessman, S. P. (1968). *Biochem. Med.* **2**, 145.
Fox, A. M. (1965). *Comp. Biochem. Physiol.* **14**, 553.
Fox, C. F., Carter, J. R., and Kennedy, E. P. (1966). *Fed. Proc.* **25**, 591.
Fox, C F. and Kennedy, E. P. (1965). *Proc. Nat. Acad. Sci.* **54**, 891.
Fox, M., Thier, S., Rosenberg, L., Kiser, W., and Segal, S. (1964*a*). *New Engl. J. Med.* **270**, 556.
Fox, M., Thier, S., Rosenberg, L., and Segal, S. (1964*b*). *Biochim. Biophys. Acta* **79**, 167.
Fozzard, H. A. and Kipnis, D. M. (1967). *Science* **156**, 1257.
Frankenhaeuser, B. (1957). *J. Physiol.* **137**, 245.
Franz, T. J., Galey, W. R. and Van Bruggen, J. T. (1968). *J. Gen. Physiol.* **51**, 1.

Franzini–Armstrong, C. and Porter, K. R. (1964). *J. Cell Biol.* **22**, 675.

Fraser, A. and Frey, A. H. (1968). *Biophys. J.* **8**, 731.

Frazier, H. S. (1962). *J. Gen. Physiol.* **45**, 515.

Frazier, H. S., Dempsey, E. F., and Leaf, A. (1962). *J. Gen. Physiol.* **45**, 529.

Frazier, H. S. and Leaf, A. (1963). *J. Gen. Physiol.* **46**, 491.

Freeman, S. and Jacobsen, A. B. (1957). *Amer. J. Physiol.* **191**, 388.

Frick, A., Rumrich, G., Ullrich, K. J., and Lassiter, W. E. (1965). *Pflügers Arch. ges. Physiol.* **286**, 109.

Fried, M. and Shapiro, R. E. (1961). *Ann. Rev. Plant Physiol.* **12**, 91.

Fridhandler, L. and Quastel, J. H. (1955). *Arch. Biochem. Biophys.* **56**, 424.

Frömter, E. and Hegel, U. (1966). *Pflügers Arch. ges. Physiol.* **291**, 107.

Frumento, A. S. (1965). *J. Theoret. Biol.* **9**, 253.

Fuhrman, F. A. and Ussing, H. H. (1951). *J. Cell. Comp. Physiol.* **38**, 109.

Fuhrman, G. F. and Rothstein, A. (1968). *Biochim. Biophys. Acta* **163**, 325.

Gamble, J. M. (1957). *J. Biol. Chem.* **228**, 955.

Gans, H., Sobramanian, V., and Tan, B. H. (1968). *Science* **159**, 107.

Gárdos, G. (1954). *Acta Physiol. Acad. Sci. Hung.* **6**, 191.

Gárdos, G. (1966). *Acta Biochim. Biophys. Acad. Sci. Hung.* **1**, 139.

Garrahan, P. J. and Glynn, I. M. (1967). *J. Physiol.* **192**, 237.

Gatzy, J. T. and Berndt, W. O. (1968). *J. Gen. Physiol.* **51**, 770.

Geduldig, D. (1968). *J. Theoret. Biol.* **19**, 67.

Gertz, K. H. (1963). *Pflügers Arch. ges. Physiol.* **276**, 336.

Ghosh, S. and Ghosh, D. (1968). *Ind. J. Biochem.* **5**, 49.

Gibbs, J. W. (1899). *Collected Works, Vol. I* (ed. by Longmans), p. 429; quoted by Guggenheim (1967).

Giebel, F. and Passow, H. (1960). *Arch. ges. Physiol.* **271**, 378.

Giebisch, G. (1958). *J. Cell. Comp. Physiol.* **51**, 221.

Giebisch, G. (1961). *J. Gen. Physiol.* **44**, 659.

Gilles–Baillien, M. and Schoffeniels, E. (1965). *Arch. Int. Physiol. Biochem.* **73**, 355.

Gilles–Baillien, M. and Schoffeniels, E. (1967). *Comp. Biochem. Physiol.* **23**, 95.

Gilman, A. and Koelle, E. S. (1960). *Amer. J. Physiol.* **199**, 1025.

Ginetzinsky, A. G. (1958). *Nature* **182**, 1218.

Ginsburg, H. and Ginzburg, B. Z. (1968). *Israel J. Bot.* **17**, 127.

Ginzburg, B. Z. and Katchalsky, A. (1963). *J. Gen. Physiol.* **47**, 403.

Gits, J. J. and Grenson, M. (1967). *Biochim. Biophys. Acta* **135**, 507.

Glasser, J. E., Weiner, I. M., and Lach, L. (1965). *Amer. J. Physiol.* **208**, 359.

Goldman, D. E. (1943). *J. Gen. Physiol.* **27**, 37.

Goldner, A. M., Hajjar, J. J., and Curran, P. F. (1969a). *Biochim. Biophys. Acta* **173**, 572.

Goldner, A. M., Schultz, S. G., and Curran, P. F. (1969b). *J. Gen. Physiol.* **53**, 362.

Goldstein, D. A. and Solomon, A. K. (1961). *J. Gen. Physiol.* **44**, 1.

Goodford, P. J. and Hermansen, K. (1961). *J. Physiol.* **158**, 426.

Gordon, S. G. and Kern, F., Jr. (1968). *Biochim. Biophys. Acta* **152**, 372.

Graepel, P. (1966). Thesis, University of Bern.

Green, D. E., Allmann, D. W., Bachmann, E., Baum, H., Kopaczyk, K., Korman, E. F., Lipton, S., MacLennan, D. H., McConnell, D. C., Perdue, J. F., Rieske, J. S., and Tzagoloff, A. (1967). *Arch. Biochem. Biophys.* **119**, 312.

Green, K. and Matty, A. J. (1963). *Gen. Comp. Endocrinol.* **3**, 244.

Green, K., Seshadri, B., and Matty, A. J. (1962). *Nature* **196**, 1322.

Green, P. B. and Stanton, F. W. (1967). *Science* **155**, 1675.

Grenson, M. (1966). *Biochim. Biophys. Acta* **127**, 339.

Grenson, M., Mousset, M., Wiame, J. M., and Bechet, J. (1966). *Biochim. Biophys. Acta* **127**, 325.

Guggenheim, E. A. (1967). *Thermodynamics*, p. 300. North-Holland Publishing Company, Amsterdam.

Gutknecht, J. (1965). *Biol. Bull.* **129**, 495.

Gutknecht, J. (1966). *Biol. Bull.* **130**, 331.

Gutknecht, J. (1967). *J. Gen. Physiol.* **50**, 1821.

Gutknecht, J. (1968). *Biochim. Biophys. Acta* **163**, 20.

Günther, Th. and Dorn, F. (1966). *Z. Naturforsch.* **21b**, 1082.

Hagihira, H., Lin, E. C. C., and Wilson, T. H. (1961). *Biochem. Biophys. Res. Comm.* **4**, 478.

Hagiwara, S., Gruener, R., Hayashi, A., Sakota, H., and Grinnell, A. D. (1968). *J. Gen. Physiol.* **52**, 773.

Haissinsky, M. and Adloff, J. P. (1965). *Radiochemical Survey of the Elements.* Elsevier Publ. Co., Amsterdam–London–New York.

Halpern, Y. S. and Even–Shoshan, A. (1967). *J. Bacter.* **93**, 1009.

Halpern, Y. S., and Lupo, M. (1966). *Biochim. Biophys. Acta* **126**, 163.

Halvorson, H. O. and Cowie, D. B. (1961). In: *Membrane Transport and Metabolism* (ed. by Kleinzeller, A. and Kotyk, A.), p. 479. Academic Press, New York–London.

Hamilton, R. D., Morgan, K. M., and Strickland, J. D. H. (1966). *Can. J. Biochem.* **12**, 995.

Hanahan, J., Gurd, F. R. N., and Zibin, I. (1960). *Lipid Chemistry.* John Wiley and Sons, Inc., New York.

Hanai, T., Haydon, D., and Taylor, J. (1965). *J. Theoret. Biol.* **9**, 433.

Handler, J. S., Preston, A. S., and Rogulski, J. (1968). *J. Biol. Chem.* **243**, 1376.

Hansen, H. H. and Zerahn, K. (1964). *Acta Physiol. Scand.* **60**, 189.

Hanson, T. E. and Anderson, R. L. (1968). *Proc. Nat. Acad. Sci.* **61**, 269.

Harold, F. M. and Baarda, J. R. (1966). *J. Bacter.* **91**, 2257.

Harold, F. M. and Baarda, J. R. (1967). *J. Bacter.* **94**, 53.

Harold, F. M., Harold, R. L., Baarda, J. R., and Abrams, A. (1967). *Biochemistry* **6**, 1777.

Harris, E. J. (1954). *J. Physiol.* **124**, 248.

Harris, E. J. (1958). *J. Physiol.* **141**, 351.

Harris, E. J. (1960). *Transport and Accumulation in Biological systems*, 2nd edition. Butterworth, London.

Harris, E. J. (1963). *J. Physiol.* **166**, 87.

Harris, E. J. (1965a). *J. Physiol.* **177**, 355.

Harris, E. J. (1965b). *J. Physiol.* **176**, 123.

Harris, E. J., Catlin, G., and Pressman, B. C. (1967a). *Biochemistry* **6**, 1360.

Harris, E. J., Cockrell, R. S., and Pressman, B. C. (1966a). *Biochem. J.* **99**, 200.

Harris, E. J., Höfer, M., and Pressman, B. C. (1967b). *Biochemistry* **6**, 1348.

Harris, E. J., Höfer, M., and Pressman, B. C. (1968). Vth FEBS Meeting, Prague. Abstr. 363.

Harris, E. J., Judah, J. D. and Ahmed, K. (1966b). In: *Current Topics in Bioenergetics, Vol. I* (ed. by Sanadi, D. R.), p. 255. Academic Press, New York–London.

Harris, E. J. and van Dam, K. (1968). *Biochem. J.* **106**, 759.

Harris, E. J., van Dam, K., and Pressman, B. C. (1967c). *Nature* **213**, 1126.

Harris, J. R. (1968). *Biochim. Biophys. Acta* **150**, 534.

Harrison, H. E. and Harrison, H. C. (1951). *J. Biol. Chem.* **188**, 83.

Hartley, G. S. and Crank, J. (1949). *Trans. Faraday Soc.* **45**, 801.

Hashimoto, Y. (1965). *Kumamoto Med. J.* **18**, 23.

Haškovec, C. and Kotyk, A. (1969). *Europ. J. Biochem.* **9**, 343.

Hauser, G. (1969a). *Biochim. Biophys. Acta* **173**, 257.

Hauser, G. (1969b). *Biochim. Biophys. Acta* **173**, 267.

Haydon, D. A. and Taylor, J. (1963). *J. Theoret. Biol.* **4**, 281.

Hays, R. M. and Leaf, A. (1962). *J. Gen. Physiol.* **45**, 905.

Heckmann, K. (1965). In: *Mechanism of Hormone Action*, p. 41. Georg Thieme Verlag, Stuttgart.

Heinz, E., Pichler, A. G., and Pfeifer, B. (1965). *Biochem. Z.* **342**, 542.

Heinz, E. and Walsh, P. M. (1958). *J. Biol. Chem.* **233**, 1488.

Helbock, H. J., Forte, J. G., and Saltman, P. (1966). *Biochim. Biophys. Acta* **126**, 81.

Henderson, P. (1907). *Z. physik. Chem.* **59**, 118.

Henderson, P. (1908). *Z. physik. Chem.* **63**, 325.

Hendrix, J. Z., Alcock, N. W., and Archibald, R. M. (1963). *Clin. Chem.* **9**, 734.

Hengstenberg, W., Egan, J. B., and Morse, M. L. (1967). *Proc. Nat. Acad. Sci.* **58**, 2741.

Hengstenberg, W., Egan, J. B., and Morse, M. L. (1968). *J. Biol. Chem.* **243**, 1881.

Henn, F. A. and Thompson, T. E. (1968). *J. Mol. Biol.* **31**, 227.

Heppel, L. A. (1967). *Science* **156**, 1451.

Herrera, F. C. (1965). *Amer. J. Physiol.* **209**, 819.

Herrera, F. C. (1968). *J. Gen. Physiol.* **51**, 261s.

Hierholzer, K., Wiederholt, M., and Stolte, H. (1965). *Pflügers Arch. ges. Physiol.* **283**, R71.

Higinbotham, N., Etherton, B., and Foster, R. J. (1967). *Plant Physiol.* **42**, 32.

Hill, A. V. (1910). *J. Physiol.* **40**, IV.

Hill, A. V. (1928). *Proc. Roy. Soc. B.* **104**, 39.

Hillman, R. E., Albrecht, I., and Rosenberg, L. E. (1968). *Biochim. Biophys. Acta* **150** 528.

Hinke, J. A. M. (1961). *J. Physiol.* **156**, 314.

Hodgkin, A. L. (1958). *Proc. Roy. Soc. B.* **148**, 1.

Hodgkin, A. L. (1964). *The Conduction of the Nervous Impulse*. Liverpool University Press, Liverpool.

Hodkgin, A. L. and Huxley, A. F. (1939). *Nature* **144**, 710.

Hodgkin, A. L. and Huxley, A. F. (1952a). *J. Physiol.* **116**, 449.

Hodgkin, A. L. and Huxley, A. F. (1952b). *J. Physiol.* **116**, 473.

Hodgkin, A. L. and Huxley, A. F. (1952c). *J. Physiol.* **117**, 500.

Hodgkin, A. L., Huxley, A. F., and Katz, B. (1949). *Arch. Sci. Physiol.* **3**, 129.

Hodgkin, A. L. and Katz, B. (1949). *J. Physiol.* **108**, 37.

Hodgkin, A. L. and Keynes, R. D. (1953). *J. Physiol.* **119**, 513.

Hodgkin, A. L. and Keynes, R. D. (1955a). *J. Physiol.* **128**, 28.

Hodgkin, A. L. and Keynes, R. D. (1955b). *J. Physiol.* **128**, 61.

Hodgkin, A. L. and Keynes, R. D. (1957). *J. Physiol.* **138**, 253.

Hodgkin, A. L. and Horowicz, P. (1959). *J. Physiol.* **145**, 405.

Hodgkin, A. L. and Rushton, W. A. H. (1946). *Proc. Roy. Soc. B.* **133**, 444.

Hoffee, P., Englesberg, E., and Lamy, F. (1964). *Biochim. Biophys. Acta* **79**, 337.

Hofmann, A. F. and Small, D. M. (1967). *Ann. Rev. Med.* **18**, 333.

Hofstee, B. H. J. (1959). *Nature* **184**, 1296.

Hogben, C. A. M. (1960). *Ann. Rev. Physiol.* **22**, 381.

Hogg, J., Williams, E. J., and Johnston, R. J. (1968). *Biochim. Biophys. Acta* **150**, 640.

Hogg, R. and Englesberg, E. (1969). *J. Bacter.* **100**, 423.

Holden, J. T. and Holman, J. (1959). *J. Biol. Chem.* **234**, 865.

Holden, J. T. and Utech, N. M. (1967). *Biochim. Biophys. Acta* **135**, 351.

Holt, P. R. (1964). *Amer. J. Physiol.* **207**, 1.

Holter, H. (1964). In: *Intracellular Membraneous Structures.*, *Symp. Soc. Cell. Chem.* **14**, *Suppl.*, 451.

Hope, A. B. (1965). *Austr. J. Biol. Sci.* **18**, 789.

Hope, A. B., Simpson, A., and Walker, N. A. (1966). *Austr. J. Biol. Sci.* **19**, 355.

Hope, A. B. and Walker, N. A. (1960). *Austr. J. Biol. Sci.* **13**, 277.

Hope, A. B. and Walker, N. A. (1961). *Austr. J. Biol. Sci.* **14**, 26.

Horák, J. and Kotyk, A. (1969). *Folia Microbiol.* **14**, 291.

Horecker, B. L., Osborn, M. J., McLellan, W. L. Avigad, G., and Asensio, C. (1961). In: *Membrane Transport and Metabolism* (ed. by Kleinzeller, A. and Kotyk, A.), p. 378. Academic Press, New York–London.

Horecker, B. L., Thomas, J., and Monod, J. (1960). *J. Biol. Chem.* **235**, 1580.

Horowicz, P. and Gerber, C. (1965). *J. Gen. Physiol.* **48**, 489.

Hoshiko, T. and Lindley, B. D. (1964). *Biochim. Biophys. Acta* **79**, 301.

Hoshiko, T., Lindley, B. D., and Edwards, C. (1964). *Nature* **201**, 932.

House, C. R. (1964). *Biophys. J.* **4**, 401.

House, C. R. (1968). *Arch. Sci. Biol.* (*Bologna*) **52**, 209.

House, C. R. (1969). *Biochim. Biophys. Acta* **173**, 344.

Howard, R. E. and Burton, R. M. (1968). *J. Amer. Oil Chemist's Soc.* **45**, 202.

Höfer, M. and Kleinzeller, A. (1963*a*). *Physiol. Bohemoslov.* **12**, 405.

Höfer, M. and Kleinzeller, A. (1963*b*). *Physiol. Bohemoslov.* **12**, 417.

Höfer, M. and Kleinzeller, A. (1963*c*). *Physiol. Bohemoslov.* **12**, 425.

Höfer, M. and Pressman, B. C. (1966). *Biochemistry* **5**, 3919.

Huang, K. C. (1961). *Fed. Proc.* **20**, 246.

Huang, K. C. (1965). *Life Sci.* **4**, 1201.

Huang, K. C. and Woosley, R. L. (1968). *Amer. J. Physiol.* **214**, 342.

Huf, E. G., Parrish, J., and Watherford, C. (1951). *Amer. J. Physiol.* **164**, 137.

Hummel, J. P. and Dreyer, W. J. (1962). *Biochim. Biophys. Acta* **63**, 530.

Hunter, A. and Downs, C. E. (1945). *J. Biol. Chem.* **157**, 427.

Hutter, O. F. and Noble, D. (1960). *J. Physiol.* **151**, 89.

Huxley, A. F. (1960). In: *Mineral Metabolism* (ed. by Comar, C. L. and Bronner, F.) Vol. I, Pt. A, p. 163. Academic Press, New York–London.

Huxley, A. F. and Stämpfli, R. (1949). *J. Physiol.* **108**, 315.

Huxley, A. F. and Stämpfli, R. (1951). *J. Physiol.* **112**, 476.

Huxley, H. E. (1964). *Nature*, **202**, 1067.

Inui, Y. and Akedo, H. (1965). *Biochim. Biophys. Acta* **94**, 143.

Inui, Y. and Christensen, H. N. (1966). *J. Gen. Physiol.* **50**, 203.

Ivanov, V. V. and Natochin, Yu. V. (1968). (In Russian.) *Fiziol. zhur. SSSR im. I. M. Sechenova* **54**, 122.

Ivanova, L. N. and Natochin, Yu. V. (1968). (In Russian.) *Dokl. Akad. Nauk SSSR* **178**, 489.

Izawa, S. (1965). *Biochim. Biophys. Acta* **102**, 373.

Jacobson, E. S. and Metzenberge, R. L. (1968). *Biochim. Biophys. Acta* **156**, 140.

Jacquez, J. A. and Sherman, J. H. (1965). *Biochim. Biophys. Acta* **109**, 128.

Jagendorf, A. T. and Uribe, E. (1966). *Proc. Nat. Acad. Sci.* **55**, 170.

Janáček, K. (1962). *Biochim. Biophys. Acta* **56**, 42.

Janáček, K. (1963). *Physiol. Bohemoslov.* **12**, 349.

Janáček, K. (1967). (In Czech.) *Československá fysiologie* **16**, 513.

Janáček, K., Morel, F., and Bourguet, J. (1968). *J. Physiol. (Paris)* **60**, 51.

Janáček, K. and Rybová, R. (1966). *Cytologia (Japan)* **31**, 199.

Janáček, K. and Rybová, R. (1967). *Nature* **215**, 992.

Janda, S. (1969a). *Physiol. Bohemoslov.* **18**.

Janda, S. (1969b). *Physiol. Bohemoslov.* **18**.

Jard, S., Bourguet, J., Carasso, N. and Favard, P. (1966). *J. de Microscopie* **5**, 31.

Jennings, D. H. (1963). *The Absorption of Solutes by Plant Cells*. Oliver and Boyd, Edinburgh.

Jervis, E. L. and Smyth, D. H. (1960). *J. Physiol.* **151**, 51.

Jirsová, V., Koldovský, O., Heringová, A., Hošková, J., Jirásek, J., and Uher, J. (1966). *Biologia Neonat.* **9**, 44.

Johnson, C. F. (1967). *Science* **155**, 1670.

Johnson, C. F. (1969). *Fed. Proc.* **28**, 26.

Johnston, J. H. and Borgström, B. (1964). *Biochim. Biophys. Acta* **84**, 412.

Johnstone, R. M. and Scholefield, P. G. (1961). *J. Biol. Chem.* **236**, 1419.

Jones, T. H. D. and Kennedy, E. P. (1968). *Fed. Proc.* **27**, 644.

Jørgensen, C. B., Levi, H. and Zerahn, K. (1954). *Acta Physiol. Scand.* **30**, 178.

Jos, J., Frézal, J., Ruy, J., and Lamy, M. (1967). *Nature* **213**, 516.

Judah, J. D., Ahmed, K., McLean, A. E. M., and Christie, G. S. (1965). *Biochim. Biophys. Acta* **94**, 452.

Ju Lin, H. and Geyer, R. P. (1963). *Biochim. Biophys. Acta* **75**, 444.

Kaback, H. R. (1968). *Fed. Proc.* **27**, 644.

Kaback, H. R. and Stadtman, E. R. (1968). *J. Biol. Chem.* **243**, 1390.

Kamiya, N. and Tazawa, M. (1956). *Protoplasma* **46**, 394.

Kaplan, J. G. (1965). *J. Gen. Physiol.* **48**, 873.

Kaplan, J. G. and Tercreiter, W. (1966). *J. Gen. Physiol.* **50**, 9.

Karlish, S. J. D. and Avron, M. (1968a). *FEBS Letters* **1**, 21.

Karlish, S. J. D. and Avron, M. (1968b). *Biochim. Biophys. Acta* **153**, 878.

Karpenko, V., Kalous, V., and Pavlíček, Z. (1968). *Coll. Czechosl. Chem. Comm.* **33**, 3457.

Kashgarian, M., Stöckle, H., Gottschalk, C. W., and Ullrich, K. J. (1963). *Pflügers Arch. ges. Physiol.* **277**, 89.

Katchalsky, A. (1961). In: *Membrane Transport and Metabolism* (ed. by Kleinzeller, A. and Kotyk, A.), p. 69. Academic Press, New York–London.

Katchalsky, A. and Curran, P. F. (1965). *Nonequilibrium Thermodynamics in Biophysics*. Harvard University Press, Cambridge, Massachusetts.

Katz, B. (1966). *Nerve, Muscle, and Synapse.* McGraw–Hill, Inc., New York–St. Louis–San Francisco–Toronto–London–Sydney.

Kedem, O. (1961). In: *Membrane Transport and Metabolism* (ed. by Kleinzeller, A. and Kotyk, A.), p. 87. Academic Press, New York–London.

Kedem, O. and Essig, A. (1965). *J, Gen. Physiol.* **48**, 1047.

Kedem, O. and Katchalsky, A. (1958). *Biochim. Biophys. Acta* **27**, 229.

Kedem, O. and Katchalsky, A. (1961). *J. Gen. Physiol.* **45**, 143.

Kelly, R. B., Kohn, P. G. and Dainty, J. (1962). *Trans. Bot. Soc. Edinburgh,* **39**, 373.

Kepes, A. (1960). *Biochim. Biophys. Acta* **40**, 70.

Kepes, A. (1964). In: *The Cellular Function of Membrane Transport* (ed. by Hoffman, J. F.), p. 155. Prentice-Hall, Inc., Englewood Cliffs.

Kepes, A. and Cohen, G. N. (1962). In: *The Bacteria* (ed. by Gunsalus, I. C. and Stanier, R. Y.), p. 179. Academic Press, New York–London.

Kernan, R. P. (1960). *Nature* **185**, 471.

Kernan, R. P. (1962). *Nature* **193**, 986.

Kernan, R. P. (1963). *Nature* **200**, 474.

Kernan, R. P. (1965). *Cell K.* Butterworths, London.

Keynes, R. D. (1951). *J. Physiol.* **114**, 119.

Keynes, R. D. (1954). *Proc. Roy. Soc. B,* **142**, 359.

Keynes, R. D. (1962). *J. Physiol.* **166**, 16P.

Keynes, R. D. (1963). *J. Physiol,* **169**, 690.

Keynes, R. D. and Lewis, P. R. (1951*a*). *J. Physiol.* **113**, 73.

Keynes, R. D. and Lewis, P. R. (1951*b*). *J. Physiol.* **114**, 151.

Keynes, R. D. and Maisel, G. W. (1954). *Proc. Roy. Soc. B* **142**, 383.

Keynes, R. D. and Rybová, R. (1963). *J. Physiol.* **168**, 58P.

Keynes, R. D. and Swan, R. C. (1959). *J. Physiol.* **147**, 591.

Khuri, R. N., Flanigan, W. J., Oken, D. E., and Solomon, A. K. (1966). *Fed. Proc.* **25**, 899.

Kinter, W. B. (1961). In: *Proc. 12th Ann. Conf. Nephrotic Syndrome* (ed. by Metcoff, J.), p. 59. National Kidney Disease Foundation.

Kinter, W. B. and Wilson, T. H. (1965). *J. Cell. Biol.* **25**, 19.

Kipnis, D. M. (1959). *Ann. N.Y. Acad. Sci.* **82**, 354.

Kipnis, D. M. and Cori, C. F. (1959). *J. Biol. Chem.* **234**, 171.

Kipnis, D. M. and Parrish, J. E. (1965). *Fed. Proc.* **24**, 1051.

Kirkwood, J. G. (1954). In: *Ion Transport across Membranes* (ed. by Clarke, T.), p. 119. Academic Press, New York.

Kishimoto, U. (1965). *J. Cell. Comp. Physiol.* **66**, Suppl. 2, 43.

Kishimoto, U. (1966). *Plant Cell Physiol.* **7**, 429.

Kishimoto, U., Nagai, R., and Tazawa, M. (1965). *Plant Cell Physiol.* **6**, 519.

Kitasato, H. (1968). *J. Gen. Phys.* **52**, 60.

Kleinzeller, A. (1961). In: *Membrane Transport and Metabolism* (ed. by Kleinzeller, A. and Kotyk, A.), p. 527. Academic Press, New York-London.

Kleinzeller, A. (1965). *Arch. Biol. (Liège)* **76**, 217.

Kleinzeller, A. and co-workers (1965). *Manometrische Methoden und ihre Anwendung in Biologie und Biochemie.* VEB Gustav–Fischer–Verlag, Jena.

Kleinzeller, A., Janáček, K., and Knotková, A. (1962). *Biochim. Biophys. Acta* **59**, 239.

Kleinzeller, A., Kolínská, J., and Beneš, I. (1967*a*). *Biochem. J.* **104**, 843.

Kleinzeller, A., Kolínská, J., and Beneš, I. (1967*b*). *Biochem. J.* **104**, 852.

Kleinzeller, A., Kostyuk, P. G., Kotyk, A., and Lev., A. A. (1968). In: *Laboratory Tech-*

niques in Membrane Biophysics (ed. by Passow, H. and Stämpfli, R.), p. 69. Springer-Verlag, Heidelberg.

Kleinzeller, A. and Kotyk, A. (1961). *Biochim. Biophys. Acta* **54**, 367.

Klingenberg, M. and Pfaff, E. (1966). In: *Regulation of Metabolic Processes in Mitochondria* (ed. by Tager, J. M., Papa, S., Quagliariello, E., and Slater, E. C.), p. 180. Elsevier Publ. Comp., Amsterdam–London–New York.

Klingmüller, W. (1967). *Z. Naturforsch.* **22b**, 327.

Klotz, I. M. (1953). In: *The Proteins* (ed. by Neurath, H. and Bailey, A.) Vol. IB, chapter 8. Academic Press, New York.

Koefoed–Johnsen, V. and Ussing, H. H. (1953). *Acta Physiol. Scand.* **28**, 60.

Koefoed–Johnsen, V. and Ussing, H. H. (1958). *Acta Physiol. Scand.* **42**, 298.

Koefoed–Johnsen, V., Ussing, H. H., and Zerahn, K. (1952). *Acta Physiol. Scand.* **27**, 38.

Kolber, A. R. and LeFevre, P. G. (1967). *J. Gen. Physiol.* **50**, 1907.

Kolber, A. R. and Stein, W. D. (1966). *Nature* **209**, 691.

Kolber, A. R. and Stein, W. D. (1967). *Curr. Mod. Biol.* **1**, 244.

Kolínská, J. (1968). Vth FEBS Congress Abstr. no. 877. Prague.

Kolínská, J. and Semenza, G. (1967). *Biochim. Biophys. Acta* **146**, 181.

Kono, T. and Colowick, S. P. (1961). *Arch. Biochem. Biophys.* **93**, 520.

Korn, E. D. (1968). *J. Gen. Physiol.* **52**, Suppl., 257.

Kostyuk, P. G. and Sorokina, Z. A. (1961). In: *Membrane Transport and Metabolism* (ed. by Kleinzeller, A. and Kotyk, A.), p. 193. Academic Press, New York.

Kotyk, A. (1959). *Folia microbiol.* **4**, 363.

Kotyk, A. (1963). *Folia microbiol.* **8**, 27.

Kotyk, A. (1966). *Biochim. Biophys. Acta* **135**, 112.

Kotyk, A. (1967). *Folia microbiol.* **12**, 121.

Kotyk, A. and Haškovec, C. (1968). *Folia microbiol.* **13**, 12.

Kotyk, A. and Höfer, M. (1965). *Biochim. Biophys. Acta* **102**, 410.

Kotyk, A. and Kleinzeller, A. (1966). *Biochim. Biophys. Acta* **135**, 106.

Kotyk, A., Kolínská, J., Vereš, K., and Szammer, J. (1965). *Biochem. Z.* **342**, 129.

Kramer, P. J. (1956). Atomic Energy Commission Report TID-7512, p. 287.

Krane, S. M. and Crane, R. K. (1959). *J. Biol. Chem.* **234**, 211.

Krebs, H. A., Bennett, D. A. H., de Gasquet, P., Gascoyne, T., and Yoshida, T. (1963). *Biochem. J.* **86**, 22.

Krebs, H. A. and Lund, P. (1966). *Biochem. J.* **98**, 210.

Krolenko, S. A. and Nikolsky, N. N. (1967). *Tsitologiya* **9**, 273.

Kundig, W., Ghosh, S., and Roseman, S. (1964). *Proc. Nat. Acad. Sci.* **52**, 1067.

Kundig, W., Kundig, F. D., Anderson, B., and Roseman, S. (1965). *Fed. Proc.* **24**, 658.

Kundig, W., Kundig, F. D., Anderson, B., and Roseman, S. (1966). *J. Biol. Chem.* **241**, 3243.

Lack, L. and Weiner, I. M. (1961). *Amer. J. Physiol.* **200**, 313.

Lack, L. and Weiner, I. M. (1966). *Amer. J. Physiol.* **210**, 1142.

Lacko, L. and Burger, M. (1962). *Biochem. J.* **83**, 622.

Lacko, L. and Burger, M. (1963). *J. Biol. Chem.* **238**, 3478.

Langdon, R. G. and Sloan, H. R. (1967). *Proc. Nat. Acad. Sci.* **57**, 401.

Lassen, U. V. (1962). *Biochim. Biophys. Acta* **57**, 123.

Lassen, U. V. (1964). *Exp. Cell Res.* **34**, 54.

Lassen, U. V. and Overgaard–Hansen, K. (1962a). *Biochim. Biophys. Acta* **57**, 111.
Lassen, U. V. and Overgaard–Hansen, K. (1962b). *Biochim. Biophys. Acta* **57**, 118.
Lassiter, W. E., Mylle, M., and Gottschalk, C. W. (1966). *Amer. J. Physiol.* **210**, 965.
Laties, G. G. (1959a). *Ann. Rev. Plant Physiol.* **10**, 87.
Laties, G. G. (1959b). *Proc. Nat. Acad. Sci.* **45**, 163.
Laue, P. and MacDonald, R. E. (1968). *J. Biol. Chem.* **243**, 680.
Läuger, P., Lesslauer, W., Marti, E., and Richter, J. (1967). *Biochim. Biophys. Acta* **135**, 20.
Leaf, A. (1965). *Rev. Physiol.* **56**, 216.
Leaf, A., Anderson, J., and Page, L. B. (1958). *J. Gen. Physiol.* **41**, 657.
Leaf, A. and Hays, R. M. (1962). *J. Gen. Physiol.* **45**, 921.
Leaf, A. and Renshaw, A. (1957a). *Biochem. J.* **65**, 82.
Leaf, A. and Renshaw, A. (1957b). *Biochem. J.* **65**, 90.
Lee, Y.-P., Sowokinos, J. R., and Erwin, M. J. (1967). *J. Biol. Chem.* **242**, 2264.
LeFevre, P. G. (1948). *J. Gen. Physiol.* **31**, 505.
LeFevre, P. G. (1959). *Science* **130**, 104.
LeFevre, P. G. (1961a). *Pharmacol. Rev.* **13**, 39.
LeFevre, P. G. (1961b). *Fed. Proc.* **20**, 139.
LeFevre, P. G. (1962). *Amer. J. Physiol.* **203**, 286.
LeFevre, P. G. (1963). *J. Gen. Physiol.* **46**, 721.
LeFevre, P. G. (1966). *Biochim. Biophys. Acta* **120**, 395.
LeFevre, P. G. (1967). *Science* **158**, 274.
LeFevre, P. G. and LeFevre, M. E. (1952). *J. Gen. Physiol.* **35**, 891.
LeFevre, P. G. and Marshall, J. K. (1958). *Amer. J. Physiol.* **194**, 333.
Leggett, J. E. (1968). *Ann. Rev. Plant Physiol.* **19**, 333.
Lehninger, A. L. (1965). *The Mitochondrion.* W. A. Benjamin Inc., New York.
Lenard, J. and Singer, S. J. (1966). *Proc. Nat. Acad. Sci.* **56**, 1828.
Lengeman, F. W. (1963). In: *The Transfer of Calcium and Strontium across Biological Membranes* (ed. by Wasserman, R.H.), p. 197. Academic Press, New York.
Leontyev, V. G. and Natochin, J. V. (1964). (In Russian.) *Zh. Obshch. Biol.* **25**, 210.
Lerner, J. and Taylor, M. W. (1967). *Biochim. Biophys. Acta* **135**, 991.
Lev, A. A. (1964a). (In Russian.) *Biofizika* **9**, 686.
Lev, A. A. (1964b). *Nature* **201**, 1132.
Lev, A. A. and Buzhinsky, E. P. (1961). (In Russian.) *Tsitologiya* **3**, 614.
Lev, A. A. and Buzhinsky, E. P. (1967). *Tsitologiya* **9**, 102.
Levi, H. and Ussing, H. H. (1948). *Acta Physiol. Scand.* **16**, 232.
Levin, R. J. (1967). *Brit. Med. Bull.* **23**, 209.
Levine, R., Goldstein, M., Klein, S., and Huddlestun, B. (1949). *J. Biol. Chem.* **179**, 985.
Levine, M. and Stein, W. D. (1966). *Biochim. Biophys. Acta* **127**, 179.
Levinson, C. (1967). *Biochim. Biophys. Acta* **135**, 921.
Liberman, E. A. and Topaly, V. P. (1968). *Biochim. Biophys. Acta* **163**, 125.
Lichtenstein, N. S. and Leaf, A. (1965). *J. Clin. Invest.* **44**, 1328.
Lin, E. C. C., Hagihira, H., and Wilson, T. H. (1962). *Amer. J. Physiol.* **202**, 919.
Lindeman, B. and Thorns, U. (1967). *Science* **158**, 1473.
Lineweaver, H. and Burk, D. (1934). *J. Amer. Chem. Soc.* **56**, 658.
Ling, G. N. (1962). *A Physical Theory of the Living State: The Association–Induction Hypothesis.* Blaisdell Publ. Co., New York.
Ling, G. N. and Cope, F. W. (1969). *Science* **163**, 1335.

Ling, G. N. and Gerard, R. W. (1949). *J. Cell. Comp. Physiol.* **34**, 382.

Ling, G. N. and Ochsenfeld, M. M. (1966). *J. Gen. Physiol.* **49**, 819.

Lodin, Z., Janáček, K., and Müller, J. (1963). *J. Cell. Comp. Physiol.* **62**, 215.

Loeschke, K. and Baumann, K. (1969). *Pflügers Arch.* **305**, 139.

Loeschke, K., Baumann, K., Renschler, H., and Ullrich, K. J. (1969). *Pflügers Arch.* **305**, 118.

Lojda, Z. (1965). *Histochemie* **5**, 339.

Longley, R. P., Rose, A. H., and Knights, B. A. (1968). *Biochem. J.* **108**, 401.

Lotspeich, W. D. (1961). *Harvey Lect.* **56**, 63.

Lotspeich, W. D. and Woronkow, S. (1958). *Amer. J. Physiol.* **195**, 331.

Lowenstein, L. M., Smith, I., and Segal, S. (1968). *Biochim. Biophys. Acta* **150**, 73.

Lubin, M. and Kessel, G. (1960). *Biochem. Biophys. Res. Comm.* **2**, 249.

Lucy, J. A. (1964). *J. Theoret. Biol.* **7**, 360.

Lucy, J. A. and Glauert, A. M. (1964). *J. Mol. Biol.* **8**, 277.

Ludvík, J., Munk, V., and Dostálek, M. (1968). *Experientia* **24**, 1066.

Lundegårdh, H. (1955). *Ann. Rev. Plant Physiol.* **6**, 1.

Lundegårdh, H. (1961). *Nature* **192**, 243.

Lüttge, U. (1965). *Planta* **66**, 331.

Lüttge, U. (1966). *Planta* **68**, 269.

Lüttge, U. and Laties, G. G. (1967). *Plant Physiol.* **42**, 181.

Luzzati, V. and Husson, F. (1962). *J. Cell. Biol.* **12**, 207.

Lyon, I. (1968). *Biochim. Biophys. Acta* **163**, 75.

Macey, R I. and Meyers, S. (1963). *Amer. J. Physiol.* **204**, 1095.

MacInnes, D. A. (1961). *The Principles of Electrochemistry.* Dover Publications.

MacRobbie, E. A. C. (1962). *J. Gen. Physiol.* **45**, 861.

MacRobbie, E. A. C. (1964). *J. Gen. Physiol.* **47**, 859.

MacRobbie, E. A. C. (1965). *Biochim. Biophys. Acta* **94**, 64.

MacRobbie, E. A. C. (1966a). *Austral. J. Biol. Sci.* **19**, 363.

MacRobbie, E. A. C. (1966b). *Austral. J. Biol. Sci.* **19**, 371.

MacRobbie, E. A. C. and Dainty, J. (1958a). *J. Gen. Physiol.* **42**, 335.

MacRobbie, E. A. C. and Dainty, J. (1958b). *Physiol. Plant.* **11**, 782.

MacRobbie, E. A. C. and Ussing, H. H. (1961). *Acta Physiol. Scand.* **53**, 348.

Malathi, P. and Crane, R. K. (1968). *Biochim. Biophys. Acta* **163**, 275.

Malathi, P. and Crane, R. K. (1969). *Biochim. Biophys. Acta* **173**, 245.

Malm, M. (1950). *Physiol. Plant.* **3**, 376.

Manchester, K. L. and Young, F. G. (1961). *Vitamins Hormones* **19**, 95.

Marsh, D. J. and Solomon, S. (1964). *Nature* **201**, 714.

Marsh, D. J., Ullrich, K. J., and Rumrich, G. (1963). *Pflügers Arch. ges. Physiol.* **277**, 107.

Marshall, E. K. and Crane, M. R. (1924). *Amer. J. Physiol.* **70**, 465.

Marshall, J. M. (1959). *Amer. J. Physiol.* **197**, 935.

Martin, D. W. and Curran, P. F. (1966). *J. Cell. Physiol.* **67**, 367.

Matile, Ph., Moor, H., and Mühlethaler, K. (1967). *Arch. Mikrobiol.* **58**, 201.

Matthews, D. M. (1967). *Brit. Med. Bull.* **23**, 258.

Mauro, A. (1960). *Circulation* **21**, 845.

Maw, G. A. (1963). *Appl. Chem.* **7**, 665.

Mawe, R. C. and Hempling, H. G. (1965). *J. Cell. Comp. Physiol.* **66**, 95.

McDougal, D. B., Jr., Little, K. S., and Crane, R. K. (1960). *Biochim. Biophys. Acta* **45**, 483.

Medzihradsky, F., Kline, M. H., and Hokin, L. E. (1967). *Arch. Biochem. Biophys.* **121**, 311.

Mendoza, S. A., Handler, J. S., and Orloff, J. (1967). *Amer. J. Physiol.* **213**, 1263.

Metlička, R. and Rybová, R. (1967). *Biochim. Biophys. Acta* **135**, 563.

Miller, D. M. (1965*a*). *Biophys. J.* **5**, 407.

Miller, D. M. (1965*b*). *Biophys. J.* **5**, 417.

Miller, D. M. (1966). *Biochim. Biophys. Acta* **120**, 156.

Miller, D. M. and Crane, R. K. (1961). *Biochim. Biophys. Acta* **52**, 281.

Mitchell, P. (1957). *Nature* **180**, 134.

Mitchell, P. (1966). *Biol. Rev.* **41**, 445.

Monnier, A. M. (1968). *J. Gen. Physiol.* **51**, 26S.

Monnier, A. M., Monnier, A., Goudeau, H., and Rebuffel–Reynier, A. M. (1965). *J. Cell. Comp. Physiol.* **66**, 147.

Monod, J., Changeux, J.-P., and Jacob, F. (1963). *J. Mol. Biol.* **6**, 306.

Monod, J., Wyman, J., and Changeux, J.-P. (1965). *J. Mol. Biol.* **12**, 88.

Moor, H. and Mühlethaler, K. (1963). *J. Cell. Biol.* **17**, 609.

Moore, J. H. and Tyler, C. (1955). *Brit. J. Nutr.* **9**, 81.

Moore, J. W. and Cole, K. S. (1960). *J. Gen. Physiol.* **43**, 961.

Mora, J. and Snell, E. E. (1963). *Biochemistry* **2**, 136.

Morel, F. (1959). In: *The Method of Isotopic Tracers Applied to the Study of Active Ion Transport*, p. 155. Pergamon Press, New York.

Morel, F. (1964). In: *Water and Electrolyte Metabolism, Vol. II* (ed. by de Graeff, J. and Leijmse, B.), p. 91. Elsevier Publ. Co., Amsterdam.

Morel, F. (1967). In: *IIIrd International Congr. of Nephrology. I. Progress in Renal Physiology* (ed. by Schreiner, G. E.), p. 1, S. Karger, Basel–New York.

Morel, F. and Bastide, F. (1964). In: *Oxytocin, Vasopressin and Their Structural Analogues*, Proc. IInd Internat. Pharmacol. Meet. Prague, p. 47. Pergamon Press, Oxford.

Morel, F. and Bastide, F. (1965). *Biochim. Biophys. Acta* **94**, 609.

Morel, F. and Bastide, F. (1967). *Protoplasma* **63**, 58.

Morel, F. and Boudiak, C. (1962). Helv. Physiol. Acta **20**, 173.

Morel, F., Odier, M. and Lucarain, C. (1961). *J. Physiol. (Paris)* **53**, 757.

Morgan, H. E., Henderson, M. J., Regen, D. M., and Park, C. R. (1959). *Ann. N.Y. Acad. Sci.* **82**, 387.

Morgan, H. E., Henderson, M. J., Regen, D. M., and Park, C. R. (1961*a*). *J. Biol. Chem.* **236**, 253.

Morgan, H. E., Post, R. L., and Park, C. R. (1961*b*). In: *Membrane Transport and Metabolism* (ed. by Kleinzeller, A. and Kotyk, A.), p. 423. Academic Press, New York–London.

Morgan, H. E., Regen, D. M., and Park, C. R. (1964). *J. Biol. Chem.* **239**, 369.

Mueller, P., Rudin, D. O., Ti Tien, H., and Wescott, W. C. (1962). *Nature* **194**, 979.

Mueller, P. and Rudin, D. O. (1963). *J. Theor. Biol.* **4**, 268.

Mueller, P., Rudin, D. O., Ti Tien, H., and Wescott, W. C. (1963). *J. Phys. Chem.* **67**, 534.

Mueller, P., Rudin, D. O., Ti Tien, H., and Wescott, W. C. (1964). In: *Recent Progress in Surface Science,* (ed. by Danielli, J. F., Pankhurst, K. G. A., and Riddiford, A. C.) Vol. I, p. 379. Academic Press, New York–London.

Mueller, P. and Rudin, D. O. (1967). *Nature* **213**, 603.

Mueller, P. and Rudin, D. O. (1968). *Nature* **217**, 713.

Mullins, L. J. (1962). *Nature* **196**, 986.

Mullins, L. J. (editor) (1965). *J. Gen. Physiol.* **48**, Pt. II.

Mullins, L. J. and Awad, M. Z. (1965). *J. Gen. Physiol.* **48**, 761.

Munck, B. G. (1966). *Biochim. Biophys. Acta* **120**, 97.

Munck, B. G. (1968). *Biochim. Biophys. Acta* **150**, 82.

Munck, B. G. and Schultz, S. G. (1969). *J. Gen. Physiol.* **53**, 157.

Muñoz, E., Freer, J. H., Ellar, D. J., and Salton, M. R. J. (1968). *Biochim. Biophys. Acta* **150**, 531.

Nagai, R. and Kishimoto, U. (1964). *Plant Cell Physiol.* **5**, 21.

Nagai, R. and Tazawa, M. (1962). *Plant Cell Physiol.* **3**, 323.

Nakamura, Y., Nakajima, S., and Grundfest, H. (1965). *J. Gen. Physiol.* **48**, 985.

Naora, H., Naora, H., Izawa, M., Allfrey, V. G., and Mirsky, A. E. (1962). *Proc. Nat. Acad. Sci.* **48**, 853.

Narahara, H. T., Özand, P. and Cori, C. F. (1960). *J. Biol. Chem.* **235**, 3370.

Nasonov, D. N. and Aleksandrov, V. Ya. (1943). (In Russian.) *Usp. Sovr. Biol.* **16**, 577.

Nasonov, D. N. and Aleksandrov, V. Ya. (1944). (In Russian.) *Usp. Sovr. Biol.* **17**, 1.

Nastuk, W. L. (1964). *Physical Techniques in Biological Research, Vol. V, VI (Electrophysiological Methods, Part A and B)* Academic Press, New York.

Nathans, D., Tapley, D. F., and Ross, J. E. (1960). *Biochim. Biophys. Acta* **41**, 271.

Natochin, J. V., Janáček, K., and Rybová, R. (1965). *J. Endocrin.* **33**, 171.

Neale, R. J. and Wiseman, G. (1968). *Nature* **218**, 473.

Nemček, O., Sigler, K., and Kleinzeller, A. (1966). *Biochim. Biophys. Acta* **126**, 73.

Netter, H. (1961). In: *Biochemie des Aktiven Transports*, p. 15. Springer–Verlag, Berlin.

Newey, H., Sanford, P. A., and Smyth, D. H. (1963). *J. Physiol.* **168**, 423.

Newey, H., Sanford, P. A., and Smyth, D. H. (1965). *Nature* **205**, 389.

Newey, H., Sanford, P. A., and Smyth, D. H. (1966). *J. Physiol.* **186**, 493.

Newey, H. and Smyth, D. H. (1959). *J. Physiol.* **145**, 48.

Newey, H. and Smyth, D. H. (1964). *Nature* **202**, 400.

Nicolaysen, R. (1951). *Acta Physiol. Scand.* **22**, 260.

Nicolle, J. and Walle, J. (1963). *C.R. Acad. Sci. Paris* **257**, 2043.

Nikolsky, B. P. and Shults, M. M. (1963). (In Russian.) *Vestn. Leningrad. Univ.* **4**, 73.

Nims, L. F. (1962). *Science* **137**, 130.

Nirenberg, M. W. and Hogg, J. F. (1958). *J. Amer. Chem. Soc.* **80**, 4407.

Nishi, Y., Yoshida, T. O., and Takesue, Y. (1968). *J. Mol. Biol.* **37**, 441.

Nishizaki, Y. (1963). *Plant Cell Physiol.* **4**, 353.

Nobel, P. S. and Packer, L. (1965). *Plant Physiol.* **40**, 633.

Noble, D. (1966). *Physiol. Rev.* **46**, 1.

Noble, D. and Tsien, R. W. (1969). *J. Physiol.* **200**, 233.

Norman, A. W., Bieber, L. L., Lindberg, O., and Boyer, P. D. (1965). *J. Biol. Chem.* **240**, 2855.

Norton, J. E., Bulmer, G. S., and Sokatch, J. R. (1963). *Biochim. Biophys. Acta* **78**, 136.

Novotny, Ch. P. and Englesberg, E. (1966). *Biochim. Biophys. Acta* **117**, 217.

Nunn, A. S., Baker, R. D., and Searle, G. W. (1963). *Life Science* **9**, 646.

Nutbourne, D. M. (1968). *J. Physiol.* **195**, 1.

O'Connor, C. M. and Wolstenholme, G. E. W. (1960). *Regulation of the inorganic ion content of cells. CIBA Foundation Study Group no. 5.*

Ogilvie, J. T., McIntosh, J. R. and Curran, P. F. (1963). *Biochim. Biophys. Acta* **66**, 441.

Oikawa, T., Spyropoulos, C. S., Tasaki, I., and Teorell, T. (1961). *Acta Physiol. Scand.* **52**, 195.

Okada, H. and Halvorson, H. O. (1964). *Biochim. Biophys. Acta* **82**, 538, 547.

Olson, E. J. and Cazort, R. J. (1969). *J. Gen. Physiol.* **53**, 311.

Orloff, J. and Handler, J. S. (1964). *Amer. J. Med.* **36**, 686.

Orloff, J. and Handler, J. S. (1967). *Amer. J. Med.* **42**, 757.

Orloff, J., Handler, J. S., and Preston, A. S. (1962). *J. Clin. Invest.* **41**, 702.

Ottosen, D., Sjöstrand, F., Stenström, S., and Swaetichin, G. (1953). *Acta Physiol. Scand.* **106**, Suppl., 611.

Overton, E. (1902). *Pflügers Arch. ges. Physiol.* **92**, 346.

Overton, J., Eichholz, A., and Crane, R. K. (1965). *J. Cell. Biol.* **26**, 693.

Oxender, D. L. and Christensen, H. N. (1963). *J. Biol. Chem.* **238**, 3686.

Ørskov, S. L. (1935). *Biochem. Z.* **279**, 241.

Packer, L., Deamer, D. W., and Crofts, A. R. (1966). Brookhaven Symp. Biol. **19**, 281.

Padien, P. (1965). *Compt. Rend.* **260**, 5119.

Palay, S. L. and Karlin, L. J. (1959). *J. Biophys. Biochem. Cytol.* **5**, 373.

Pall, M. L. (1969). *Biochim. Biophys. Acta* **173**, 113.

Pallaghy, C. K. (1968). *Planta* **80**, 147.

Pappenheimer, J. R., Renkin, E. M., and Borrero, L. M. (1951). *Amer. J. Physiol.* **167**, 13.

Pardee, A. B. (1967). *Science* **156**, 1627.

Pardee, A. B. (1968). *Science* **162**, 632.

Pardee, A. B. and Prestidge, L. S. (1966). *Proc. Nat. Acad. Sci.* **55**, 189.

Pardee, A. B., Prestidge, L. S., Whipple, M. B., and Dreyfuss, J. (1966). *J. Biol. Chem.* **241**, 3962.

Pardee, A. B. and Watanabe, K. (1968). *J. Bacter.* **96**, 1049.

Park, C. R., Post, R. L., Kalman, C. F., Wright, J. H., Jr., Johnson, J. H., and Morgan, H. E. (1956). *CIBA Colloquia Endocrin.* **9**, 240.

Park, R. B. (1966). *Internat. Rev. Cytol.* **20**, 67.

Parkinson, T. M. and Olson, J. A. (1963). *Life Science* **2**, 393.

Parsons, D. S. (1967). *Brit. Med. Bull.* **23**, 252.

Passow, H. (1961). In: *Biochemie des Aktiven Transports*, p. 54. Springer–Verlag, Berlin.

Passow, H. (1964). In: *The Red Blood Cell* (ed. by Bishop, Ch. and Surgenor, D. M.), p. 71. Academic Press, New York.

Patlak, C. S. (1957). *Bull. Math. Biophys.* **19**, 209.

Patlak, C. S., Goldstein, D. A., and Hoffman, J. F. (1963). *J. Theoret. Biol.* **5**, 426.

Pavlasová, E. and Harold, F. M. (1969). *J. Bacter.* **98**, 198.

Payne, J. W. and Gilvarg, Ch. (1968). *J. Biol. Chem.* **243**, 335.

Peachey, L. D. (1965). *J. Cell. Biol.* **25**, 209.

Penrose, W. R., Nichoalds, G. E., Piperno, J. R., and Oxender, D. L. (1968). *J. Biol. Chem.* **243**, 5921.

Pessoa, V. C., Kim, K. S., and Ivy, A. C. (1953). *Amer. J. Physiol.* **174**, 209.

Petersen, M. J. and Edelman, I. S. (1964). *J. Clin. Invest.* **43**, 583.

Pfaff, E. (1967). In: *Mitochondrial Structure and Compartmentation* (ed. by Quagliariello, E., Papa, S., Slater, E. C., and Tager, J. M.), p. 165. Adriatica Editrice, Bari.

Phibbs, P. V. and Eagon, R. G. (1969). *Bacter. Proc.*, p. 143.

Piperno, J. R. and Oxender, D. L. (1966). *J. Biol. Chem.* **241**, 5732.

Piperno, J. R. and Oxender, D. L. (1968). *J. Biol. Chem.* **243**, 5914.

Planck, M. (1890). *Ann. Physik* **40**, 561.

Playoust, M. R. and Isselbacher, K. J. (1964). *J. Clin. Invest.* **43**, 467.

Podolsky, R. J. (1965). *Fed. Proc.* **24**, 1112.

Polissar, M. J. (1954). In: *The Kinetic Basis of Molecular Biology* (by Johnson, F. H., Eyring, H., and Polissar, M. J.), p. 515. John Wiley and Sons, Inc., New York.

Poncová, M. and Kotyk, A. (1967). Curr. Mod. Biol. **1**, 189.

Porteous, J. W. (1969). In: *Subcellular Components* (*Preparation and Fractionation*) (ed. by Birnie, G. D. and Fox, S. M.), p. 57. Butterworths, London.

Porter, K. R. (1969). *Fed. Proc.* **28**, 35.

Porter, K. R. and Palade, G. E. (1957). *J. Biophys. Biochem. Cytol.* **3**, 269.

Post, R. L. and Albright, C. D. (1961). In: *Membrane Transport and Metabolism* (ed. by Kleinzeller, A. and Kotyk, A.), p. 219. Academic Press, New York.

Post, R. L., Albright, C. D., and Dayani, K. (1967). *J. Gen. Physiol.* **50**, 1201.

Post, R. L., Morgan, H. E. and Park, C. R. (1961). *J. Biol. Chem.* **236**, 269.

Poulos, P. P. (1957). *J. Lab. Clin. Med.* **49**, 253.

Pressman, B. C. (1964). *Biochem. Biophys. Res. Comm.* **15**, 556.

Pressman, B. C. (1965). *Proc. Nat. Acad. Sci.* **53**, 1076.

Pressman, B. C. (1967). In: *Methods of Enzymology, Vol. X* (ed. by Colowick, S. P. and Kaplan, N. O.), p. 714. Academic Press, New York.

Pressman, B. C., Harris, E. J., Jagger, W. S., and Johnson, J. H. (1967). *Proc. Nat. Acad. Sci.* **58**, 1949.

Probine, M. C. and Preston, R. D. (1960). *J. Exptl. Bot.* **12**, 261.

Quinn, P. J. and White, I. G. (1967). *Res. Vet. Sci.* **8**, 58.

Randall, H. G. and Evered, D. F. (1964). *Biochim. Biophys. Acta* **93**, 98.

Randle, P. J. and Smith, G. H. (1958). *Biochem. J.* **70**, 490.

Rang, H. P. and Ritchie, J. M. (1968). *J. Physiol.* **196**, 183.

Raven, J. A. (1967*a*). *J. Gen. Physiol.* **50**, 1607.

Raven, J. A. (1967*b*). *J. Gen. Physiol.* **50**, 1627.

Razin, S., Gottfried, L., and Rottem, S. (1968). *J. Bacter.* **95**, 1685.

Rector, F. C., Jr., Carter, N. W., Seldin, D. W., and Nunn, A. C. (1965). *J. Clin. Invest.* **44**, 278.

Rector, F. C., Jr., Martinez–Maldonado, M., Brunner, F. P., and Seldin, W. D. (1966). *J. Clin. Invest.* **45**, 1060.

Reeves, R. E. (1951). *Adv. Carbohydrate Chem.* **6**, 107.

Regen, D. M. and Morgan, H. E. (1964). *Biochim. Biophys. Acta* **79**, 151.

Renkin, E. M. (1954). *J. Gen. Physiol.* **38**, 225.

Rhodes, J. B., Eichholz, A., and Crane, R. K. (1967). *Biochim. Biophys. Acta* **135**, 959.

Rhodes, M. E. and Payne, W. J. (1968). *Antonie van Leeuwenhoek* **34**, 298.

Ribierre, J. R. M., Job, J. C., and Girault, M. (1958). *Arch. Franc. Pédiatr.* **15**, 374.

Rickenberg, H. V., Cohen, G. N., Buttin, G., and Monod, J. (1956). *Ann. Inst. Pasteur* **91**, 829.

Ried, A. (1968). *Biochim. Biophys. Acta* **153**, 653.

Ring, K., Gross, W., and Heinz, E. (1967). *Ber. Bunsenges. Physik. Chem.* **71**, 893.

Ring, K. and Heinz, E. (1966). *Biochem. Z.* **344**, 446.

Robertson, J. D. (1955). *J. Biophys. Biochem. Cytol.* **1**, 271.

Robertson, J. D. (1959). *Biochem. Soc. Symp.* **16**, 3.

Robertson, J. D. (1964). *Symp. Soc. Cell. Chem. Suppl.* **14**, 379.

Robertson, J. S. (1957). *Physiol. Rev.* **37**, 133.

Robinson, J. W. L. (1966). Thesis, University of Lausanne.

Robinson, J. W. L. (1968). *Europ. J. Biochem.* **7**, 78.

Robinson, R. P. and Schmidt–Nielsen, B. (1964). *J. Cell. Comp. Physiol.* **62**, 147.

Robson, A. M., Srivastava, P. L., and Bricker, N. S. (1968). *J. Clin. Invest.* **47**, 329.

Roess, W. R. (1968). *J. Gen. Microbiol.* **52**, 421.

Rogers, A. W. (1967). *Techniques of Autoradiography.* Elsevier Publ. Co., Amsterdam.

Rogers, D. and Yu, S. H. (1962). *J. Bacter.* **84**, 877.

Rose, I. A., O'Connell, E. L., and Langdon, R. (1968). *Arch. Biochem. Biophys.* **126**, 727.

Rosenberg, L. E., Albrecht, I., and Segal, S. (1967). *Science* **155**, 1426.

Rosenberg, L. E., Bergman, M., and Segal, S. (1963). *Biochim. Biophys. Acta* **71**, 664.

Rosenberg, L. E., Blair, A., and Segal, S. (1961). *Biochim. Biophys. Acta* **54**, 479.

Rosenberg, L. E., Downing, S., Durant, J. L., and Segal, S. (1966). *J. Clin. Invest.* **45**, 365.

Rosenberg, Th. (1954). *Symp. Soc. Exptl. Biol.* **8**, 27.

Rosenberg, Th. and Wilbrandt, W. (1955). *Exptl. Cell. Res.* **9**, 49.

Rosenberg, Th. and Wilbrandt, W. (1957*a*). *Helv. Physiol. Pharmacol. Acta* **15**, 168.

Rosenberg, Th. and Wilbrandt, W. (1957*b*). *J. Gen. Physiol.* **41**, 289.

Rosenberg, Th. and Wilbrandt, W. (1962). *Exptl. Cell Res.* **27**, 100.

Roskoski, R. and Steiner, D. F. (1967*a*). *Biochim. Biophys. Acta* **135**, 717.

Roskoski, R. and Steiner, D. F. (1967*b*). *Biochim. Biophys. Acta* **135**, 727.

Rossi, C. S., Carafoli, E., Drahota, Z., and Lehninger, A. L. (1966). In: *Regulation of Metabolic Processes in Mitochondria* (ed. by Tager, J. M., Papa, S., Quagliariello, E., and Slater, E. C.), p. 317. Elsevier Publ. Co., Amsterdam.

Rothenberg, M. A. (1950). *Biochim. Biophys. Acta* **4**, 96.

Rothstein, A. (1954). *Symp. Soc. Exptl. Biol.* **8**, 165.

Rothstein, A. and Donovan, K. (1963). *J. Gen. Physiol.* **46**, 1075.

Rothstein, A., Frenkel, A., and Larrabee, C. (1948). *J. Cell. Comp. Physiol.* **32**, 261.

Rotman, B., Ganesan, A. K., and Guzman, R. (1968). *J. Mol. Biol.* **36**, 247.

Rotman, B. and Radojikovic, J. (1964). *J. Biol. Chem.* **239**, 3153.

Rotunno, C. A., Kowalewski, V., and Cereijido, M. (1967). *Biochim. Biophys. Acta* **135**, 170.

Roush, A. H. and Shieh, T. R. (1962). *Biochim. Biophys. Acta* **61**, 255.

Russel, R. S. and Barber, D. A. (1960). *Ann. Rev. Plant Physiol.* **11**, 127.

Rybová, R. (1965). *Physiol. Bohemoslov.* **14**, 412.

Rybová, R. and Janáček, K. (1968). Vth FEBS Meeting, abstr. no. 405. Prague.

Saha, J. and Coe, E. L. (1967). *Biochem. Biophys. Res. Comm.* **26**, 441.

Salton, M. R. J. and Freer, J. H. (1965). *Biochim. Biophys. Acta* **107**, 531.

Sanford, P. A., Smyth, D. H., and Watling, M. (1965). *J. Physiol.* **179**, 72P.

Scatchard, G. (1949). *Ann. N.Y. Acad. Sci.* **51**, 660.

Scatchard, G., Scheinberg, I. H. and Armstrong, S. H., Jr. (1950). *J. Amer. Chem. Soc.* **72**, 535.

Schachter, D. (1963). In: *The Transfer of Calcium and Strontium across Biological Membranes* (ed. by Wasserman, R. H.), p. 197. Academic Press, New York and London.

Schachter, D., Johnson, N. and Kirkpatrick, M. A. (1966). *Biochem. Biophys. Res. Comm.* **25**, 603.

Schachter, D. and Rosen, S. M. (1959). *Amer. J. Physiol.* **196**, 357.

Schafer, J. A. and Jacquez, J. A. (1967). *Biochim. Biophys. Acta* **135**, 1081.

Schatzmann, H. J. (1953). *Helv. Physiol. Acta* **11**, 346.

Schatzmann, H. J. (1966). *Experientia* **22**, 364.

Scheer, B. T. and Mumbach, M. W. (1960). *J. Cell. Comp. Physiol.* **55**, 259.

Schilde, C. (1968a). *Fortschr. Bot.* **30**, 43.

Schilde, C. (1968b). *Z. Naturforsch.* **236**, 1369.

Schlegel, H. G. and Lafferty, R. (1961). *Arch. Mikrobiol.* **38**, 52.

Schleif, R. (1969). *J. Mol. Biol.* **46**, 185.

Schliephake, W., Junge, W., and Witt, H. T. (1968). *Z. Naturforsch.* **23b**, 1571.

Schnebli, H. P. and Abrams, A. (1969). *Fed. Proc.* **28**, 464.

Schönheimer, R. (1946). *The Dynamic State of Body Constituents.* Harvard Univ. Press, Boston.

Schultz, S. G., Curran, P. F., Chez, R. A., and Fuisz, R. E. (1967). *J. Gen. Physiol.* **50**, 1241.

Schultz, S. G., Epstein, W., and Goldstein, D. (1964). *J. Gen. Physiol.* **46**, 343.

Schultz, S. G., Fuisz, R. E., and Curran, P. F. (1966). *J. Gen. Physiol.* **49**, 849.

Schultz, S. G. and Solomon, A. K. (1961). *J. Gen. Physiol.* **45**, 355.

Schultz, S. G., Wilson, N. L., and Epstein, W. (1962). *J. Gen. Physiol.* **46**, 159.

Schultz, S. G. and Zalusky, R. (1963). *Biochim. Biophys. Acta* **71**, 503.

Schultz, S. G. and Zalusky, R. (1964a). *J. Gen. Physiol.* **47**, 567.

Schultz, S. G. and Zalusky, R. (1964b). *J. Gen. Physiol.* **47**, 1043.

Schultz, S. G. and Zalusky, R. (1965). *Nature* **204**, 292.

Schwartzman, L., Blair, A., and Segal, S. (1966). *Biochem. Biophys. Res. Comm.* **23**, 220.

Scriver, C. R. (1965). *New Engl. J. Med.* **273**, 530.

Scriver, C. R. and Goldman, H. (1966). *J. Clin. Invest.* **45**, 1357.

Scriver, C. R., Pueschel, S., and Davies, E. (1966). *New Engl. J. Med.* **274**, 636.

Scriver, C. R. and Wilson, O. H. (1967). *Science* **155**, 1428.

Segal, H. L., Kachmar, J. F., and Boyer, P. D. (1952). *Enzymologia* **15**, 187.

Segal, J. R. (1968). *Biophys. J.* **8**, 470.

Segal, S., Blair, A., and Rosenberg, L. E. (1963). *Biochim. Biophys. Acta* **71**, 676.

Segal, S., Schwartzman, L., Blair, A., and Bertoli, D. (1967). *Biochim. Biophys. Acta* **135**, 127.

Segal, S., Thier, S., Fox, M., and Rosenberg, L. (1962). *Biochim. Biophys. Acta* **65**, 567.

Semenza, G. (1967). *J. Theoret. Biol.* **15**, 145.

Semenza, G. (1968). In: *Handbook of Physiology* (*Alimentary Canal*), p. 2543, Amer. Physiol. Soc., Washington.

Semenza, G. (1969a). *Biochim. Biophys. Acta* **173**, 104.

Semenza, G. (1969b). VIth FEBS Meeting, Abstr., no. 55. Madrid.

Semenza, G., Tosi, R., Vallotton–Delachaux, M. C., and Mülhaupt, E. (1964). *Biochim. Biophys. Acta* **89**, 109.

Sen, A. K. and Post, R. L. (1964). *J. Biol. Chem.* **239**, 345.

Sen, A. K. and Post, R. L. (1966). *Abstr. Biophys. Soc. 10th Ann. Meeting*, p. 152.

Sen, A. K. and Widdas, W. F. (1962). *J. Physiol.* **160**, 392.

Shanes, A. M. (1958a). *Pharmacol. Rev.* **10**, 59.
Shanes, A. M. (1958b). *Pharmacol. Rev.* **10**, 165.
Shannon, J. A., Farber, S., and Troast, L. (1941). *Amer. J. Physiol.* **133**, 752.
Sharp, G. W. G. and Leaf, A. (1966). *Physiol. Rev.* **46**, 593.
Sibaoka, T. (1966). *Symp. Soc. Exptl. Biol.* **20**, 49.
Siekevitz, P. (1961). In: *Membrane Transport and Metabolism* (ed. by Kleinzeller, A. and Kotyk, A.), p. 598. Academic Press, New York–London.
Silverman, M. and Goresky, C. A. (1965). *Biophys. J.* **5**, 487.
Simoni, R. D., Levinthal, M., Kundig, F. D., Kundig, W., Anderson, B., Hartman, Ph. E., and Roseman, S. (1967). *Proc. Nat. Acad. Sci.* **58**, 1963.
Singer, I. and Tasaki, I. (1968). In: *Biological Membranes* (ed. by Chapman, D.), p. 347. Academic Press, London.
Sjöstrand, F. S. (1963a). *J. Ultrastruct. Res.* **8**, 517.
Sjöstrand, F. S. (1963b). *J. Ultrastruct. Res.* **9**, 340.
Sjöstrand, F. S. (1964). *Symp. Soc. Chem. Biol. Suppl.* **14**, 103.
Skadhauge, E. (1967). *Comp. Biochem. Physiol.* **23**, 483.
Skou, J. C. (1965). *Physiol. Rev.* **45**, 596.
Slayman, C. L. (1965a). *J. Gen. Physiol.* **49**, 69.
Slayman, C. L. (1965b). *J. Gen. Physiol.* **49**, 93.
Slayman, C. W. and Tatum, E. L. (1964). *Biochim. Biophys. Acta* **88**, 578.
Slayman, C. W. and Tatum, E. L. (1965). *Biochim. Biophys. Acta* **109**, 184.
Smith, F. A. (1966). *Biochim. Biophys. Acta* **126**, 94.
Smith, H. W. (1951). *The Kidney*. Oxford Univ. Press, London.
Smyth, D. H. (1964). *Symp. Soc. Exptl. Biol.* **19**, 307.
Smyth, D. H. (1966). *Excerpta Med. Monographs Nucl. Med. Biol.* **1**, 195.
Smyth, D. H. (1967) (editor): *Intestinal Absorption. Brit. Med. Bull.* **23**, 205.
Smyth, D. H. and Taylor, C. B. (1957). *J. Physiol.* **136**, 632.
Sollner, K. (1955). In: *Electrochemistry in Biology and Medicine* (ed. by Shedlovsky, Th.), p. 33. John Wiley and Sons, Inc., New York.
Solomon, A. K. (1960a). In: *Mineral Metabolism, Vol. IA* (ed. by Comar, C. L. and Bronner, F.), p. 119. Academic Press, New York.
Solomon, A. K. (1960b). *J. Gen. Physiol.* **43**, Suppl. 1, 1.
Solomon, A. K. (1961). In: *Membrane Transport and Metabolism* (ed. by Kleinzeller, A. and Kotyk, A.), p. 94. Academic Press, New York.
Sols, A. and de la Fuente, G. (1961). In: *Membrane Transport and Metabolism* (ed. by Kleinzeller, A. and Kotyk, A.), p. 361. Academic Press, New York.
Sorokina, Z. A. (1964). (In Russian.) *Tsitologiya* **6**, 152.
Spanswick, R. M., Stolarek, J. and Williams, E. J. (1967). *J. Exptl. Bot.* **18**, 1.
Spanswick, R. M. and Williams, E. J. (1964). *J. Exptl. Bot.* **15**, 193.
Spanswick, R. M. and Williams, E. J. (1965). *J. Exptl. Bot.* **16**, 463.
Spencer, R. P. and Brody, K. R. (1964). *Biochim. Biophys. Acta* **88**, 400.
Spiegler, K. S. and Wyllie, M. R. J. (1956). In: *Physical Techniques in Biological Research*, *Vol.* 2 (ed. by Oster, G. and Pollister, A. W.), p. 301. Academic Press, New York.
Stämpfli, R. (1959). *Ann. N.Y. Acad. Sci.* **81**, 265.
Stein, W. D. (1962a). *Biochim. Biophys. Acta* **59**, 47.
Stein, W. D. (1962b). *Biochim. Biophys. Acta* **59**, 66.
Stein, W. D. (1964). In: *The Structure and Activity of Enzymes* (ed. by Goodwin, T. W., Hartley, B. S., and Harris, J. I.), p. 133. Academic Press, New York.

Stein, W. D. and Danielli, J. F. (1956). *Disc. Faraday Soc.* **21**, 238.

Steinbach, H. B. (1933). *J. Cell. Comp. Physiol.* **3**, 1.

Steinbach, H. B. (1941). *J. Cell. Comp. Physiol.* **17**, 57.

Steward, F. C. and Sutcliffe, J. E. (1961). In: *Plant Physiology, Vol. II* (ed. by Steward, F. C.), p. 253. Academic Press, New York.

Stirling, C. E. (1967). *J. Cell. Biol.* **35**, 605.

Stoeckenius, W. (1962). *J. Cell Biol.* **12**, 221.

Straub, F. B. (1953). *Acta Physiol. Hung.* **4**, 235.

Strauss, E. W. (1963). *J. Cell Biol.* **17**, 597.

Suomalainen, H. (1968). *Suomen Kemistilehti* **41**, 239.

Sutcliffe, J. E. (1960). *Nature* **188**, 294.

Tanaka, S., Fraenkel, D. G., and Lin, E. C. C. (1967). *Biochem. Biophys. Res. Comm.* **27**, 63.

Tasaki, I., Lerman, L., and Watanabe, A. (1969). *Amer. J. Physiol.* **216**, 130.

Tasaki, I. and Spyropoulos, C. S. (1961). *Amer. J. Physiol.* **201**, 413.

Tasaki, I., Singer, I., and Takenaka, T. (1965). *J. Gen. Physiol.* **48**, 1095.

Tasaki, I. and Singer, I. (1968). *Ann. N.Y. Acad. Sci.* **148**, 36.

Tasaki, I., Watanabe, A., and Takenaka, T. (1962). *Proc. Nat. Acad. Sci.* **48**, 1177.

Taylor, A. E., Wright, E. M., Schultz, S. G., and Curran, P. F. (1968). *Amer. J. Physiol.* **214**, 836.

Taylor, R. E. (1963). In: *Physical Techniques in Biological Research, Vol. VI B* (ed. by Nastuk, W. L.), p. 219. Academic Press, New York.

Tazawa, M. (1964). *Plant Cell Physiol.* **5**, 33.

Tazawa, M. and Kamiya, N. (1965). *Ann. Rep. Biol. Works, Fac. Sci., Osaka Univ.* **13**, 123.

Tazawa, M. and Kishimoto, U. (1964). *Plant Cell Physiol.* **5**, 45.

Tempest, D. W. and Meers, J. L. (1968). *J. Gen. Microbiol.* **54**, 319.

Teorell, T. (1949). *Arch. Scand. Physiol.* **3**, 205.

Teorell, T. (1953). In: *Progress in Biophysics and Biophysical Chemistry, Vol. III* (ed. by Butler, J. A. V. and Randall, J. T.), p. 305. Academic Press, New York.

Thier, S., Blair, A., Fox, M., and Segal, S. (1967). *Biochim. Biophys. Acta* **135**, 300.

Thier, S., Fox, M., Rosenberg, L., and Segal, S. (1964). *Biochim. Biophys. Acta* **93**, 106.

Thorn, M. B. (1953). *Biochem. J.* **54**, 540.

Ti Tien, H., Huemoeller, W. A., and Ping Ting, H. (1968). *Biochem. Biophys. Res. Comm.* **33**, 207.

Tosteson, D. C. (1968). *Fed. Proc.* **27**, 1269.

Tosteson, D. C. and Hoffman, J. F. (1960). *J. Gen. Physiol.* **44**, 169.

Troshin, A. S. (1958). *Das Problem der Zellpermeabilität.* VEB Gustav–Fischer–Verlag, Jena.

Tyler, D. D. (1969). *Biochem. J.* **111**, 665.

Tyree, M. T. (1968). *Canad. J. Biochem.* **46**, 317.

Ugolev, A. M., Jesuitova, N. N., and De Lacy, P. (1964). *Nature* **203**, 879.

Ullrich, K. J. (1967). In: *IIIrd Internat. Congr. Nephrol.* (ed. by Schreiner, G. E.), p. 48. S. Karger, Basel–New York.

Ullrich, K. J., Kramer, K., and Boylan, J. W. (1961). *Prog. Cardiovascular Dis.* **3**, 395.

Ullrich, K. J., Rumrich, G., and Fuchs, G. (1964). *Pflügers Arch. ges. Physiol.* **280**, 99.

Ullrich, K. J., Rumrich, G., and Schmidt–Nielsen, B. (1967). *Pflügers Arch. ges. Physiol.* **295**, 147.

Ulrich, F. (1959). *Amer. J. Physiol.* **197**, 997.

Ussing, H. H. (1949). *Acta Physiol. Scand.* **19**, 43.

Ussing, H. H. (1960). In: *The Alkali Metal Ions in Biology* (*Handbuch der experimentellen Pharmakologie*). Springer–Verlag, Berlin.

Ussing, H. H. (1961). *J. Gen. Physiol.* **43**, Suppl. 1, 135.

Ussing, H. H. (1965). *Acta Physiol. Scand.* **63**, 141.

Ussing, H. H. (1966). *Ann. N.Y. Acad. Sci.* **137**, 543.

Ussing, H. H. (1967). *Protoplasma* **63**, 292.

Ussing, H. H. and Windhager, E. E. (1964). *Acta Physiol. Scand.* **61**, 484.

Ussing, H. H. and Zerahn, K. (1951). *Acta Physiol. Scand.* **23**, 110.

Vallée, M. and Jeanjean, R. (1968). *Biochim. Biophys. Acta* **150**, 599.

Van Bruggen, J. T. and Zerahn, K. (1960). *Nature* **188**, 499.

van Dam, K. and Slater, E. C. (1967). *Proc. Nat. Acad. Sci.* **58**, 2015.

Vandenheuvel, F. A., quoted in: Warner, D. T. (1966). *Science* **153**, 324.

van Steveninck, J. (1968). *Biochim. Biophys. Acta* **163**, 386.

van Steveninck, J. and Dawson, E. C. (1968). *Biochim. Biophys. Acta* **150**, 47.

van Steveninck, J. and Rothstein, A. (1965). *J. Gen. Physiol.* **49**, 235.

van Steveninck, J., Weed, R. I., and Rothstein, A. (1965). *J. Gen. Physiol.* **48**, 617.

Vicenzi, F. F. and Schatzmann, H. J. (1967). *Helv. Physiol. Pharmacol. Acta* **25**, 233.

Vidaver, G. A. (1966). *J. Theoret. Biol.* **10**, 301.

Vidaver, G. A. and Shepherd, S. L. (1968). *J. Biol. Chem.* **243**, 6140.

Villegas, L. (1963). *Biochim. Biophys. Acta* **75**, 131.

Villegas, R., Blei, M., and Villegas, G. (1965). *J. Gen. Physiol.* **48**, 35.

Vogel, G. (1966). *Pflügers Arch. ges. Physiol.* **288**, 342.

Vogel, G. and Kürten, M. (1967). *Pflügers Arch.* **295**, 42.

Vogel, G., Lauterbach, F., and Kröger, W. (1965). *Pflügers Arch.* **283**, 151.

Vorobyev, L. N. (1967). *Nature* **216**, 1325.

Vorobyev, L. N. and Kurella, G. A. (1965). (In Russian.) *Biofizika* **10**, 788.

Vorobyev, L. N., Radenovich, T. N., Khitrov, Y. A., and Yavlova, L. G. (1967). (In Russian.) *Biofizika* **12**, 1016.

Waddell, W. J. and Butler, T. C. (1959). *J. Clin. Invest.* **38**, 720.

Walker, A. M. and Hudson, C. L. (1937). *Amer. J. Physiol.* **118**, 130.

Walker, N. A. (1960). *Austral. J. Biol. Sci.* **13**, 468.

Walser, M., and Robinson, B. H. B. (1963). In: *The Transfer of Calcium and Strontium across Biological Membranes* (ed. by Wasserman, R. H.), p. 305. Academic Press, New York.

Wasserman, R. H., Corradino, R. A. and Taylor, A. N. (1968). *J. Biol. Chem.* **243**, 3978.

Watlington, C. O., Campbell, A. D., and Huf, E. G. (1964). *J. Cell. Comp. Physiol.* **64**, 389.

Webber, W. A. (1963). *Canad. J. Biochem. Physiol.* **41**, 131.

Webber, W. A. (1965). *Canad. J. Physiol. Pharmacol.* **43**, 79.

Wei, L. Y. (1969). *Science* **163**, 280.

Weidman, S. (1955). *J. Physiol.* (*London*) **129**, 568.

Weigl, J. (1967). *Planta* **75**, 327.

Weigl, J. (1969). *Planta* **84**, 311.

Weinstein, S. W. and Hempling, H. G. (1964). *Biochim. Biophys. Acta* **79**, 329.

Werkheiser, W. C. and Bartley, W. (1957). *Biochem. J.* **66**, 79.

Wheeler, K. P. and Christensen, H. N. (1967). *J. Biol. Chem.* **242**, 3782.

Whittam, R. and Ager, M. E. (1965). *Biochem. J.* **97**, 214.

Whittembury, G. (1960). *J. Gen. Physiol.* **43**, 43.

Whittembury, G. (1963). *Amer. J. Physiol.* **204**, 401.

Whittembury, G. (1964). *J. Gen. Physiol.* **47**, 795.

Whittembury, G. (1967). *Acta Cient. Venezolana, Suppl.* 3, 71.

Whittembury, G., Sugino, N., and Solomon, A. K. (1960). *Nature* **187**, 699.

Widdas, W. F. (1952). *J. Physiol.* **118**, 23.

Widdas, W. F. (1954). *J. Physiol.* **125**, 163.

Wilbrandt, W. (1948). *Pflügers Arch. ges. Physiol.* **250**, 569.

Wilbrandt, W. (1954). *Symp. Soc. Exptl. Biol.* **8**, 136.

Wilbrandt, W. (1961a). In: *Biochemie des Aktiven Transports*, p. 112. Springer–Verlag Berlin–Göttingen–Heidelberg.

Wilbrandt, W. (1961b). *Pharmacol. Rev.* **13**, 109.

Wilbrandt, W., Frei, S., and Rosenberg, Th. (1956). *Exptl. Cell Res.* **11**, 59.

Wilbrandt, W. and Kotyk, A. (1964). *Naunyn-Schmiedebergs Arch. Expt. Pathol. Pharmakol.* **249**, 279.

Wilbrandt, W. and Rosenberg, Th. (1950). *Helv. Physiol. Acta* **8**, C82.

Wiley, W. R. and Matchett, W. H. (1966). *J. Bacter.* **92**, 1705.

Wilkins, P. O. and O'Kane, D. J. (1964). *J. Gen. Microbiol.* **34**, 389.

Williams, E. J. and Bradley, J. (1968a). *Biophys. J.* **8**, 145.

Williams, E. J. and Bradley, J. (1968b). *Biochim. Biophys. Acta* **150**, 626.

Williams, E. J., Johnston, R. J., and Dainty, J. (1964). *J. Exptl. Bot.* **15**, 1.

Williams, H. H., Erickson, B. N., and Macy, I. G. (1941). *Quart. Rev. Biol.* **16**, 80.

Wilson, O. H. and Scriver, C. R. (1967). *Amer. J. Physiol.* **213**, 185.

Wilson, T. H. (1962). *Intestinal Absorption.* W. B. Saunders and Co., Philadelphia.

Wilzbach, K. E. (1956). *J. Amer. Chem. Soc.* **79**, 1013.

Winkler, H. H. (1966). *Biochim. Biophys. Acta* **117**, 231.

Winkler, H. H. and Wilson, T. H. (1966). *J. Biol. Chem.* **241**, 2200.

Winn, P. M., LaPrade, N. S., Tolbert, W. R., and Huf, E. G. (1966). *MCV Quarterly* **2**, 116.

Winn, P. M., Smith, T. E., Campbell, A. D., and Huf, E. G. (1964). *J. Cell. Comp. Physiol.* **64**, 371.

Winter, C. G. and Christensen, H. N. (1964). *J. Biol. Chem.* **239**, 872.

Winter, C. G. and Christensen, H. N. (1965). *J. Biol. Chem.* **240**, 3594.

Wiseman, G. (1951). *J. Physiol.* (*London*) **114**, 7P.

Wiseman, G. (1953). *J. Physiol.* **120**, 63.

Wiseman, G. (1964). *Absorption from the Intestine.* Academic Press, London.

Wood, R. E., Wirth, F. P., and Morgan, H. E. (1968). *Biochim. Biophys. Acta* **163**, 171.

Wool, I. G. (1965). *Fed. Proc.* **24**, 1060.

Woolf, B. (1932). Quoted in: Haldane, J. B. S. and Stern, K. G.: *Allgemeine Chemie der Enzyme*, p. 119. Steinkopff–Verlag, Dresden.

Woosley, R. L. and Huang, K. C. (1967). *Proc. Soc. Exptl. Biol. Med.* **124**, 20.

Wright, E. M. (1966). *J. Physiol.* **185**, 486.

Yabu, K. (1967). *Biochim. Biophys. Acta* **135**, 181.
Yang, T. T. (1950). *Schweiz. Mediz. Wochenschr.* **80**, 1157.
Yudkin, M. (1966). *Biochem. J.* **98**, 923.

Zadunaisky, J. A. (1966). *J. Cell Biol.* **31**, C11.
Zadunaisky, J. A., Candia, O. A., and Chiardini, D. J. (1963). *J. Gen. Physiol.* **47**, 393.
Zadunaisky, J. A. and DeFisch, F. W. (1964). *Amer. J. Physiol.* **207**, 1010.
Zadunaisky, J. A., Gennaro, J. F., Jr., Bashirelahi, N., and Hilton, M. (1968). *J. Gen. Physiol.* **51**, 290S.
Zerahn, K. (1955). *Acta Physiol. Scand.* **33**, 347.
Zerahn, K. (1956). *Acta Physiol. Scand.* **36**, 300.
Zerahn, K. (1958). *Oxygen Consumption and Active Sodium Transport.* Universitetsforlaget, Aarhus.
Zerahn, K. (1961). In: *Membrane Transport and Metabolism* (ed. by Kleinzeller, A. and Kotyk, A.), p. 237. Academic Press, New York.

SUBJECT INDEX